TODAY'S TECHNICIAN™

SHOP MANUAL

For Basic Automotive Service & Systems

SIXTH EDITION

TODAY'S TECHNICIAN™

SHOP
MANUAL

For Basic Automotive Service & Systems

SIXTH EDITION

Chris Hadfield

John Witthauer

Australia • Brazil • Canada • Mexico • Singapore • United Kingdom • United States

***Today's Technician: Shop Manual for Basic
Automotive Service & Systems,***
Sixth Edition
Chris Hadfield
John Witthauer

SVP, Higher Education & Skills Product: Erin Joyner

Product Director: Matthew Seeley

Senior Product Manager: Katie McGuire

Product Assistant: Kimberly Klotz

Director, Learning Design: Rebecca von Gillern

Senior Manager, Learning Design: Leigh Hefferon

Learning Designer: Mary Clyne

Marketing Director: Sean Chamberland

Marketing Manager: Andrew Ouimet

Director, Content Creation: Juliet Steiner

Content Creation Manager: Alexis Ferraro

Senior Content Manager: Meaghan Tomaso

Digital Delivery Lead: Amanda Ryan

Art Designer: Erin Griffin

Text Designer: Angela Sheehan

Cover Designer: Angela Sheehan

Cover image(s): Umberto Shtanzman/Shutterstock.com

For product information and technology assistance, contact us at
**Cengage Customer & Sales Support, 1-800-354-9706
or support.cengage.com.**

For permission to use material from this text or product, submit all requests online at **www.copyright.com.**

Library of Congress Control Number: 2020901746

Book Only ISBN: 978-1-337-79569-2
Package ISBN: 978-1-337-79566-1

Cengage
5191 Natorp Boulevard
Mason, OH 45040
USA

Cengage is a leading provider of customized learning solutions. Our employees reside in nearly 40 different countries and serve digital learners in 165 countries around the world. Find your local representative at: **www.cengage.com.**

To learn more about Cengage platforms and services, register or access your online learning solution, or purchase materials for your course, visit **www.cengage.com.**

Notice to the Reader

Publisher does not warrant or guarantee any of the products described herein or perform any independent analysis in connection with any of the product information contained herein. Publisher does not assume, and expressly disclaims, any obligation to obtain and include information other than that provided to it by the manufacturer. The reader is expressly warned to consider and adopt all safety precautions that might be indicated by the activities described herein and to avoid all potential hazards. By following the instructions contained herein, the reader willingly assumes all risks in connection with such instructions. The publisher makes no representations or warranties of any kind, including but not limited to, the warranties of fitness for particular purpose or merchantability, nor are any such representations implied with respect to the material set forth herein, and the publisher takes no responsibility with respect to such material. The publisher shall not be liable for any special, consequential, or exemplary damages resulting, in whole or part, from the readers' use of, or reliance upon, this material.

Printed in the United States of America
Print Number: 03   Print Year: 2024

CONTENTS

PHOTO SEQUENCES

JOB SHEETS

Thanks to the support the *Today's Technician* series has received from those who teach automotive technology, Cengage, the leader in automotive-related textbooks, is able to live up to its promise to regularly provide new editions of texts of this series. We have listened and responded to our critics and our fans and present this new updated and revised sixth edition. By revising this series on a regular basis, we can respond to changes in the industry, changes in technology, changes in the accreditation process, and to the ever-changing needs of those who teach automotive technology.

The *Today's Technician* series features textbooks that cover all mechanical and electrical systems of automobiles and light trucks. Principally, the individual titles correspond to the areas for National Institute for Automotive Service Excellence (ASE) certification.

All titles in the *Today's Technician* series include remedial skills and theories common to all of the certification areas and advanced or specific subject areas that reflect the latest technological trends.

Today's Technician: Basic Automotive Service & Systems, 6e is designed to give students a chance to develop the same skills and gain the same knowledge that today's successful technician has. This edition also reflects the changes in the guidelines established by the ASE Education Foundation in 2017.

The purpose of the ASE Education Foundation is to evaluate technician training programs against standards developed by the automotive industry and recommend qualifying programs for accreditation by ASE. Programs can earn ASE accreditation upon the recommendation of the ASE Education Foundation. ASE Education Foundation's national standards reflect the skills that students must master. ASE Education Foundation accreditation ensures that the training programs meet or exceed industry-recognized, uniform standards of excellence.

The technician of today and for the future must know the underlying theory of all automotive systems and be able to service and maintain those systems. Dividing the material into two volumes, a Classroom Manual and a Shop Manual, provides the reader with the information needed to begin a successful career as an automotive technician without interrupting the learning process by mixing cognitive and performance learning objectives into one volume.

The design of Cengage Learning's *Today's Technician* series was based on features known to promote improved student learning. The design was further enhanced by a careful study of survey results in which the respondents were asked to value particular features. Some of these features can be found in other textbooks, whereas others are unique to this series.

Each Classroom Manual contains the principles of operation for each system and subsystem. The Classroom Manual also discusses design variations in key components used by the different vehicle manufacturers and considers emerging technologies that will be standard or optional features in the near future. This volume is organized to build upon basic facts and theories. Its primary objective is to help the reader gain an understanding of how each system and subsystem operates. This understanding is necessary to diagnose the complex automobiles of today and tomorrow. Although the basics contained in the Classroom Manual provide the knowledge needed for diagnostics, diagnostic procedures appear only in the Shop Manual. An understanding of the underlying theories is also a requirement for competence in the skill areas covered in the Shop Manual.

A spiral–bound Shop Manual delivers hands-on learning experiences with step-by-step instructions for diagnostic and repair procedures. Photo Sequences are used to illustrate some of the common service procedures. Other common procedures are listed and accompanied with fine-line drawings and photos that let the reader visualize and conceptualize the finest details of the procedure. This volume explains the reasons for performing the procedures, as well as the circumstances when each particular service is appropriate.

The two volumes are designed to be used together and are arranged in corresponding chapters. Not only are the chapters in the volumes linked together, but the contents of the chapters are also linked. The linked content is indicated by marginal callouts that refer the reader to the chapter and page where the same topic is addressed in the companion volume. This valuable feature saves users the time and trouble of searching the index or table of contents to locate supporting information in the other volume. Instructors will find this feature especially helpful when planning the presentation of material and when making reading assignments.

Both volumes contain clear and thoughtfully selected illustrations, many of which are original drawings or photos specially prepared for inclusion in this series. This means that the art is a vital part of each textbook and not merely inserted to increase the number of illustrations.

The page design of this series uses available margin space to deliver helpful information efficiently without interrupting the pedagogical lesson material. This information includes examples of concepts just introduced in the text, explanations or definitions of terms that are not defined in the text, examples of common trade jargon used to describe a part or operation, and unique applications of the system or service described in the text. Many textbooks also include this information but insert it in the main body of text; this tends to interrupt the reader's thought process. By placing this information to the side of the main text, students can read through the text uninterrupted and refer to the additional information when it is best for them.

Jack Erjavec
Series Editor

HIGHLIGHTS OF THIS EDITION—CLASSROOM MANUAL

The new sixth edition has been fully updated to align with the new ASE Education Foundation accreditation model introduced in 2017.

This edition's full-color design with new photographs and updated line art helps readers better visualize processes and procedures. All chapters have been updated and arranged for easier learning progression. This edition is designed to build skills and knowledge as the student advances through the chapters. The entire text was updated with information about new tools, equipment, and vehicle systems in use today and expected in the near future.. Coverage of hybrid vehicles and emerging technologies, along with revised and updated coverage of electronic engine controls, prepares students for today's workplace.

What's New to This Edition:

- All chapters have been updated and arranged for easier learning progression. This edition is designed to build skills and knowledge as the student advances through the chapters.
- New and updated photos to show modern industry.
- New Chapter 1, "Introduction to the Automotive Industry." This chapter highlights the different types of businesses, careers, training, and ethics needed to operate a successful operation.

- New Chapter 2, "Introduction to the automobile." This chapter explains the complex systems and the purpose for today's automobiles.
- New Chapter 4, "Fasteners and measurements." This chapter combines the 5th edition Chapters 4 and 5 to consolidate and prioritize basic fasteners and measurement tools for the entry-level technician.
- New Chapter 5, "Electrical and electronic systems." This chapter corresponds with the ASE Maintenance and Light repair requirements needed for the importance of basic electrical fundamentals.
- New Chapter 6, "Accessory Lighting and Electrical Systems." This chapter provides a deeper dive onto electrical systems.
- New Chapter 7, "Engine Systems." This chapter provides a better and basic understanding for engine operation.
- New Chapter 11, "Heating and Air Conditioning Systems." This chapter focuses on operation and theory of the HVAC systems working together.
- New Chapter 12, "Engine Performance Theories." This chapter is designed to align with ASE and MLR required material.

HIGHLIGHTS OF THIS EDITION—SHOP MANUAL

Along with the Classroom Manual, the Shop Manual was revised and updated in both text and figures to emphasize the changing technology in tools, equipment, and procedures. The chapters parallel with and complement the Classroom Manual's layout and closely match a typical automotive school's sequence of instruction. The sixth edition includes even more all-new Job Sheets that are focused on skill building.

New content cross-references current industry practices to help the student understand and practice real-world workplace competencies, and employer-focused soft skills are integrated into each topic. The revised edition features a new chapter on Safety Inspection and Preventive Maintenance which focuses on the importance on complete safety inspections and proper documentation.

What's New to This Edition

- All chapters have been updated and arranged for easier learning progression. This edition is designed to build skills and knowledge as the student advances through the chapters.
- All chapters include new task sheets that include "Required Supplemental Tasks" (RST) that align with the ASE standards required for Entry-Level Technician Certification.
- Task sheets also have new grading rubric with explanation of the rubric terms.
- Task sheets also added "instructor check points" these are designed to stop the students and let the instructor quiz and or inspect the progress of each student or groups of students.
- All chapters align and support the Classroom manual and all changes are followed in the Shop manuals.
- New Chapter 12, Safety Inspection and Preventive maintenance." This chapter focuses on the importance of a complete safety inspection to provide the customer an overview of the current condition of their vehicle.
- This chapter also stresses the importance of proper documentation of the safety inspections finding and suggesting options for preventive maintenance.

CLASSROOM MANUAL

Features of the Classroom Manual include the following:

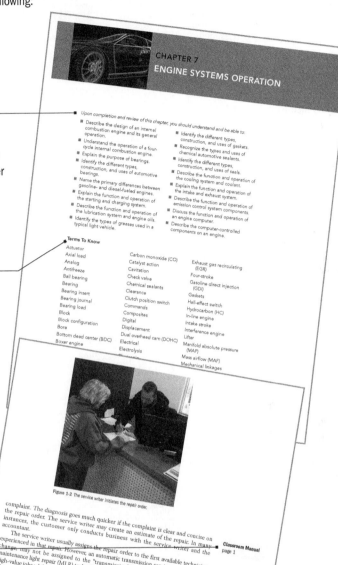

Cognitive Objectives

These objectives outline the chapter's contents and identify what students should know and be able to do upon completion of the chapter. Each topic is divided into small units to promote easier understanding and learning.

Terms to Know List

A list of key terms appears immediately after the Objectives. Students will see these terms discussed in the chapter. Definitions can also be found in the Glossary at the end of the manual.

Cross-References to the Shop Manual

References to the appropriate page in the Shop Manual appear whenever necessary. Although the chapters of the two manuals are synchronized, material covered in other chapters of the Shop Manual may be fundamental to the topic discussed in the Classroom Manual.

Margin Notes

The most important terms to know are highlighted and defined in the margin. Common trade jargon also appears in the margin and gives some of the common terms used for components. This helps students understand and speak the language of the trade, especially when conversing with an experienced technician.

A Bit of History

This feature gives the student a sense of the evolution of the automobile. This feature not only contains nice-to-know information but also should spark some interest in the subject matter.

Author's Note

This feature includes simple explanations, stories, or examples of complex topics. These are included to help students understand difficult concepts.

Summary

Each chapter concludes with summary statements that contain the important topics of the chapter. These are designed to help the reader review the contents.

Review Question

Short-answer essay, fill in the blank, and multiple-choice questions follow each chapter. These questions are designed to accurately assess the student's competence in the stated objectives at the beginning of the chapter.

Figure 4-53 Common types of threaded automotive fasteners.

A BIT OF HISTORY

Threads provide a mechanical advantage to hold parts together. The geometry behind threads was developed a long time ago, around 200 BCE. They were used by the ancient Romans to press grapes for wine.

AUTHOR'S NOTE Threads act like a lever that multiplies the applied force. This is the mechanical principle behind cranes, hand-operated presses, wine presses—and the high clamping forces applied by a wrench with minimum force.

Fastener Sizing and Torquing

For a fastener to work correctly, it must have the strength and size to handle the task. Fasteners are designed in different grade (strength) ratings for specific loads and applications. Replacement fasteners should meet the same requirements as the original item.

A light-duty fastener may be replaced with a heavy-class fastener in most cases. However, some industry and vehicle parts have a certain class of fastener designed to break or shear to prevent damage to the components. A typical example would be the shear pin in a winch. To prevent overloading a cable or chain that may result in very serious injury and great damage, the shear pin is designed to shear apart when its limits are reached. Replacing a shear pin that "broke too quickly" with a stronger pin has resulted in at least one death to the author's knowledge.

Introduction to the Automobile 33

...from the remote control "fob" in the driver's hand or pocket as he or she approaches the vehicle. Other accessories include DVD and game players for the passengers. Navigation features allow the driver to access driving directions without pulling over for assistance. Like air conditioning in the southeastern United States, all of these "exotic" features will be standard on future vehicles. Each requires some degree of technical expertise for diagnosing and repairs. Although this means better pay and employment for the automotive technician, the downside is increased cost of parts and labor. The cost of the parts may eventually be lower, but the labor cost of a well-trained, motivated technician will likely increase. This further emphasizes the technician's golden rule: *Fix it right the first time*.

SUMMARY

- The electrical system has become the single most important system regarding vehicle operation.
- Starting and charging systems provide the means to start the engine and power the electrical system.
- The power system is the component that changes heat from burning or reactive chemicals to mechanical energy.
- Power systems are constantly in a state of change regarding fuels and operations.
- Alternative fuels may provide a means to reduce the use of hydrocarbon fuels while lowering harmful emissions.
- The powertrain transmits the power-system output to the drive wheels.
- The transmission or transaxle provides a means to harness the engine's output to the load based on vehicle load and environment.

- Steering allows the driver to direct the vehicle in the desired direction.
- Suspension helps protect vehicle components and aids in a better ride.
- A brake system provides the driver a way to stop and control the vehicle.
- The frame supports the body and all of its mounted components.
- The body protects the passengers and vehicle components from outside hazards.
- The climate control system, although not needed to operate the vehicle, reduces driver and passenger fatigue, and increases alertness.
- Accessories include many features or subsystems that assist the driver or passenger and provide a better interior environment.

REVIEW QUESTIONS

Short Answer Essay

1. Discuss how the vehicle electrical system has expanded in use over the last 20 years.
2. Explain how the starter circuit turns the crankshaft during engine startup.
3. Discuss the use of alternative fuels in today's and future vehicles.
4. Explain the purpose of the powertrain.
5. Explain the general usage of a computer or computers in today's vehicles.
6. List five of the major computer modules used on light vehicles.
7. List and explain the purpose of the powertrain's major components.

8. Explain the primary operational difference(s) between a manual transmission and an automatic transmission.
9. Discuss the operation of a brake system.
10. Explain the overall operation of an internal combustion engine.

Fill-in-the-Blanks

1. The largest single vehicle computer is the _____.

2. The charging system uses a belt-driven _____ to produce electrical current to operate the vehicle.

SHOP MANUAL

To stress the importance of safe work habits, the Shop Manual also dedicates one full chapter to safety. Other important features of this manual include:

Performance-Based Objectives

These objectives define the contents of the chapter and define what the student should have learned on completion of the chapter.

Terms to Know List

Terms in this list are also defined in the Glossary at the end of the manual.

Margin Notes

The most important terms to know are highlighted and defined in the margin. Common trade jargon also appears in the margins and gives some of the common terms used for components. This feature helps students understand and speak the language of the trade, especially when conversing with an experienced technician.

Basic Tools Lists

Each chapter begins with a list of the basic tools needed to perform the tasks included in the chapter.

Special Tools Lists

Whenever a special tool is required to complete a task, it is listed in the margin next to the procedure.

Author's Note

This feature includes simple explanations, stories, or examples of complex topics. These are included to help students understand difficult concepts.

Photo Sequences

Many procedures are illustrated in detailed Photo Sequences. These photographs show the students what to expect when they perform particular procedures. They also familiarize students with a system or type of equipment that the school might not have.

References to the Classroom Manual

References to the appropriate page in the Classroom Manual appear whenever necessary. Although the chapters of the two manuals are synchronized, material covered in other chapters of the Classroom Manual may be fundamental to the topic discussed in the Shop Manual.

Service Tips

Whenever a shortcut or special procedure is appropriate, it is described in the text. Generally, these tips describe common procedures used by experienced technicians.

Cautions and Warnings

Cautions appear throughout the text to alert the reader to potentially hazardous materials or unsafe conditions. Warnings advise the student of things that can go wrong if instructions are not followed or if an incorrect part or tool is used.

PHOTO SEQUENCE 1
Using a Dry Chemical Fire Extinguisher

Automotive shop and safety procedures   33

P1-1 Multipurpose dry chemical fire extinguisher.

P1-2 Hold the fire extinguisher in an upright position.

P1-3 Pull the safety pin from the handle.

P1-4 Stand 8 to 10 feet upwind of the fire. Do not go any closer to the fire.

P1-5 Free the hose from its retainer and aim it at the base of the fire.

P1-6 Squeeze the lever while sweeping the hose from side to side.

---

102   Chapter 4

**Classroom Manual
page 94**

makes it easier and quicker to use. Warpage or parallelism between the two opposing friction areas can also be checked with a micrometer or a disc parallel gauge.

SERVICE TIP In the last few years, brake rotors have an additional minimum specification: *minimum refinishing thickness*. If the rotor is at that thickness or thinner, then the rotor is replaced. The minimum refinishing limit is usually greater than the minimum or discard thickness. The refinishing specification is found in the service manual and usually not stamped on the rotor.

Brake disc thickness and parallelism can be checked on the car, a table, or a brake lathe. Find the minimum thickness specification for the disc. Many times, it is stamped on the inside or outside of the hub portion of the disc and most are listed as metric measurements (**Figure 4-25**). If the brake disc micrometer is USC, convert the specification to USC.

Place the micrometer's measuring faces on each side of the disc midway to the wear area (**Figure 4-26**). Rotate the thimble until the spindle and anvil faces are against the disc. Read and record the measurement. In almost all cases, the measurement will be less than 1 inch. Repeat the measurement at a minimum of 12 different points around the disc (**Figure 4-27**). Remember to measure at the midpoint of the friction areas at each of the measuring points. Compare your readings to the specified minimum thickness. Any measurement that is less than specified means the disc is not to be reused and should be discarded. Differences over 0.002 inch between any of

---

222   Chapter 6

In many instances, it will be necessary to have electric power to test components located in or near the airbag areas. There are two ways to do this safely:

1. Use an ohmmeter (a digital multimeter or DMM) to measure resistance of the component and compare it to manufacturer specifications after it has been disarmed and removed from the vehicle.
2. Remove the component(s) from the vehicle and test it at a test bench equipped with the proper power supply and tools.

Special Tools
Service manual
Vehicle's diagnostic routine

WARNING Do not reconnect the battery again until all the service work is completed, including reassembly of components and their mounting. A quick touch of the cable to the battery will instantly recharge the capacitor. This would supply the power needed to fire the airbag, and severe damage and injury could result.

WARNING LIGHTS

Bulb check is a program where certain dash indicator lights will illuminate for only a few seconds so that the technician can check their operation.

KOEO means ignition key on, engine off, a test condition required to performed certain tests.

KOER means ignition key on, engine running, another test condition required for certain tests.

Many of the warning lights are simple, electrical circuits. A good understanding of their operation will give the technician a head start on basic repairs. Most tests must be run when the key is on with engine off (KOEO) or key on with the engine running (KOER). You may have to turn the key on and back on several times to read and understand all the lights. A light that goes on for only a few seconds and then goes off does so because of the lights. A light that goes on for only a few seconds. This is done so that the technician (or driver) can check if the bulb is working or not. If the oil pressure warning light bulb is burnt out, then the driver would never know when or if there is a problem with the oil pressure.

KOEO and KOER running are two test conditions associated with powertrain control module (PCM) test routines, but they appear in other system test procedures as well.

Older vehicles use a mechanical switch to turn the light on when certain conditions are present. Other vehicles use a sending unit for the warning light and gauge. One basic function occurs on both types: the ground to the warning light is completed.

Caution
The following tests are for nonelectronic circuits only. Sensor circuits can be determined by observing the number of wires attached to the sensor or sending unit. True sending units will have only one wire, whereas sensors will have two or more wires. Grounding a sensor circuit could damage the computer or the circuit.

WARNING Do not grab hoses or other components under the hood if an overheated engine is suspected. Even an engine operating at normal temperature can cause serious burns.

WARNING Never remove the radiator cap on a warm or hot engine. Always let the engine cool down. Extremely serious burns and other injuries can occur when removing a hot radiator cap.

If the red engine light is lit, the problem indicated is either an overheated engine or low oil pressure. On some vehicles, there is a light for oil and one for overheating (**Figure 6-1**). Before starting work on the electrical circuit, do the obvious and check the level and condition of the coolant and oil. For a cooling problem, look for missing or damaged fan belts. The coolant level on a hot engine can be checked in the overflow reservoir. The reservoir may not contain a true representation of the coolant level, but it will be an indication. Note that, on some cooling systems, the reservoir may become pressurized. Notice the condition of the engine by checking for oil or coolant leaks. The hoses should be inspected for swelling. Do not touch hot hoses. In many cases of low oil pressure or overheating, the condition will be found during the inspection.

Case Studies

Each chapter ends with a Case Study describing a particular vehicle problem and the logical steps a technician might use to solve the problem. These studies focus on system diagnosis skills and help student gain familiarity with the process.

ASE–Style Review Questions

Each chapter contains ASE-style review questions that reflect the performance objectives listed at the beginning of the chapter. These questions can be used to review the chapter as well as to prepare for the ASE certification exam.

Customer Care

This feature highlights those little things a technician can do or say to enhance customer relations.

ASE Challenge Questions

Each technical chapter ends with five ASE challenge questions. These are not more review questions; rather, they test the students' ability to apply general knowledge to the contents of the chapter.

Accessory Lighting and Electrical Systems Inspection, Maintenance, and Testing   241

CASE STUDY

This is a problem that was national in scope, but it only concerned one brand of vehicle and only for a model year or two. The customers complained that the vent air was not cool when the air conditioning was on in MAX or NORMAL. The heater worked fine. This system used stepper motors on the blend doors. Routine electrical tests did not locate the problem. The doors could be heard moving back and forth as the controls were moved. Studying the service manual's system operation section revealed a possible cause. When the ignition key was moved to ON, the A/C computer ran a test on its circuits. Part of this test was the search for a 35-mA load on each of the motors as they moved the doors through their entire range of movement. A DMM set to amps and connected between the heat blend door motor and the computer showed 35 mA at full cold and at full hot. Supposedly the system was functioning correctly. Further physical investigation found the problem on top of the air box. The arm from the motor was held in place to the door lever with a small plastic clip. The clip had broken and was hanging down. As the motor moved the door toward the cold position, the clip would hook a ridge of the air box. The motor would try to push the door further, but the 35-mA load was reached, and the computer assumed this was the limit of door travel and stopped the motor. The end result was the door never reached its cold position even though the ampere reading was correct. Replacing the clip with a better connection solved the problem.

ASE-STYLE REVIEW QUESTIONS

1. Keyless entry system is being tested.
Technician A says that all doors and windows should always be closed when test the keyless remote.
Technician B says the driver's window should be down when testing the keyless remote.
Who is correct?
A. A only   C. Both A and B
B. B only   D. Neither A nor B

2. Lighting systems are being discussed.
Technician A says inoperative marking lights may be caused by an open circuit breaker.
Technician B says testing a lighting system without relays can normally be done with a test light.
Who is correct?
A. A only   C. Both A and B
B. B only   D. Neither A nor B

3. A customer complains that the headlights turn on and off, generally when traveling on a rough road.
Technician A says the circuit breaker may be weak or failing.
Technician B says a loose ground on the relay may be the fault.
Who is correct?
A. A only   C. Both A and B

4. A power window is stuck about halfway up and will not move up or down. There is a "clunk" each time the switch is operated.
Technician A says the drive gear is stripped.
Technician B says the window motor switch is shorting to ground.
Who is correct?
A. A only   C. Both A and B
B. B only   D. Neither A nor B

5. Instrument gauge diagnosis is being discussed.
Technician A says a test light can be used to test digital gauge systems.
Technician B says supplying 12 VDC to a gauge will cause it to read full.
Who is correct?
A. A only   C. Both A and B
B. B only   D. Neither A nor B

6. Warning light systems are being discussed.
Technician A says grounding an oil pressure warning light wire at the sending unit will cause the gauge to read low.
Technician B says grounding the engine temperature sending unit will cause the warning light to come on and the gauge to read hot.
Who is correct?
A. A only   C. Both A and B

299

dipstick hole.

CUSTOMER CARE Although a cylinder leakage test can identify the primary cause of low compression readings, remember to keep your mind and eyes open to other problems when repairing the engine. Do not automatically inform a customer that an exhaust valve is leaking and a valve job will cure the problem. Consider the engine mileage, oil condition, pressure and usage, and degree of leakage past the rings. If the engine has high miles, it is very likely that the rings are significantly worn. By replacing the faulty valve(s) and restoring the others to like-new condition, you will increase the compression and combustion pressures. The change may be significant enough that the old rings will not seal as well as they had been before the valve job. In this case, the customer would be back complaining about blue smoke from oil consumption or about the engine still lacking adequate power. This may very well be an example of when a replacement engine would be the most cost-effective repair for the customer. Evaluate the whole situation, and make your recommendation to the customer. If you offer the customer repair options, be very clear about explaining the possible consequences.

ASE CHALLENGE QUESTIONS

1. Tires are being discussed.
Technician A says that the contact patch will change due to different tire inflation.
Technician B says the contact patch is where the tire supports the weight of the vehicle.
Who is correct?
A. A only   C. Both A and B
B. B only   D. Neither A nor B

2. Tire pressure is being discussed.
Technician A says that the proper tire pressure for each tire is on the side of the tire.
Technician B says that the proper tire pressure can be found on the tire placard located on the drivers door frame.
Who is correct?
A. A only   C. Both A and B
B. B only   D. Neither A nor B

3. Tire pressure monitor system is being discussed.
Technician A says that the indirect system can identify the tire that is under inflated.
Technician B says the direct system needs to be relearned after a tire rotation.
Who is correct?
A. A only   C. Both A and B
B. B only   D. Neither A nor B

4. Two technicians are discussing coil springs.
Technician A says that when the spring compresses it is called a jounce.
Technician B says that when the spring extends it is called a rebound.
Who is correct?
A. A only   C. Both A and B
B. B only   D. Neither A nor B

5. Suspension systems are being discussed.
Technician A says that a dependent suspension is mainly used for light duty application.
Technician B says that independent suspension can mainly be found on heavy duty trucks.
Who is correct?
A. A only   C. Both A and B
B. B only   D. Neither A nor B

Engine Inspection and Preventive Maintenance   303

Name _____   Date _____

JOB SHEET
27

PERFORMING AN ENGINE OIL AND FILTER CHANGE

Upon completion and review of this job sheet, you should be able to perform a routine engine oil and filter change service.

ASE Education Foundation Correlation

This job sheet addresses the following **MLR** tasks:

I.A.3   Inspect engine assembly for fuel, oil, coolant, and other leaks; determine necessary action. **(P-1)**

I.C.5   Perform engine oil and filter change; use proper fluid type per manufacturer specification; reset maintenance reminder as required. **(P-1)**

This job sheet addresses the following **AST/MAST** tasks:

I.A.4   Inspect engine assembly for fuel, oil, coolant, and other leaks; determine needed action. **(P-1)**

I.D.10   Perform engine oil and filter change; use proper fluid type per manufacturer specification. **(P-1)**

We Support
ASE | Education Foundation

Tools and Materials

Drain pan
Filter wrenches
Box-end wrenches
Service manual

Protective Clothing

Goggles or safety glasses with side shields

Describe the vehicle being worked on:

Year _____ Make _____ Model _____

VIN# _____ Engine type and size _____

Procedure   **Task Completed**

1. Move the vehicle to the lift. Record any noise or obvious defects found during the movement. _____
    _____
    _____
    _____

2. Engine Identification:
    Displacement _____ Block _____ Valvetrain _____

3. Determine the type and amount of oil and the filter:
    Oil _____   Amount _____
    Filter/customer preferred brand

4. Raise the vehicle and place the drain pan under the oil pan plug.   ☐

5. Remove the plug and drain the oil. Note the condition of the oil.
    _____

6. Check the plug gasket and replace if necessary.   ☐

Job Sheets

Located at the end of each chapter, the Job Sheets provide a format for students to perform procedures covered in the chapter. A reference to the ASE Education Foundation Tasks addressed by the procedure is referenced on the Job Sheet.

SUPPLEMENTS

Instructor Resources

The *Today's Technician* series offers a robust set of instructor resources, available online at Cengage's Instructor Resource Center (http://login.cengage.com) and on DVD. The following tools have been provided to meet any instructor's classroom preparation needs:

- Lesson Plans provide lecture outlines, teaching tips, and complete answers to end-of-chapter questions.
- PowerPoint presentations include images, videos, and animations that coincide with each chapter's content coverage.
- Cengage Learning Testing Powered by Cognero® delivers hundreds of test questions in a flexible, online system. You can choose to author, edit, and manage test bank content from multiple Cengage Learning solutions and deliver tests from your LMS, or you can simply download editable Word documents from the DVD or Instructor Resource Center.
- An Image Gallery includes photos and illustrations from the text.
- The Job Sheets from the Shop Manual are provided in Word format.
- End-of-Chapter Review Questions are also provided in Word format, with a separate set of text rejoinders available for instructors' reference.
- To complete this powerful suite of planning tools, a pair of correlation guides map this edition's content to the ASE Education Foundation tasks and to the previous edition.

MindTap for Today's Technician: Basic Automotive Service and Systems, 6e

MindTap is a personalized teaching experience with relevant assignments that guide students to analyze, apply, and improve thinking, allowing you to measure skills and outcomes with ease.

- Personalized Teaching: Becomes yours with a Learning Path built with key student objectives. Control what students see and when they see it. Use it as-is or match to your syllabus exactly—hide, rearrange, add, and create your own content.
- Guide Students: A unique learning path of relevant readings, multimedia, and activities that move students up the learning taxonomy from basic knowledge and comprehension to analysis and application.
- Promote Better Outcomes: Empower instructors and motivate students with analytics and reports that provide a snapshot of class progress, time in course, engagement, and completion rates.

REVIEWERS

The author and publisher would like to extend a special thanks to the instructors who reviewed this text and offered invaluable feedback:

Rodney Batch
University of Northwestern Ohio
Lima, OH

Matthew Bockenfeld
Vatterott College, Joplin Campus
Joplin, MO

Shih-Wei "Jack" Chen
Utah State University Eastern
Price, UT

Cory Peck
Vatterott College St.
Louis, MO

Todd Richardson
Utah State University Eastern
Price, UT

Steve Roessner
University of Northwestern Ohio
Lima, OH

Joshua Summey
Richmond High School / Ivy Tech
Richmond, IN

CHAPTER 1

WORKING IN THE AUTOMOTIVE INDUSTRY

Upon completion and review of this chapter, you should understand and be able to:

- Describe the basic duties of key shop personnel.
- Understand the technician's technical responsibilities.
- Describe how technicians are paid for the work they perform.
- Explain the shop's and technician's legal concerns.
- Describe the differences between warranties and recalls.
- Perform the process for completion of a service repair order.

Terms To Know

Accountant

Commission

Entry-level technician

Flat rate

Hourly rate

Labor guide

Labor time

Manager

Recalls

Repair order

Service manager

Service writer

Technical Service Bulletin (TSB)

The 3 Cs

Vehicle identification number (VIN)

Warranties

INTRODUCTION

A shop employs personnel in several key positions. As stated in the Classroom Manual, the service manager is responsible for the service department. To make the operation run smoothly, he or she hires service writers, accountants, cashiers, parts managers, personnel, and technicians. We will discuss shop operations that would be typical of dealerships and large independents.

MANAGERS AND ADMINISTRATIVE OFFICES

There are several levels of management in the larger shops, especially at dealerships. In smaller shops, the owner may be the only manager and may also cover billing and writing service. The **owner** or **manager** may help by performing repairs when necessary. In this section, the operation of a typical local dealership will be used to outline the common positions and responsibilities of each. A typical manager usually has training and experience in business operations, whereas persons in other positions have the knowledge for their particular job such as the accountant's knowledge of debits and credits. Each person is assigned a specific job, including technicians, but there are times, particularly in a smaller shop, when the assignments may have to overlap to satisfy a sudden short-term increase in customers.

The **owner** or **manager** is responsible for the daily operation of a segment of a business and reports to the general manager.

Without a concerted effort by all members of an organization, a customer may buy a vehicle or parts from one business, then go elsewhere to have it installed or maintained. This reduces profit and essentially loses a customer.

The managers of each department must work together to ensure the customer receives first-class service from any employee they may contact. Some dealerships introduce their new-car customers to the service department within a short time of purchase. The customer's perception of receiving individual, personalized treatment boosts the prestige of the company and its employees. At many dealerships' service departments, you are greeted as soon as you get out of the car. Some dealerships even have the customers drive their vehicle into the service department so that they are not outside in the weather.

Independents, on the other hand, are usually smaller and locally owned operations. Most of the employees and the owner are from the local area and know their customers firsthand. Many times, independents will work with local dealerships to handle each other's work and capabilities. For example, a local independent may specialize in transmission repair. A dealership that has an older vehicle or a trade-in vehicle with a bad transmission may subcontract some of the repair of the transmission after it has been removed, and vice versa: The independent may send a customer to the dealership when the computer needs reprogramming, which can be an expensive and not profitable repair for an independent. Both operations profit from this arrangement in monetary terms and customer trust.

Service Departments

The service writer is usually the first person to greet the customer and the **accountant** is the last. Both are essential to the shop's operation.

Usually, a **service manager**, who is responsible for the **service writer**, technicians, department support staff, and parts, heads the service department. The service manager works with vehicle sales in a dealership to bring new (and preowned) car customers back for service when needed (**Figure 1-1**). The parts department has a manager who reports to the service manager. In an independent shop, the number of people involved in these types of duties and responsibilities is often much less than that of a dealership.

Service Writers and Accountants

The service writer greets the customer and initiates the **repair order** (**Figure 1-2**). The service writer must be able to communicate with a wide variety of customers and must understand enough of the vehicle's operating systems to interpret the customer's

> **Accountant** A person whose job is to keep track or inspect financial account records.

> The **service manager** controls the operation of the service department, including parts, technicians, support staff, and the training of all department employees.

> The **service writer** meets the customer, writes the repair order, and assigns a technician based on company policy.

> The **repair order** is a document that tracks the vehicle from its arrival at the shop for repairs to the time the repairs are paid by the customer. This is considered a legal document. This document helps protect the customer and the business from any problems that may arise during or after the repairs.

GENERAL MANAGERIAL DUTIES

General Manager	Sales Manager	Service Manager
Reports to owner	Reports to general manager	Reports to general manager
Delegates authority	Supervises sales staff	Supervises service staff
Supervises department managers		
General budget reports	Sales budget reports	Service budget reports
Hires, fires, and trains managers and general staff personnel	Hires, fires, and trains sales and staff personnel	Hires, fires, and trains parts personnel, staff, and technicians
Supervises interactions between department	Coordinates with general manager and service to gain and keep customers for the business	Coordinates with general manager and sales to gain and keep customers for the business

Figure 1-1 Managers share some common duties, so the various departments can work together to build and maintain the business.

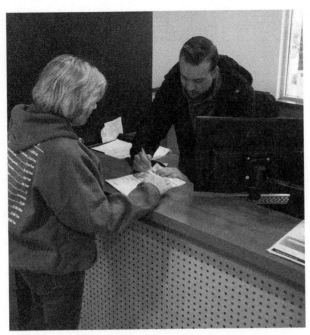

Figure 1-2 The service writer initiates the repair order.

complaint. The diagnosis goes much quicker if the complaint is clear and concise on the repair order. The service writer may create an estimate of the repair. In many instances, the customer only conducts business with the service writer and the accountant.

Classroom Manual page 1

The service writer usually assigns the repair order to the first available technician experienced in that repair. However, an automatic transmission service, such as a filter change, may not be assigned to the "transmission" tech but to the entry-level or maintenance light repair (MLR) tech. This allows the higher-paid technician to perform high-value jobs while the hourly person performs the low-profit, quick-turnaround jobs. This results in an efficient use of personnel while providing excellent training for the less experienced tech.

Express service is a term used to emphasize the speed of basic and regularly schedule maintenance. As mentioned in the Classroom Manual, express service gives the customer a quick turnaround and a detailed inspection of the vehicle. To speed things up more, some dealerships have service attendants. These are people who assist the service writer. The attendant has several duties. One is greeting the customer if the writer is busy. One protocol is to never let the customer stand around waiting to be noticed. A quick greeting and "How may we help you?" immediately upon the customer's arrival indicate that person is recognized and welcomed. A second duty consists of placing disposable floor mats and seat covers in the car and retrieving the odometer mileage and the **vehicle identification number (VIN)**. The third duty, but not necessarily the last, is to begin the vehicle inspection. Dealership service attendants check all lighting, tire wear and air pressure, parking brakes, horn, and warning lights, and perform a walk-around visual check for damage to the vehicle body. The attendant starts the inspection sheet and marks the result of this first phase. This is done while the service order is being prepared and will save time in the express service bay. In addition, any defects found can now be shown immediately to the customer by the writer. This express service job may become a bigger repair job for items such as tires or a lighting problem. Part of this third duty is to remove the ignition key from the customer's key ring and return all other keys to the customers. Explain, if questioned, that the shop can replace a lost ignition key easily, but

The **vehicle identification number (VIN)** is a number assigned to a specific vehicle. Information on the engine size and type, transmission, year of production, and other vehicle data is encoded in the VIN.

Figure 1-3 Good customer relations is important for all who work in the automotive industry.

Figure 1-4 The parts department stocks parts and materials required on a routine basis.

personal ones will present a problem—for example, the customer would be locked out of his or her home. This shows that the business takes care of the customer. With that said, the single most important duty is still greeting the customer and making that customer feel appreciated.

The accountant may also act as the cashier for the service department (**Figure 1-3**). The customer will pick up his or her vehicle keys and pay the repair cost at the cashier's window. However, the accountant does much more. The information on the customer, the vehicle, the repair accomplished, and so on is usually collected and entered into a computerized database. The information is used to remind customers of scheduled maintenance and may be used to resolve a customer's complaint. In larger shops, there may be several individuals working in a service department's accounting office.

Parts

The parts manager usually reports to the service manager. The parts department is responsible for ensuring that enough routine maintenance items are present (**Figure 1-4**). Also, parts personnel must order, deliver, pick up, and enter parts on the repair order. The department operates a pickup window for walk-in customers. A running inventory is always kept to prevent shortages of routine items. In addition, dealerships are linked by telephone and satellite communications to other dealerships. In this manner, it may be possible to get an out-of-stock part from a dealer on the other side of town instead of waiting several days to get one from the manufacturer's warehouse. The parts are charged against the repair order or a separate invoice, which is sent to the accountant for payment collection.

TECHNICIAN TECHNICAL RESPONSIBILITIES AND PAY

Classroom Manual
page 6

The technician performs the basic profit-making portion of the shop's operations: vehicle repairs. The technician may also have the most personal money invested in the position. Most shops will not hire a technician who does not have personal tools (**Figure 1-5**). A basic tool set may cost more than a thousand dollars, and it is not unusual for an experienced technician to have more than $20,000 worth of tools collected through his or her career. Remember that a quality tool set is a good way to make sure that you can make money. Good tools don't break and get the job done faster and with high quality. This is the recipe for success and a good paycheck.

Figure 1-5 This is a typical tool set for an entry-level or MLR technician. An experienced technician's toolbox is far more robust than a student's tool set.

The technician must follow all regulations concerning automotive environmental hazards and safety. The repair should be performed correctly the first time. The best customer relations possible are quick, fair, and accurate repairs to the vehicle.

Entry-level technicians are hired to perform basic services such as oil and filter changes, tires, brakes, and other routine maintenance (**Figure 1-6**). As technicians gain experience, they gain more and more responsibilities. To stay abreast of vehicle operations, it is necessary for the technician to attend training at the manufacturer school, vendor-sponsored training, or a technical school.

Technicians are paid through several methods. The most common are hourly, **commission,** and **flat rate.** Hourly is usually reserved for entry-level technicians and is paid by the clock hours that the technician is at the shop, regardless of the amount of repair business. This will give the person a certain pay that is consistent week to week and provides time to become proficient at the job. Technicians drawing commission will usually be guaranteed a minimum hourly pay plus a percentage of the shop's charges for labor and parts. In most cases, although the pay can be excellent, the technician is only considered to have worked specific hours per week.

Entry-level technicians are usually recent graduates or may still be attending school. This is their first official job as automotive technicians. An entry-level technician may have different titles when he or she is employed. Such titles often are lube technician, MLR technician, express service technician, or general service technician.

Commission pay is when the technician receives a percentage of the labor and part sales as part or all of his or her pay.

Flat rate is the time it takes to perform a job. It is an average based on research and practice. Some technicians are paid by flat rate.

Figure 1-6 Although this is the type of entry-level task performed by new technicians, it is only the first step to a successful career.

A **labor guide** is a book or guide that the service writer uses to estimate the amount of time a repair will take. This rate is then used to calculate the amount that the customer is charged.

Flat rate is a means to motivate technicians' speed and proficiency by paying them a per-hour rate based on a **labor guide.** A labor guide is a publication, paper or computerized, that shows the amount of time an average technician will take to complete the repair. Another name for the per-hour rate found in the labor guide is flat rate. A trained, motivated technician can beat the time allowed in many instances. The guide may state it takes 2 hours to complete a repair, but the trained and experienced technician only takes 1 hour, thus earning 2 hours of pay. Also, it allows an additional clock hour that day to do another job. It is not unusual for an experienced technician to turn in 60 to 80 labor hours for a 40-clock-hour week. However, if a "comeback" (shop warranty) repair comes in, the technician isn't paid for that repair. The lack of pay for a comeback provides an excellent incentive to fix the vehicle right the first time. Many shops offer the best technicians a guaranteed weekly check even when business is slow. This also applies to those working on commission.

AUTHOR'S NOTE Flat rate and commission pay also provide an incentive for managing money. There are several times during the year, such as the weeks between Thanksgiving and Christmas and right after the school session ends, when repair business drops drastically. Recall, warranty, and preventive services will still be performed, but many customers will be out buying gifts or going on vacation. They tend to put off any repairs they deem not immediately necessary. Then there are times, such as tax refund season, when business is booming. There are also times when the weather changes drastically, and business is busy from the drop in temperature. A sudden drop in temperature can bring in vehicles with dead batteries, flooded engines, flat tires, and more. Auto collision repair facilities are busiest just after a weather storm passes through. A technician who is on commission or working flat rate quickly learns to put aside a little pay to cover those times when work can be seasonally slower than normal.

SHOP AND TECHNICIAN LEGAL CONCERNS

Regardless of the bad publicity about crooked shops and technicians sometimes seen on television, probably at least 99 percent of repair shops are honest and work hard to satisfy all customers. But like all of life, sometimes mistakes are made. Mistakes in automotive repair can range from a simple broken wire to mistakes in brake or wheel repairs that could be life threatening. Sometimes a vehicle inspection reveals a very unsafe condition that the customer, for whatever reason, does not want the shop to fix. This could be due to lack of funds on the customer's part. This places the shop and the technician in a vulnerable position. The problem vehicle could be unsafe when operated. Some courts have held the shop responsible for damage and injury because the shop and its employees are considered professionals in their jobs and should have known not to release the vehicle for use on public roads. This problem should be handled by the shop's management and the customer; it is not a decision for the technician to make. Fortunately, this would only happen in an extreme case. However, the shop and the technician do have some legal obligations in this area. State and local laws dictate those obligations, and moral guidance should do the rest.

With regard to the situation noted in the previous paragraph, the author is not a legal expert, but it is recommended that the shop and technician do at least one thing: Write a detailed comment concerning the safety situation on the repair order and have the customer sign it in acknowledgement of the problem, the shop's position, and the customer's refusal to allow the repair work. This may not have any bearing in a trial situation, but it will at least show the shop owner's state of mind at the time of disagreement. There are few cases overall that require court or arbitration actions, but it is the customer's vehicle, and he or she is paying the bill and has a right to expect good work. Many shops will voluntarily lose money for repairs on warranty work, even if the complaint may be unfair, to gain good public relations and the good word-of-mouth advertising they receive from a satisfied customer. Experienced service managers know how to gain customer confidence and prevent or minimize incidents from occurring in the first place. With time and a good mentor, you will learn techniques to deal with, and hopefully avoid, many of the difficult situations.

Environmental Protection

Most laws and regulations on environmental protection are set by the local, state, and federal statutes. Hazardous materials and waste are controlled or regulated by federal environmental laws and supplemented by lower levels of government. The shop must also consider damaged parts, such as vehicle computers, as hazardous waste because of the chemicals contained in the circuit boards. Most computers and many other damaged parts are recycled or rebuilt by domestic and foreign companies. At a minimum, removed parts should not just be thrown in the trash bin. They should be sent to a scrap yard for recycling.

Repairs

Both the shop and its employees at all levels should strive to repair the vehicle right the first time and charge a fair price. If a mistake should happen, it should be repaired at the shop's expense. Charging to fix something you broke is not a good business tactic. This could cause lawsuits, legal costs, and generally poor public relations. Prime examples of absorbing the cost of a mistake are recalls issued by a vehicle manufacturer. A mistake was made either by the engineering and design team or by the part manufacturer. Although it may seem to take a long time to get a government-mandated recall in place, most manufacturers issue recalls without outside pressure.

At the shop level, local arbitration boards or committees consisting of members from outside the industry review and settle disputes between a shop and an unhappy customer. The board's decision is usually binding on the shop, but the customer has the right to appeal the decision or even go to court for a settlement. Most states have a "lemon" law stating that after a certain number or types of repairs are unsuccessful on a given vehicle within a reasonable time frame, the shop or the manufacturer is required to properly fix the vehicle or replace the vehicle at minimum cost to the owner. In almost all cases, the lemon law applies to the business seller of the vehicle. Violations of this law are not usually considered a criminal act unless there is compelling evidence of intentional fraud. Over the years, this law has been applied to new and used vehicle sales.

WARRANTIES AND RECALLS

Warranties A written guarantee, issued to the purchaser a repair by the business, promising to repair or replace if necessary within a specified period of time.

Warranties are offered to the customer either as part of the business or at cost, like life insurance. Warranties are issued by the vehicle manufacturer for a specified time and mileage. The one that seems best is 10 years or 100,000 miles from the date of sale. Based on the driver's lifestyle and commuting distance, the mileage limit may expire before the time limit. In large cities and suburbs where the job may be up to 70 miles from home, 100,000 miles is reached in 2 or 3 years or less. Compared to the old days when a warranty on a new vehicle was 12/12 (12 months or 12,000 miles), modern-day vehicle warranties are generous. Many warranties are bumper-to-bumper for 36 months or 36,000 miles, so that any repair is covered. Bumper-to-bumper means every component between the front and rear bumpers—other than routine maintenance such as oil changes, belts, and brakes—is fixed free. Many new car sales, shops, and used vehicle sales offer some type of extended warranty that covers specific repairs for a given time or mileage period. Most extended warranties require the customer to pay the warranty premium and a fixed fee or deductible for the repair. Most repair shops guarantee or warrant their labor and parts for a fixed time or mileage. This has led to a change in the business practice of many shops. At one time a repair shop would overhaul or rebuild an engine or transmission. However, with the growth of Jasper Engine and Transmission Rebuilders and similar operations, it is more profitable for the shop to replace the engine or transmission with a rebuilt unit. If the component fails within its warranty period, the vendor replaces the component free of cost and usually reimburses the shop for some, if not all, of the labor costs. This allows the shop to offer excellent guarantees without having to absorb the full cost of an expensive replacement. In turn, it frees that bay and technician quickly for other work.

Recalls A safety-related defect that does not comply with federal safety standards. The automaker must alert and repair the problem at no charge to the vehicle owner.

Recalls are issued only by the vehicle manufacturer, are usually safety or emission related, and apply to any new vehicle sold in the United States and Canada. However, the manufacturer usually attempts to locate any recalled vehicle worldwide. At times, the manufacturer is forced to issue a recall by the National Safety and Transportation Board (NSTB), a federal agency. Other times, a recall may be issued when a problem is found through recurring defects or by the engineering or quality control sections of the manufacturer. Recalls may be necessary on a certain batch of vehicles because of an error at the parts manufacturer or at the vehicle assembly plant. A single recall usually only concerns a certain brand, year, and model of vehicle. Recalls can also be to update the software program of a vehicle computer. Recall repairs are performed by manufacturer-certified technicians, generally at a dealership. To accomplish a recall, the manufacturer must notify the recorded owner of the vehicle. If the recall is for vehicles that are several years old, the manufacturer's records may not be current. The person receiving the mailed notice is supposed to either take the vehicle in for repairs or complete the recall form with respect to the present owner of the vehicle. In many cases, the vehicle was traded for another or sold, and the present owner is unknown to the original owner. The notified owner is expected to provide as much information as possible on the vehicle and return the card to the manufacturer. Unfortunately, if the vehicle is no longer owned by the original individual, the card often goes in the trash, and thus the recall is not completed. The manufacturer is protected to some extent provided it made a reasonable effort to locate and fix the vehicle.

Recall listings may be found at the dealerships, on paper manuals, or on the Internet site of the NSTB: http://www.nstb.gov/cars/problems.

SERVICE REPAIR ORDERS

A completed repair order lists the customer's complaint, the diagnosis, repair parts, labor, and costs of each. Some may include a history of the most recent services performed on the vehicle (**Figure 1-7**). Although the individual form may differ from shop to shop, the same information is normally used.

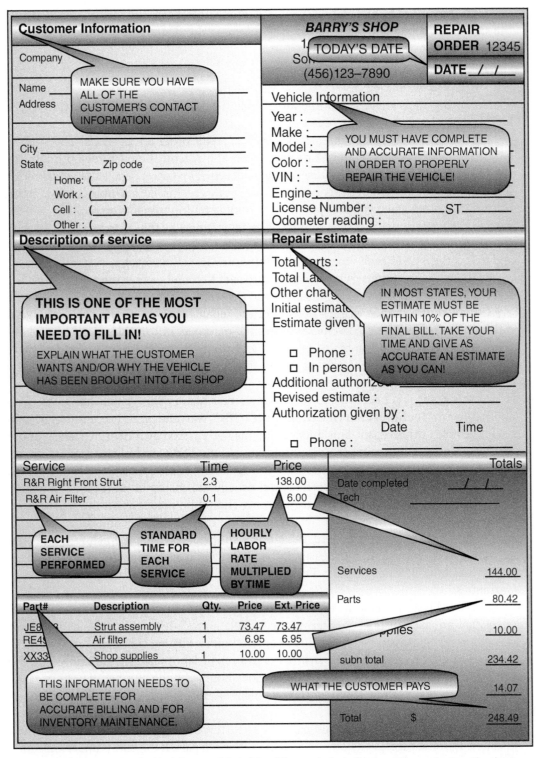

Figure 1-7 A service repair order blank form used for training. It is very similar to those used by most automotive shops.

Upon receipt of the repair order, the technician retrieves the vehicle and performs the diagnosis. Assuming that parts are available, the technician performs the repairs and conducts a test drive as needed. Parts personnel charge the parts to the repair order. The technician includes the **labor time** and other information about the repairs required by the shop. This may include the method or type of testing performed and the diagnosis.

The repair order is routed to the accountant, who computes the bill based on information supplied by the technician, parts, and shop policy. Warranty repairs do not cost the customer, but the same information is entered on the order and the customer receives a copy. The repair order information is entered into the database, and the order is filed after the customer pays. The manufacturer reimburses the dealership for warranty recall repairs.

Completing a Repair Order

The following steps are based on the form in Figure 1-7. Most of the information will be included on almost any shop's repair order.

One of the most important parts of a service repair order are the **3 Cs.** The 3 Cs are concern, cause, and correction. The concern is the reason the customer brought the vehicle in. The cause is the diagnosis of the concern. The correction is the actual repair that was made as a result of the concern. Although this seems like a simple concept, it can be easily forgotten when you are busy and performing multiple repairs on multiple vehicles each day. Always remember that your job is to determine why the customer brought their vehicle to you, what it will take to bring the vehicle back to correct operating conditions, and to link the 3 Cs to each other. Another important part is having the customer sign the repair order to allow the services to be performed. Remember that the repair order is a legal document that protects the business and the customer.

Complete the service writer's portion of the repair order by entering the date, VIN, mileage, make and type of vehicle, customer personal information, and complaint (**Figure 1-8**). Enter the complaint as clearly and concisely as possible. Be sure to secure the customer's signature authorizing the repairs. If requested, provide an estimate of the cost and time of completion to the customer.

At this point it is necessary to record any repair history this vehicle had in the past and locate any **Technical Service Bulletins (TSBs),** warranties, and recalls that pertain to this vehicle or this repair. This information may be entered by the service writer or the technician after confirming the customer complaint. TSBs are bulletins issued by the manufacturer based on problems found on the same model vehicle. Repairs performed under TSBs may be covered by warranty but usually the customer must pay. The title or code listings for any recalls, TSBs, or warranty should be recorded on the repair order before the actual repair is started. This information may give the technician a head start on the diagnosis. During the actual diagnosis and repairs, each step should be listed according to shop rules. In most cases, if the technician does not list the diagnostic steps taken, that time will not be paid. This is particularly true when dealing with recalls or warranties.

Usually, the shop has some type of computer software or manual that has numerical codes that are used instead of entering the complete procedures. Most dealerships have manufacturer-approved codes for this purpose. Before turning the repair order over to the accountant or service writer, ensure that *all* repair labor and parts are noted on the repair order. The number of labor, guide, or clock hours and the number of parts, such as eight spark plugs, must be listed for the technician to receive pay and the shop to collect the final cost from the customer.

Many shops offer the removed part to the customer as "proof" that a new part was installed. In some cases, a customer will demand the old part. If the part is under warranty, the dealership or parts vendor may hold it in an area of the shop until a representative can

labor time Time devoted to the work to complete a repair. A measure of value for a given effort.

The **3 Cs** of automotive repair are concern, cause, and correction. The correction (actual repair the technician does) needs to match the concern (the reason it was brought in for service in the first place).

Technical Service Bulletins (TSBs) Recommended procedures for an unanticipated problem that the automaker has had several occurrences. They are a form of diagnosis.

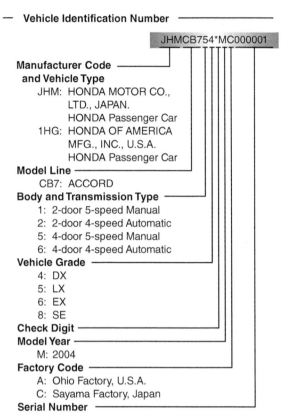

—— **Vehicle Identification Number** ——

JHMCB754*MC000001

Manufacturer Code and Vehicle Type
    JHM: HONDA MOTOR CO., LTD., JAPAN.
            HONDA Passenger Car
    1HG: HONDA OF AMERICA MFG., INC., U.S.A.
            HONDA Passenger Car

Model Line
    CB7: ACCORD

Body and Transmission Type
    1: 2-door 5-speed Manual
    2: 2-door 4-speed Automatic
    5: 4-door 5-speed Manual
    6: 4-door 4-speed Automatic

Vehicle Grade
    4: DX
    5: LX
    6: EX
    8: SE

Check Digit

Model Year
    M: 2004

Factory Code
    A: Ohio Factory, U.S.A.
    C: Sayama Factory, Japan

Serial Number

Figure 1-8 The VIN is a 13- to 17-digit number that identifies the vehicle and certain components of that vehicle. Character interpretation of a typical VIN is shown here.

look at it and verify that it was bad. If the new part has a core charge, it can be passed to the customer if he or she desires to take the old part. A core is a component that can be rebuilt for future sales. The core charge is the cost of replacing the base part, such as the housing. This is a common practice for items such as engines, starters, alternators, and brake shoes, which can be refurnished much more cheaply and quickly than recasting and manufacturing the entire component.

Enter the technician's diagnosis based on the complaint. Most shops require the technician to enter a starting and ending time on the repair order to track the actual time the vehicle was in the shop. Enter the labor time and, if not already provided, the **hourly rate.** Once the repair is complete, provide a short, concise statement about the repair that will give the customer an idea of the work performed. The information may be handwritten or printed by a computer terminal based on numerical codes entered by the technician. The technician must enter the employee identification in order to receive credit and payment for the repair. Many shops use computer-generated forms that allow the service writer, technician, parts personnel, and accountant to process a complete repair order by entering codes into a terminal without any paper used.

Complete the parts department information by entering part numbers, quality, cost per unit, and total cost of parts. The repair order is given to the accountant for final calculations including taxes before the customer pays.

Some shops require a foreman or supervisor to sign orders for major repairs. The supervisor may choose to test drive the vehicle. The repair order should not go to the accountant until the required signature is obtained. The accountant is the last stop for the repair order before filing. With the repair order complete, notify the service writer that the vehicle is ready for the customer to pick up.

Hourly rate is the amount paid by clock hour to the technician regardless of the amount or type of work done in that hour. Entry level technicians are usually paid by the hour.

ADDITIONAL VEHICLE DATA PLATES

There are several places on the vehicle with plates or tags containing specific information about that component or assembly. The engine normally has a casting number forged into the engine block. This number can be used to correctly identify the engine size, manufacture date, and location of the manufacturing plant. There should be a plate or tag on the transmission or transaxle identifying that assembly. The same applies to the differential of a rear-wheel-drive vehicle. Some older Fords had a calibration tag in the glove box and on the radiator support bracket showing the calibration for that vehicle's emission system. A label is sometimes placed on the radiator support or under the hood listing which emission devices are installed on the vehicle. This same label lists the vehicle's emission certifications and other information related to the engine's performance.

Within the engine compartment, there is usually a decal listing the type and amount of refrigerant used in the climate control system, one showing the routing of the serpentine belt, one showing the remote jump points for the concealed battery, and any one of several others showing information the manufacturer or technician deem necessary for the maintenance and operation of the vehicle.

Other data plates and tags are located on the driver door pillar listing gross vehicle weight, individual axle weight, tire size, and pressure. Recently, the VIN number has been added to major components, such as the transmission and the unibody, to assist in identifying the vehicle if stolen, used in a crime, or involved in an accident. Most service data will list the location of specific tags, plates, or decals for a specific component or assembly.

The major problem with tags as opposed to plates is the loss during repair procedures. It is not uncommon to find that the metal tag on the differential or transmission has disappeared because it was not reinstalled after a former service. Most plates are fastened in such a way that they cannot be easily lost. Decals, however, can be soiled with grease and other material that covers or erases the ink.

> **AUTHOR'S NOTE** My students were having a problem determining how much refrigerant to be installed in an air-conditioning system. The decal was present, but the lettering could not be read because of grime. Being the brilliant person that I am, I simply grabbed a rag and wiped off the grime. It disappeared, as the printing on the decal. What was left was a nice, clean, perfectly blank white decal. Fortunately, a service manual was available. Moral of the story: Don't waste your time trying to clean a dirty decal. Just get the service manual.

SUMMARY

- Automotive repair shops have personnel ranging from janitorial services to managers and owners.
- Service managers, service writers, and accountants generally provide the direct customer contact.
- The parts manager is responsible for storing, ordering, issuing, and charging parts.
- The technician most directly influences the profits for the shop.
- Technicians should be trained to the basic level and then attend continuous additional training to keep current with changing automotive technology.
- Technicians can receive certifications in specific systems through several national agencies.
- Technicians may receive a salary based on an hourly, commission, or flat rate.
- Technicians usually are required to have their own tool set.

■ All shop employees and the owners share responsibility for environmental protection.
■ Warranties cover the parts and services required that may be due to errors in manufacturing, design, or previous repairs.
■ Recalls are issued by the vehicle manufacturers to correct some deficiency in design or manufacturing.

■ A repair order is the first step in the diagnostic and repair procedure.
■ The VIN is a number that identifies a particular vehicle and will contain information pertinent to that vehicle's manufacture.
■ Additional data plates and tags are placed in various locations on the vehicle.

ASE-STYLE REVIEW QUESTIONS

1. The duties of the service writer are being discussed.
   Technician A says the service writer completes the repair order.
   Technician B says the service writer may provide a repair estimate to the customer.
   Who is correct?
   A. A only
   B. B only
   C. Both A and B
   D. Neither A nor B

2. *Technician A* says a service writer usually has only business administration training.
   Technician B says most technicians have a background in business.
   Who is correct?
   A. A only
   B. B only
   C. Both A and B
   D. Neither A nor B

3. The parts department is being discussed.
   Technician A says all parts are sold on the shop's repair orders.
   Technician B says the technician performing the service must charge the parts to the order.
   Who is correct?
   A. A only
   B. B only
   C. Both A and B
   D. Neither A nor B

4. *Technician A* says, based on a labor guide, a technician may receive up to 80 hours of pay per 40-hour week under the commission pay scale.
   Technician B says flat-rate pay does not strictly rely on actual clock-hours worked.
   Who is correct?
   A. A only
   B. B only
   C. Both A and B
   D. Neither A nor B

5. The position of an entry-level technician is being discussed.
   Technician A says this individual is usually a graduate or student of a postsecondary automotive program.
   Technician B says most individuals in this position are expected to diagnose a vehicle using a scan tool.
   Who is correct?
   A. A only
   B. B only
   C. Both A and B
   D. Neither A nor B

6. Service repair orders are being discussed.
   Technician A says recall information is entered only by the technician assigned to the order.
   Technician B says it is the responsibility of the parts department to make sure all parts are entered onto the order.
   Who is correct?
   A. A only
   B. B only
   C. Both A and B
   D. Neither A nor B

7. Arbitration boards are being discussed.
   Technician A says the board's decision is only binding for the customer filing the complaint.
   Technician B says the board's decision is binding upon the shop.
   Who is correct?
   A. A only
   B. B only
   C. Both A and B
   D. Neither A nor B

8. Repair orders are being discussed.

   Technician A says the technician is responsible for computing the labor cost.

   Technician B says the accountant usually enters the labor time on the order.

   Who is correct?

   A. A only

   B. B only

   C. Both A and B

   D. Neither A nor B

9. *Technician A* says the service manager is responsible for the company's accounting office.

   Technician B says the parts manager usually reports to the service manager.

   Who is correct?

   A. A only

   B. B only

   C. Both A and B

   D. Neither A nor B

10. *Technician A* says the service writer initiates the repair order.

    Technician B says the technician enters the VIN onto the repair order.

    Who is correct?

    A. A only

    B. B only

    C. Both A and B

    D. Neither A nor B

Name _____   **Date** _____

PREPARING A SERVICE REPAIR ORDER

Upon completion and review of this job sheet, you should be able to prepare a service repair order based on customer input, vehicle information, and service history. Demonstrate proper use of fender covers, floor mats, and steering wheel covers.

ASE Education Foundation Correlation

This job sheet addresses the following (**RST**) tasks:

3-1. Identify information needed and the service requested on a repair order.

3-2. Identify purpose and demonstrate proper use of fender covers and mats.

3-3. Demonstrate use of the three Cs (concern, cause, and correction).

3-4. Review vehicle service history.

3-5. Complete work order to include customer information, vehicle identifying information, customer concern, related service history, cause, and correction.

We Support

ASE | Education Foundation

Tools and Materials

Goggles or safety glasses with side shield

An assigned vehicle or the vehicle of your choice

Service repair order or computer-based shop management package

Parts and labor guide

Seat covers

Protective floor mats

Steering wheel covers

Repair Order Source

Work Order Source

Describe the system used to complete the work order. If a paper repair order is being used, describe the source. Install seat covers, floor mats, and steering wheel cover if needed.

_____

_____

Procedure   **Task Completed**

1. Prepare the shop management software for entering a new work order or obtain a   ☐
   blank paper work order.

2. Enter customer information, including name, address, and phone numbers on the   ☐
   work order.

3. Locate and record the vehicle's VIN.   ☐

4. Enter the necessary vehicle information, including year, make, model, engine type and   ☐
   size, transmission type, license number, and odometer reading.

5. Does the VIN verify that the information about the vehicle is correct? ☐ Yes ☐ No

6. Normally, you would interview the customer to identify his or her concerns. However,   ☐
   to complete this job sheet, assume the only concern is that the valve (cam) cover is
   leaking oil. This concern should be added to the repair order.

Task Completed

7. The history of service to the vehicle can often help diagnose problems as well as ☐
   indicate possible premature part failure. Gathering this information from the customer
   can provide some of this information. For this job sheet, assume the vehicle has not
   had a similar problem and was not recently involved in a collision. Service history is
   further obtained by searching files based on customer name, VIN, and license number.
   Check the files for any related service work.

8. Search for TSBs on this vehicle that may relate to the customer's concern. ☐

9. Based on the customer's concern, service history, TSBs, and your knowledge, what is
   the likely cause of this concern?

   _____

   _____

10. Enter this information onto the work order. ☐

11. Prepare to make a repair cost estimate for the customer. Identify all parts that may be
    needed to be replaced to correct the concern. List them here.

    _____

    _____

12. Describe the task(s) that will be necessary to replace the part.

    _____

    _____

13. Using the parts and labor guide, locate the cost of the parts that will be replaced and ☐
    enter the cost of each item onto the work order at the appropriate place for creating an
    estimate.

14. Now, locate the flat-rate time for work required to correct the concern. List each task
    and its flat-rate time.

    _____

    _____

15. Multiply the time for each task by the shop's hourly rate and enter the cost of each ☐
    item onto the work order at the appropriate place for creating an estimate.

16. Many shops have a standard amount they charge each customer for shop supplies and ☐
    waste disposal. For this job sheet, use an amount of $10 for shop supplies.

17. Add the total costs and insert the sum as the subtotal of the estimate. ☐

18. Taxes must be included in the estimate. What is the sales tax rate in your area, and
    does it apply to both parts and labor, or just one of these?

    _____

    _____

19. Enter the appropriate amount of taxes to the estimate, then add this to the subtotal. ☐
    The result is the estimate to give to the customer.

20. By law, how accurate must your estimate be?

_____

21. The repair order is complete and is ready for the customer's signature. However, some businesses require additional information; make sure you enter that information on the work order. On the work order there is a legal statement that defines what the customer is agreeing to. Briefly describe the contents of that statement.

_____

_____

Instructor's Response

_____

_____

_____

Name _____   Date _____

DECODE VEHICLE IDENTIFICATION NUMBER

Upon completion of this activity, you will have demonstrated the ability to locate and identify the vehicle identification numbers. You will be assigned a school vehicle for this activity.

ASE Education Foundation Correlation

This job sheet addresses the following **RST** tasks:

3-1. Identify information needed and the service requested on a repair order.

3-2. Identify purpose and demonstrate proper use of fender covers and mats.

3-3. Demonstrate use of the three Cs (concern, cause, and correction).

3-4. Review vehicle service history.

3-5. Complete work order to include customer information, vehicle identifying information, customer concern, related service history, cause, and correction.

We Support

ASE | **Education Foundation**

Tools and Materials

Goggles or safety glasses with side shield

An assigned vehicle or the vehicle of your choice

Service repair order or computer-based shop management package

Describe the vehicle being worked on:

Vehicle year _____ Make _____ Model _____

Procedure

1. What does the abbreviation "VIN" stand for? _____ _____
    _____

2. How many characters are in a VIN? _____

3. Locate and record the VIN: _____

4. Describe where the federal government mandates the VIN # is located.
    _____

5. Is there more than one place to find the VIN? _____

6. Describe where the Vehicle Emission Control Information (VECI) label is located.
    _____

7. Describe where the belt routing map label is located. _____

8. What type of information is contained in the VIN?
    _____
    _____

9. Why is the VIN # located in more than one spot? _____

10. Using the VIN # from earlier, determine the location and enter the letter or number:
    a) Model year _____
    b) Country of manufacture _____
    c) Engine size _____
    d) Manufacturer/division _____
    e) Body/platform _____
    f) Serial number _____
    g) Check digit _____
    h) Plant code _____

11. What is the eighth digit of the vehicle you are working on, and what does it represent?

    _____

    _____

12. What is the 10th digit of the vehicle you are working on, and what does it represent?

    _____

13. What is the first digit of the vehicle you are working on, and what does it represent?

    _____

14. What is the fourth digit of the vehicle you are working on, and what does it represent?

    _____

15. What is the engine code for the vehicle you are working on, and what is the cubic inch displacement (engine size)?

    _____

Instructor's Response

_____

_____

_____

Name _____   Date _____

COLLECTING SERVICE INFORMATION

Upon completion and review of this job sheet, you should be able to collect service information based on a customer complaint.

ASE Education Foundation Correlation ————————————————————

This job sheet addresses the following **RST** tasks:

3-1. Identify information needed and the service requested on a repair order.

3-2. Identify purpose and demonstrate proper use of fender covers and mats.

3-3. Demonstrate use of the three Cs (concern, cause, and correction).

3-4. Review vehicle service history.

4-1. Ensure vehicle is prepared to return to customer per school/ company policy (floor mats, steering wheel cover, etc.).

We Support
ASE | Education Foundation

Tools and Materials

Goggles or safety glasses with side shields
Service manuals, computerized or paper
Labor and parts guide
Blank repair order if available

Procedure

This job sheet is based on the following information provided by the customer or from the vehicle.

Customer complaint: The vehicle pulls to the right when the brakes are applied. It has no unusual noise, and other than the brake pull, it operates fine. He stated the vehicle had the brakes repaired at an out-of-town shop about 3 weeks ago.

VIN: 1B46P54R7TB3XXXXXX. The vehicle has antilock brakes on rear wheels only.

NOTE: Labor fee for this shop is $65.00 per hour.

Note: Information gathered during this job sheet may be recorded in the spaces provided, but the use of a blank repair order is preferred.

1. Use the VIN to determine the year, make, model, and engine for this vehicle, and list below.

   _____

   _____

2. Locate and record any recall or TSBs that pertain to the customer's complaint. If none, state none.

   _____

   _____

3. Record the vehicle's repair history as stated by the customer or shop records.

   _____

   _____

Task Completed

4. Locate and record a brief outline of the diagnostic procedures for the vehicle fault as stated by the customer.

_____

_____

5. Locate and record a brief outline of the repair procedures for this problem. Note any cautions or warnings.

_____

_____

6. Use the parts guide to locate and price the parts needed to complete the repair. Record both the suggested retail price and the shop's cost.

_____

_____

7. Use the labor guide to calculate and record the number of hours needed to complete a total repair.

_____

8. Calculate the total cost of this repair, and record below. Tax is 7 percent on parts; no tax on labor.

_____

9. Were any problems encountered during the completion of this job sheet? What type of service manual was used? If you used a different source (e.g., paper instead of computer) before this job sheet, which appears to be the easiest to use?

_____

_____

10. Complete the repair order if used and present it or this job sheet to the instructor.   ☐

Instructor's Response

_____

_____

_____

CHAPTER 2

AUTOMOTIVE SHOP AND SAFETY PROCEDURES

Upon completion and review of this chapter, you should understand and be able to:

- Understand personal safety and work practices.
- Define how personal protection equipment applies to you.
- Identify common safety hazards in an automotive repair business.
- Explain the technical and legal requirements and expectations of the business and the technician.
- Understand the legal requirements and practices for control of hazardous material and waste.
- Explain the purposes of the Environmental Protection Agency (EPA)

and Occupational Safety and Health Administration (OSHA).

- Identify the information contained on Safety Data Sheets (SDS) and container labels.
- Inspect fire extinguishers and understand operating procedures.
- Perform basic first aid steps.
- Understand hybrid vehicle safety.
- Explain safety layout of a shop.
- Work safely on and around a vehicle.
- Disarm an airbag system.

Terms To Know

Capacitor	Professionalism	Upwind
Logo	Safety Data Sheet (SDS)	
PASS	Tag	

INTRODUCTION

Accidents cost employers and employees millions of dollars and work hours each year.

An accident affects the individual employee, the family, the employer, and possibly all coworkers. Paying attention to the task, practicing safe habits, and applying common sense will lessen the chance of being injured. Knowing the environment—the workplace—also helps prevent loss of time from mechanical, chemical, and burn accidents.

Many repair facilities and car dealerships have raised insurance rates because of employee accidents. This added cost must be either passed onto the consumer or taken out of the profit margin. As a prospective employee (or maybe even an owner), it is important to understand safety because a serious accident can make a significant difference at the workplace.

Governmental agencies at all levels provide guidance and legal requirements to assist in accident prevention. However, because all technicians are human, there may be a time when some first aid is needed. Understanding some basic steps may reduce injury and possibly save a life.

PERSONAL AND COWORKER SAFETY

The way you and your coworkers dress and act in the shop reflects your **professionalism**. Wearing professional clothing and acting in a safe manner helps keep you healthy and makes your customers feel happier about hiring you to repair their vehicles.

Clothing, Hair, and Jewelry

Professional clothing ranges from lab coats to jump suits to clean "civilian" work clothes (**Figure 2-1**). To present a better image, most shops supply or require you to wear uniforms with the business **logo**. Uniforms are fitted to make sure no loose sleeves or shirttails can become tangled. If uniforms are not required, follow the same procedures with civilian work clothes. Extra looseness is not beneficial in the automotive shop environment.

Most uniforms have polyester as the main material. This material is more resistant to battery acid and cleans easier. Cotton clothing does not go well with battery acid.

Long hair in a shop can be dangerous. If the hair is long enough to get tangled in rotating engine or drivetrain parts, it must be pinned or fastened securely out of the way. Never assume it is short enough. Under some circumstances, hair that is slightly more than shoulder length can cause problems.

Jewelry is one item that technicians do not like to remove, particularly wedding rings and religious images. Do not wear rings when working in an automotive shop. Rings are made of metal, and they tend to conduct electricity and react in strange ways when exposed to chemicals. Typical of personal religious symbols is a long necklace with the jewelry hanging inside the shirt or blouse. Long necklaces should be removed before beginning work. Ask the supervisor if there are any storage areas that may be used for storage of personal valuable jewelry.

Eye and Face Protection

A person can go to almost any shop and see technicians working without eye or face protection. Most of them feel the protection is uncomfortable or cuts visibility. Most of

professionalism
A person who shows consideration and respect for others and demonstrates loyalty and exceeds expectations.

A **logo** is artwork used by a business to promote its name and product more easily. The "Golden Arches" is a visual logo for McDonald's restaurants. The "blue seal" (ASE) is a visual logo for technicians who are certified.

Hearing protection

Respirator

Safety goggles

Welding shield

Welding gloves

Gloves

Working clothes

Safety shoes

Figure 2-1 Work clothes should fit neatly without binding or looseness.

Figure 2-2 The only way to protect the eyes during work is to wear proper eye or face protection.

them also have not seen serious eye or face injuries firsthand. The wearing of eye protection during work is the safest method for protecting your sight (**Figure 2-2**). Proper safety glasses are rated by the **American National Standards Institute** (ANSI) and have a stamp on the frames or lens. This stamp recognizes that the safety lens will not break during a "concentrated high impact." This rating is **Z.87**. Most eye injuries in the automotive shop come from a piece of rust or metal from a routine job such as exhaust work. Face injuries often happen during cleaning or grinding parts.

Although it may feel uncomfortable when first wearing eye protection, the technician will become accustomed to it with daily wear and eventually feel naked without it. Remember your first ring or your first watch? For a few days, it worried you to wear it. Now, forgetting that ring or watch throws the entire day off. The same applies to eye and face protection. The vision problem is corrected in two ways. First, buy a set of safety glasses that are worth the money, and second, wear them every day. Technicians' tools are the best investment in their career. Eye and face protection devices are part of the tool set.

Hearing and Hand Protection

With some of the loud music played on vehicle audio equipment, it is hard to imagine that an automotive shop can cause ear damage. A typical repair business does not have a great deal of **continuous noise** as in a body shop, but there are instances when hearing needs to be protected. Earmuffs or plugs are excellent for temporary protection on the job (**Figure 2-3**).

Hands are subject to cuts and abrasions by sharp tools and burning by chemicals and heat from the exhaust system. Be careful how the hands are worked into those crevices on the vehicle and use good hand tools. When cleaning parts, wear chemical protection gloves to prevent burns. One thing about hearing and hand protection is that most technicians use these types of protection devices without prompting. The pain and discomfort of exposure is immediate as opposed to eye and face protection where the danger is not clear. Lack of eye, face, hearing, or hand protection may put the technician's career on hold for some time.

 Caution

Ordinary prescription glasses are not safety glasses even though they may offer some impact protection. Order safety prescription glasses to include the frame. You are not protecting yourself if the glasses do not meet occupational safety glasses requirements.

Caution

Trying to stop a falling object can result in injury to you without regaining control of the object. Step back and allow the object to drop. If possible, you may try to control it after it hits the floor, but do not let yourself be placed in jeopardy.

Figure 2-3 Ear protection should be worn during any metal working, such as grinding or drilling. Air impact tools can also cause hearing damage. Ear protection should always be worn.

LIFTING AND CARRYING

Back pain does not normally result when a person tries to lift something above his or her capabilities. It usually happens when the body is in a position that will not accept the weight being moved. Before lifting any object, always position the body so the additional weight is placed over the legs. If lifting from the floor, bend at the knees and attempt to keep the back straight (**Figure 2-4**). This will allow the legs to perform the lifting motion. The reverse applies if an object is to be placed on the floor or a lower surface. Secure the object and lower it to the floor by bending the knees. This may be awkward and could cause a loss of control with heavier objects. Before moving heavy objects, ask another person to help distribute the weight and maintain control. If the object slips or moves off balance, the best method is to step back and allow the object to drop.

Also, if a change of direction is needed, such as moving an object from one table to another, do not twist the body. Turn the entire body with the object. Keeping feet and legs in place and twisting at the waist places the weight of the object at an angle to the muscles. Back and shoulder muscles may not be able to fully support the weight and may be overexerted.

Carrying heavier objects is dangerous to the technician, the object, and personnel working nearby. If the object cannot be carried comfortably, ask a second person to help. Do not wait until the object becomes unstable.

Figure 2-4 With the back kept straight, the load will be placed on the legs.

AUTHOR'S NOTE One of the worst back injuries I endured was the result of picking up a 2-pound motorcycle battery. Down on one knee, I twisted to the left, bent slightly, grabbed the battery, and straightened up to slide it in place on the bike. About halfway around through the straightening and twisting, I felt an extremely sharp pain near my right shoulder blade. The battery and I hit the ground about the same time. I was out of work for 2 weeks, had trouble breathing, and was not fully recovered for about 5 weeks. Remember: It's not always the weight so much as the position of the body that can cause injury.

Coworker Safety

Another safe practice is to watch out for coworkers. Each person needs to have an idea of what is happening around him or her. The person in the next bay may be busy moving something, not knowing someone is approaching from the blind side. In addition, coworkers are like everyone else. Do them a favor and remind them of eye protection and other safety violations they may have missed.

HAZARDOUS MATERIAL AND WASTE CONTROL

Right-to-know laws place a great deal of responsibility on employers to ensure that employees know about all hazardous materials they may be using. It is the employee's responsibility to utilize that training.

Environmental Protection Agency (EPA)

The Environmental Protection Agency, or EPA, is a federal agency charged with instituting and enforcing regulations that assist in protecting the environment. It was formed in the early 1970s to reduce air pollution caused by vehicle and manufacturing emissions. Inherent within that charter was the control and disposal of waste products from almost all businesses, including local automotive repair shops and individuals. The main issue with the EPA is the storage and disposal of hazardous waste from major manufacturers, plants, the local garbage dump, and everything in between. Although its formation met with much resistance, the results some 40 years later are cleaner air and less ground and water pollution. The EPA has had a very positive effect on businesses and the environment and will continue to do so in the foreseeable future. The EPA's Web site is http://www.epa.gov.

Environment Canada

Environment Canada is the Canadian version of the U.S. EPA. It has requirements that relate to Canada's more northern environment and citizens. Within Environmental Canada, there are subagencies that perform the same governmental functions as their sister agencies with the U.S. EPA. As far as the automotive industry is concerned, however, the legal and environmental control requirements are almost the same. Section 7 of the Canadian Environmental Protection Act specifically covers the Canadian automotive industry. The Web site best suited for obtaining information on this agency is http://www.ceaa-acee.gc.ca.

Occupational Safety and Health Administration (OSHA)

Occupational Safety and Health Administration (OSHA) was formed to help protect employees and ultimately employers. It has the legal authority to inspect businesses and ensure that working areas are safe for the employees. Some safety concerns of utmost importance are the control of chemicals within the workplace, the equipment and facilities to store or use those chemicals, the equipment and tools used within the facility, and the general working environment. It should be noted that, since the formation of OSHA, accidents resulting from unsafe working environments have been reduced, with an increase in production associated with lowered lost man-hours and fewer accidents. Visit OSHA's Web site at http://www.osha.gov/SLTC/index.html for more detailed information.

Canadian Center for Occupational Health and Safety (CCOHS)

This agency is like the U.S. OSHA, with a similar mandate, responsibility, and authority. It performs inspections, determines administrative fines, may file criminal charges, and directs training programs in much the same manner as the U.S. OSHA. Its Web site is http://www.ccohs.ca/.

It should be noted that each of the four agencies listed operate across borders because many pollutants tend to cross borders. Automotive manufacturing, vehicle repair, and vehicle operations are shared by the United States and Canada, and many associated problems are the result of actions in one country affecting the environment of its neighbor. Each of the listed Web sites contains information pertaining to many environmental and safety issues.

Right-to-Know

Each of these agencies enforces what are known as "right-to-know" laws or regulations. Basically, right-to-know laws require the employer to notify employees of dangerous materials housed or used on-site. They also require the initial training of new employees, annual (or more often) refresher training of all employers, and employer-designated personnel with specific authority to train, maintain records, and in some instances act as first responders to fires or accidents. Of direct interest to all employees are the three main informational documents pertaining to on-site chemicals, described next.

Safety Data Sheets

Safety Data Sheet (SDS) A document that contains information on the potential health effects of exposure to chemicals or other potentially dangerous substances.

The **Safety Data Sheet (SDS)** is important for an employee to understand. In **Figure 2-5**, the chemical listed is a common brake cleaner (Sections 1 and 2). Search the sheet for the safety hazards that may cause harm to humans and the environment (Sections 3 and 4). Using that information, it should be easy to decide what personal protection measures must be taken to protect people. Go to Section 5 on Figure 2-5 and compare those measures with the items listed there. Analyze the first aid treatment that may be needed in case of accident. Survey the shop and locate the items that may be needed to prevent injury and the items needed as first aid for this chemical. The minimum would be an eye wash station and a place to wash any flesh that may be heavily exposed to this chemical.

Container Label

Every chemical must have a label on its container listing the same items found on an SDS. A container from a manufacturer will have that label, but many times chemicals are stored in a large container and transferred to a smaller (transfer) container for ease of use. Sometimes this transfer container isn't labeled. It is required by law, and common sense, that hazardous materials not be stored in an unmarked or improperly marked container. Surprises can lead to serious injuries or damage. The label on the transfer container must list exactly the same information shown on the manufacturer's label.

MATERIAL SAFETY DATA SHEET

PRODUCT NAME: KLEAN-A-KARB (aerosol) #- HPMS 102068

PRODUCT: 5078, 5931, 6047T

(page 1 of 2)

1. Ingredients	CAS #	ACGIH TLV	OSHA PEL	OTHER LIMITS	%
Acetone	67-64-1	750 ppm	750 ppm		2–5
Xylene	1330-20-7	100 ppm	100 ppm		68–75
2-Butoxy Ethanol	111-76-2	25 ppm	25 ppm	(skin)	3–5
Methanol	67-56-1	200 ppm	200 ppm		3–5
Detergent	-	NA	NA		0–1
Propane	74-98-6	NA	1000 ppm		10–20
Isobutane	75-28-5	NA	NA	1000 ppm	10–20

2. PHYSICAL DATA : (without propellent)

Specific Gravity   : 0.865   Vapor Pressure   : ND

% Volatile   : >99

Boiling Point   : 176°F Initial   Evaporation Rate : Moderately Fast

Freezing Point   : ND   Vapor Density   : ND

Solubility: Partially soluable in water   pH   :NA

Appearance and Odor: A clear colorless liquid, aromatic odor

3. FIRE AND EXPLOSION DATA

Flashpoint   : –40°F   Method   : TCC

Flammable Limits propellent   LEL: 1.8   UEL: 9.5

Extinguishing Media: CO_2, dry chemical, foam

Unusual Hazards   : Aerosol cans may explode when heated above 120°F.

4. REACTIVITY AND STABILITY

Stability   : Stable

Hazardous decomposition products: CO_2, carbon monoxide (thermal)

5. PROTECTION INFORMATION

Ventilation : Use mechanical means to ensure vapor concentration
        is below TLV.

Respiratory: Use self-contained breathing apparatus above TLV.

Gloves   : Solvent resistant   Eye and Face: Safety Glasses

Other Protective Equipment: Not normally required for aerosol product usage

Figure 2-5 SDS sheets contain essential information about chemical use, handling, and disposal.

Hazardous Material Inventory Roster

This roster is not readily available to every employee every day, but the employee should know its location. The roster is a list of every hazardous material, mostly chemicals with their individual hazards, stored or used on the business property. It is usually stored away from the "dangerous" areas of the property so it will be available to first responders in case of accident or fire. It shows the location of each material, and several copies may be located in different places, especially in large, single-site businesses. The roster is available to every employee and may be required reading for selected employees.

Storage of Hazardous Materials and Waste

Hazardous **waste** is a different matter from hazardous **materials.** Waste is something to be discarded; it is not reusable and must be stored in marked and completely sealed containers. In most cases, oil and antifreeze are the most common hazardous wastes in an automotive shop. The EPA does not consider most used materials as hazardous waste until they are picked up for recycling or disposal.

Because hazardous materials are used almost daily, they need to be stored close to the work area, usually inside the shop. The storage device should be a closed metal cabinet of some type marked with fire warnings (**Figure 2-6**). Some materials, such as fuel, should be **stored outside the building** in approved specialty metal containers. Remember, the storage and safe use of hazardous materials is one of the safety requirements of the shop and employees. Anyone using a hazardous material must use it safely and store it properly. Any material remaining in the container after use must be returned to the storage area or disposed of properly.

Used material such as oil must be stored under certain conditions (**Figure 2-7**). Each material must have its own container. The container must be labeled with the material's name and be marked as USED.

Another type of hazardous waste is the can used to store propellant-driven material such as brake cleaner. Once empty with all its pressure released, the can must be split open before its disposal. The EPA considers a container of this type to be hazardous waste if it is not split open.

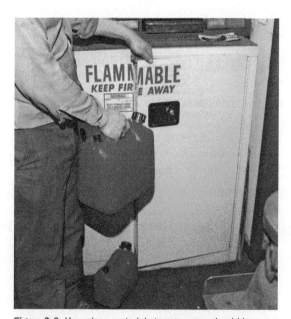

Figure 2-6 Hazardous material storage areas should be marked as flammable and should be enclosed or locked.

Figure 2-7 Used hazardous fluids must be stored and clearly labeled containers.

FIRES AND FIRE EXTINGUISHERS

The danger of fire in an automotive shop is present at most times. Fire can be caused by oily rags improperly stored, hazardous materials exposed to heat, fuel leaks, or poor storage of hazardous material or waste. There are four general classifications of fires and a type of fire extinguisher to match the burning materials (**Figure 2-8**). Each class of fire is matched with a type of fire extinguisher containing the best material for controlling or extinguishing that fire. In most cases, the type of fire in an automotive shop will not be paper or metallic such as magnesium.

	Class of Fire	Typical Fuel Involved	Type of Extinguisher
Class **A** Fires (green)	For Ordinary Combustibles Put out a class A fire by lowering its temperature or by coating the burning combustibles.	Wood Paper Cloth Rubber Plastics Rubbish Upholstery	Multipurpose dry chemical
Class **B** Fires (red)	For Flammable Liquids Put out a class B fire by smothering it. Use an extinguisher that gives a blanketing, flame-interrupting effect; cover whole flaming liquid surface.	Gasoline Oil Grease Paint Lighter fluid	Carbon dioxide Halogenated agent Standard dry chemical Purple K dry chemical Multipurpose dry chemical
Class **C** Fires (blue)	For Electrical Equipment Put out a class C fire by shutting off power as quickly as possible and by always using a nonconducting extinguishing agent to prevent electric shock.	Motors Appliances Wiring Fuse boxes Switchboards	Carbon dioxide Halogenated agent Standard dry chemical Purple K dry chemical Multipurpose dry chemical
Class **D** Fires (yellow)	For Combustible Metals Put out a class D fire of metal chips, turnings, or shavings by smothering or coating with a specially designed extinguishing agent.	Aluminum Magnesium Potassium Sodium Titanium Zirconium	Dry power extinguishers and agents only

Figure 2-8 Class B and C type fires present the greatest fire concern in an automotive shop.

AUTHOR'S NOTE Be aware that some wheels offered for sale contain magnesium. They do not usually present a great fire hazard to the shop or the employees, but there is always that possibility.

A fire in a shop has a good chance of being a Class B type, which involves flammables such as fuel, oil, and solvents. However, the technician must understand the different types of fires and know whether the fire extinguishers in the shop are the right kind and in good working order.

The fire extinguisher normally found in an automotive shop is either a Class B or a multipurpose extinguisher that can be used on several types of fires. Look for a "B" enclosed in a square on the fire extinguisher (**Figure 2-9**). It may have more than one symbol indicating that it is a multipurpose extinguisher. Obviously, inspection of a fire extinguisher cannot be done while extinguishing the fire. Fire extinguishers must be inspected periodically, preferably by the same person. There are usually several fire extinguisher points throughout the shop. Look at the **tag** and gauge of each to make sure the extinguisher is charged and ready for work. The safety clip or wire must be in place, and there must be no damage to the fire extinguisher. The inspector initials and dates the tag if the extinguisher is ready for use. Study the different types of extinguishers available; learn their locations and the proper use of each style during the inspection. Most fire codes call for an inspection of fire extinguishers on a periodic basis.

The use of a Class B fire extinguisher is discussed because this is the most common type of fire in an automotive shop. Class B fires must have their oxygen supply cut off before they go out. Water or other spreading material is not applicable on Class B fires because it spreads the fuel. When a fire is discovered, sound the alarm and locate a Class B fire extinguisher. Position yourself about 8 feet from the base of the fire, **upwind** if possible. Remove the safety clip, aim the nozzle at the base of the fire, and pull the trigger. Continue using the extinguisher until the fire is out or the extinguisher is empty. If necessary, spray the fire area to cool any hot spots that may reignite the fire. If you or any of your coworkers feel the fire is too large for the shop's fire extinguishers, leave the area and notify the local fire department.

Although Class B fires are the most common in an auto shop, it is important to know the other types of fires and the classes. A Class A fire is a combustible products fire. This includes wood, paper, or cloth. These fires can be extinguished by water. A Class C fire is

The extinguisher's **tag** is usually a small card tied directly to the extinguisher. The inspector may enter his or her initials, date, and status of the extinguisher.

Upwind means the wind is blowing on the person's back. If the wind is in the firefighter's face (downwind), the flame and the firefighting material can be blown toward the fighter.

Figure 2-9 The fire extinguisher in the center is a multipurpose fire extinguisher for use against Class A, B, and C fires.

PHOTO SEQUENCE 1
Using a Dry Chemical Fire Extinguisher

P1-1 Multipurpose dry chemical fire extinguisher.

P1-2 Hold the fire extinguisher in an upright position.

P1-3 Pull the safety pin from the handle.

P1-4 Stand 8 to 10 feet upwind of the fire. Do not go any closer to the fire.

P1-5 Free the hose from its retainer and aim it at the base of the fire.

P1-6 Squeeze the lever while sweeping the hose from side to side. Keep the hose aimed at the base of the fire.

an electrical fire. On some occasions, you may experience a Class C fire. The smell of a Class C fire is very distinct, and you will never forget it. A Class D fire is very rare in an auto shop. Class D fires are caused when pyrophoric metals explode. An example of this metal is magnesium. Most shops today have multipurpose fire extinguishers.

The easiest way to remember how to use a fire extinguisher is to remember the acronym **PASS**. This stands for Pull, Aim, Squeeze, and Sweep. Pull the trigger, aim the nozzle, squeeze the lever, and sweep the hose side to side. **Photo Sequence 1** explains the use of a dry chemical multipurpose fire extinguisher.

PASS A method to extinguish a fire using a pull, aim, squeeze, sweep (PASS) technique

FIRST AID

This section is not intended to make the reader a certified medical first responder. **First aid** is defined as actions to protect the victim and reduce the injury until trained medical personnel are on the scene. Remember that we are not discussing extreme cases but common and usually non-life-threatening injuries. There are four basic first aid steps that can be applied by any person in the shop and in whatever order is appropriate to the situation:

1. Stop the bleeding and protect the wound.
2. Treat burns.
3. Treat for shock.
4. Protect the victim.

For the most part, injuries during an automotive repair will be cuts or burns. However, shock may kill the victim faster than the injury. Some people will go into deep shock at the sight of blood—whether it's their own or another's. In event of an accident, call or have someone contact medical personnel. In the United States, this is usually accomplished by dialing "911" and giving necessary information for a quick response. Don't forget that most schools and businesses require a prefix number, usually "9" in the United States, to get an outside telephone line. Most schools also have a medical person or department that oversees administrating first aid. Schools also have a first aid procedure in their school policies or handbook. Make it a point to check the first aid procedures where you are.

Follow the appropriate step(s) listed earlier by using: sterile bandages and pressure for bleeding, burn treatment or cooling the area with clean water for burns, talking quietly to the individual for shock, and, when necessary, evacuate the person from the area in cases such as a nearby fire or chemical spill. One of the best things a first aid provider can do is remain calm and keep the injured calm.

First Aid for Eye Injury

One of the most common injuries that occurs in an automotive repair shop is injury to the eye. This is all too often caused by the failure, intentional or otherwise, to wear protective face and eye protection. Common protection devices to help prevent this were discussed earlier in this chapter. However, many technicians, smart in every other way, fail to use them for several reasons. Eye injuries are painful, distracting, and can be permanent. In almost every case of eye injury, certain steps should be followed immediately.

First, do not rub the eye(s). This will only cause more damage and pain. If necessary, guide the person (or be guided) to the eye wash station. It is very important to know the location of the eye wash area, in case you need it and not be able to see well to get there. There are two general types. One is an attachment to a water source by which low-pressure, high-volume water can be directed over the eyes (**Figure 2-10**). After

Figure 2-10 A typical eye wash station.

Figure 2-11 This type of eye wash can be taken to the injured worker. Once opened, any remaining liquid should be discarded.

flushing the eyes, the water naturally flows back off the face. Some state safety codes call for this water to be warmed for a certain amount of time that it is on. The second common type is the use of sterile water in a throwaway plastic bottle (**Figure 2-11**). The head is tilted backward, and the water is poured from the bottle over the affected eye(s). Tilting the head toward the side of the injured eye helps the water flush the chemical or debris from the eye over the cheek. Both methods work well but may require some assistance from another person. Regardless of the method used, do not rub the eye(s) and seek medical attention as soon as possible. Metal debris can cut the eye and cause infection, and chemicals can cause immediate serious burns. It is best if water is constantly flushed over the eye until medical assistance arrives or a visit is made to the nearest medical facility. Chemical intrusion into the eye should be flushed out for at least 15 minutes.

Supervisors and teachers should not, and usually don't, accept the "I'm OK, I don't need a doctor" idea. Supervisors must maintain an accident report, and more important, the technician's health and well-being are more of a concern than the medical cost or lost time. In this instance, the injured should flush the eyes, refrain from the natural tendency to rub them, and accept medical assistance as soon as possible.

HYBRID VEHICLE SAFETY AND HIGH-VOLTAGE SAFETY

With the increasing sales of hybrid vehicles, a specific safety issue has emerged for the automotive technician. Greater details on this issue are discussed in Chapter 5 of this manual and the Classroom Manual, but they basically concern the much higher available electrical current from the drive batteries or power pack. This higher available current in the vehicle's drive batteries require much larger conductors and cables. The positive cable of the drive battery will be orange in color and equal to or larger than typical 12-volt battery cables. In some cases, the voltage can be more than 300 volts. The negative cable will be the same size and black in color. Paying close attention and using common sense can prevent most, if not all, of the electrical safety concerns with a hybrid vehicle. If the cable is **big and orange**, don't touch it without proper equipment and training.

 Caution

For some repair tasks on a hybrid vehicle, it may be necessary to disconnect the drive batteries from the rest of the vehicle. This should be done while wearing insulated gloves like those worn by power company linemen. Very serious injury or death could result if proper personal protection equipment (PPE) is not worn. See Chapter 5 for further details.

SERVICE TIP Waiting time for the capacitor to discharge varies. Consult the service manual for specifics.

Some nonhybrid electric vehicles have high-voltage circuits. Some vehicles have high-intensity headlights. These headlights are brighter and provide a better focus for the driver because they illuminate the road more brightly, especially during severe weather. These headlights operate on voltage higher than the battery. They range from 14 to 24 volts. When servicing this voltage level, it is important to follow the manufacturer's suggestions and safety precautions. Just like other parts of the vehicle, there are always safety warnings and labels under the hood.

WORKING AROUND SUPPLEMENTAL RESTRAINT SYSTEMS (SRS)

Supplemental restraint systems (SRS) are commonly known as airbags. Although many lives have been saved with SRS, some lives have been lost or injuries incurred by technicians and emergency personnel working near or around the airbag. Like all computers, the one for the airbags is "dumb." Its sole purpose in life is to wait for a signal and then ignite the airbag chemicals. It does not know or care if that signal comes from an airbag arming switch or from a technician probing the wires under the dash or an emergency medical technician (EMT) trying to extract an individual from the vehicle. It just senses a signal and activates the airbag, sometimes with drastic results.

Before beginning any work inside the vehicle's passenger compartment, check to see how many airbags are installed. Naturally, there is the driver-side airbag and probably a passenger-side airbag, but there may be side or curtain bags at the outside edge of the seat and possibly in the headrests. (Some vehicles are now equipped with airbags to protect rear-seat passengers.) There are two definite steps that must be taken with any airbag system. First, disconnect the negative battery cable and place it so that accidental contact cannot be made with the battery. Second, wait a specific time for the airbag's **capacitor** to discharge.

A **capacitor** is a small electrical storage device.

If work is being done on the electrical system inside the passenger compartment, all airbag connections should be disconnected before work begins. With the capacitor discharged, look near the lower part of the steering column and disconnect all bright yellow connections. Airbags are the only systems now using bright yellow connections in this area. Check in and around the glove compartment and under the seats for additional bright yellow connections. Disconnect all of them. Consult the service manual for the location of all airbag electrical connections of this type. Once this is done, the battery can be reconnected to make electrical checks. Before reconnecting any of the airbag connectors, disconnect the battery again. Reconnect all airbag connectors and then reconnect the battery. This prevents any surges from activating the airbags.

SUMMARY

- Eye and face protection provides excellent protection against injuries.
- Gloves should always be worn when working with chemicals.
- Lifting should be done with the legs while keeping the back straight to avoid back injury.
- Most businesses, particularly automotive repair, keep hazardous material available on-site.
- Right-to-know regulations require the business owners and managers to inform employees of

the presence of hazardous materials and waste on the site.

- The EPA is a federal agency charged with regulating possible pollutants that may harm the environment.
- OSHA is a federal agency primarily concerned with safety at the work site.
- Each container must have a container label.
- Canada has environmental and safety agencies that are similar to the ones in the United States.

- Each chemical sold in the United States requires a Material Safety Data Sheet to be kept at the user's location.
- A Hazardous Material Inventory Roster is kept in a secure location and is available to all emergency first responders.
- There are four classes of fires and a matching type of fire extinguisher.
- Class B fires are most common in automotive repair shops.

- A combination fire extinguisher is the best type for an automotive shop.
- First aid is the immediate assistance given to a victim before the arrival of medical personnel.
- Hybrid vehicles usually operate on a combination of electric power and internal combustion engines.
- Hybrid vehicles have high-energy electrical systems requiring additional safety measures.

ASE-STYLE REVIEW QUESTIONS

1. Firefighting is being discussed.

   Technician A says to use a fire extinguisher that has a circle and a "B" on it to fight a cleaner solvent fire.

   Technician B says no spreading agent is to be used for fighting a gasoline fire.

   Who is correct?

   A. A only   C. Both A and B
   B. B only   D. Neither A nor B

2. Hazardous material and waste are being discussed.

   Technician A says both should be stored the same way.

   Technician B says waste is not considered to be waste until it has been picked up for disposal or thrown into a dumpster.

   Who is correct?

   A. A only   C. Both A and B
   B. B only   D. Neither A nor B

3. *Technician A* says that an oil fire is a Class B fire.

   Technician B says that you can use a water hose to extinguish a Class A fire.

   Who is correct?

   A. A only   C. Both A and B
   B. B only   D. Neither A nor B

4. Fire extinguisher use is being discussed.

   Technician A says to stand 8 feet from the base of a fire and downwind if possible.

   Technician B says to squeeze or pull the trigger and then pull out the safety clip.

   Who is correct?

   A. A only   C. Both A and B
   B. B only   D. Neither A nor B

5. The SDS and container label DOES NOT contain which of the following information?

   A. Reactivity data
   B. Transportation method
   C. Special fire control
   D. Storage procedures

6. The route of entry into a human body may be:

   A. Inhalation   C. Ingestion
   B. Skin   D. All of the above

7. *Technician A* says the container label of a cleaning material should be read before using the cleaner.

   Technician B says an empty cleaning container should be split before it is thrown away.

   Who is correct?

   A. A only   C. Both A and B
   B. B only   D. Neither A nor B

8. Class B fires are being discussed.

   Technician A says a fire extinguisher labeled with B is to be used on this type of fire.

   Technician B says a spreading agent fire extinguisher can be used to separate the burning materials.

   Who is correct?

   A. A only   C. Both A and B
   B. B only   D. Neither A nor B

9. *Technician A* says gloves should be worn when using caustic cleaning materials.

   Technician B says face and eye protection should be worn when cleaning parts.

   Who is correct?

   A. A only

   B. B only

   C. Both A and B

   D. Neither A nor B

10. Each of the following hybrid vehicle electrical characteristics is true EXCEPT:

    A. The positive conductor will be orange in color.

    B. The positive cable will be equal to or smaller than the typical 12-volt battery cable.

    C. The drive battery cables carry more current.

    D. Insulated gloves should be worn when disconnecting the drive batteries from the vehicle electrical system.

Name _____   Date _____

IDENTIFYING THE SHOP LAYOUT AND SAFETY EQUIPMENT

Upon completion of this job sheet, you will be able to identify the general shop safety rules and procedures, safety features, and location of the safety equipment and evacuation routes.

ASE Education Foundation Correlation

This job sheet addresses the following (**RST**) tasks:

1.1 Identify general shop safety rules and procedures.

1.5 Utilize proper ventilation procedures for working within the lab/shop area.

1.6 Identify marked safety areas.

1.7 Identify the location and the types of fire extinguishers and other fire safety equipment; demonstrate knowledge of the procedures for using fire extinguishers and other fire safety equipment.

1.8 Identify the location and use of eye wash stations.

1.9 Identify the location of the posted evacuation routes.

1.10 Comply with the required use of safety glasses, ear protection, gloves, and shoes during lab/shop activities.

1.15 Locate and demonstrate knowledge of material safety data sheets (MSDS).

2.1 Identify tools and their usage in automotive applications.

2.2 Identify standard and metric designation.

2.3 Demonstrate safe handling and use of appropriate tools.

2.4 Demonstrate proper cleaning, storage, and maintenance of tools and equipment.

We Support

ASE | Education Foundation

Tools and Materials

Goggles or safety glasses with side shield

Procedure

In the blank space below, make a neat line drawing of the shop. Then walk around the shop, and locate and identify items listed below. Mark the location of the item on your drawing using the corresponding number.

1. Fire extinguishers
2. Fire alarms
3. Fire evacuation route
4. Building evacuation rally point
5. First aid kits
6. Eye wash stations
7. Service write-up area
8. Identify marked safety area
9. Shop repair orders
10. Vehicle parking permits
11. MSDS manual
12. Parts department window
13. Hand washing stations
14. Tools and tool boxes
15. Locations of all vehicle hoists
16. Vehicle entrance doors
17. Air ventilation system
18. Parts washers
19. Chemical cabinets
20. Vehicle key lock box
21. Battery chargers
22. Used oil disposal containers
23. Used oil filters containers
24. Oil change carts
25. Bolts and hardware cabinets
26. Floor mats/seat covers
27. Supply room(s)
28. Shop cleaning supplies

Instructor's Response

_____

_____

_____

Name _____   Date _____

IDENTIFYING PERSONAL PROTECTION EQUIPMENT (PPE)

OBJECTIVES:

Upon completion of this activity, you will be able to identify and demonstrate the use of Personal Protective Equipment (PPE).

ASE Education Foundation Correlation

This job sheet addresses the following **RST** tasks:

1.5 Utilize proper ventilation procedures for working within the lab/shop area.

1.10 Comply with the required use of safety glasses, ear protection, gloves, and shoes during lab/shop activities.

1.11 Identify and wear appropriate clothing for lab/shop activities.

1.12 Secure hair and jewelry for lab/shop activities.

2.1 Identify tools and their usage in automotive applications.

2.2 Identify standard and metric designation.

2.3 Demonstrate safe handling and use of appropriate tools.

2.4 Demonstrate proper cleaning, storage, and maintenance of tools and equipment.

We Support
ASE | **Education Foundation**

Tools and Materials

Goggles or safety glasses with side shield

Procedure

Using the text from Chapter 2 Shop Manual, answer the following questions.

1. What does the abbreviation PPE stand for? _____

    _____

2. What equipment must always be worn when working in the shop/lab?

    _____

3. No aspect of automotive repair and service is more important than what?

    _____

4. What is the best way to protect your hands while working on a vehicle?

    _____

5. Losing your _____ is often a gradual process and may be less noticeable than other types of injuries!

6. What is the best way to prevent hearing loss?

    _____

7. What is the best way to prevent back injuries?

    _____

8. What type of footwear is recommended in the shop/lab?

    _____

9. What is the best way to protect yourself from airborne dust or chemicals?

_____

10. What does the abbreviation OSHA stand for?

_____

11. Back injuries can cause long-term _____ and can bring an early end to a technician's career.

12. Maintaining good personal _____ means keeping your hair washed and combed as well as taking a frequent shower and washing your clothes.

13. Explain the purpose of the Environmental Protection Agency (EPA).

_____

_____

14. To prevent injury and damage to the environment, _____ _____ must be handled, stored, and disposed of properly.

15. What are the four steps for using a fire extinguisher? (PASS method)

    a. _____

    b. _____

    c. _____

    d. _____

16. What exhaust gas can be dangerous to you and all people working in the shop/lab?

_____

17. Any time a vehicle is running in the shop/lab, a _____ must be connected to remove exhaust gases.

18. Why is it important to wear protective gloves when disposing of waste oils?

_____

19. Define "work ethic" and describe what having a "good work ethic" means to you.

_____

_____

20. Explain the purpose of OSHA.

_____

_____

Instructor's Response

_____

_____

_____

Name _____   Date _____

PROPERLY DISABLE AND ENABLE THE AIRBAG SYSTEM

Upon completion of this activity, you will be able to properly disable and enable the SIR/SRS. Verify indicator lamp.

ASE Education Foundation Correlation

This job sheet addresses the following **RST** tasks:

1.13 Demonstrate awareness of the safety aspects of supplemental restraint systems (SRS), electronic brake control systems, and hybrid vehicle high voltage circuits.

2.1 Identify tools and their usage in automotive applications.

2.2 Identify standard and metric designation.

2.3 Demonstrate safe handling and use of appropriate tools.

2.4 Demonstrate proper cleaning, storage, and maintenance of tools and equipment.

We Support

ASE | Education Foundation

Tools and Materials

Goggles or safety glasses with side shields

Describe the vehicle being worked on:

Year _____ Make _____ Model _____

VIN _____ Engine type and size _____

Procedure   **Task Completed**

1. Drive vehicle into the stall, and observe the indicator lamp. Is there a SIR/SRS light on? _____ If yes, see Instructor.

2. With the engine running, turn the steering wheel from right stop to the left stop. Do you hear any noise coming from the steering column? Explain.

   _____

   _____

3. With the engine running, point the wheels forward. (Center the steering wheel.) Turn the engine off and turn the key to the lock position.   ☐

4. Test the horn. Does it work properly? _____

5. Using the service manuals, **PRINT OUT AND ATTACH** the service procedure for disabling the SIR/SRS system.   ☐

6. According to the service procedure, how long must you wait after removing the battery terminals until the airbag is safe to work on?

   _____

7. Explain why you must wait for this amount of time.

   _____

   _____

8. Remove the SIR/SRS fuse and the negative battery terminal and wait the specified amount of time. Follow the service procedure for this vehicle.   ☐

9. Using the service manuals, follow the service procedure for enabling the SIR/SRS system.   ☐

10. Verify indicator lamp operation. Does the lamp turn off after the bulb check?

_____

Instructor's Response

_____

_____

_____

CHAPTER 3

USING AUTOMOTIVE TOOLS AND EQUIPMENT

Upon completion and review of this chapter, you should understand and be able to:

- Demonstrate the use of a screw extractor to remove a broken stud or screw.
- Demonstrate the use of a flat chisel to split and remove a stuck nut from a stud or bolt.
- Demonstrate the use of an oxyacetylene torch for heating.
- Collect vehicle service information.

- Connect and use an air impact wrench to remove the lug nuts on a wheel.
- Research the torque specifications for wheel lug nuts.
- Torque wheel lug nuts to the correct torque using a beam, dial readout, and break-over hinge-type torque wrench.
- Select, install, and use a gear puller to correctly remove a pulley from a shaft.

Terms To Know

Cautions	Lug nuts	Quick-disconnect coupling
Charge	Male coupling	Spring-loaded switch
Charge rate	Neutral flame	Striker
Condition seizing	Nut splitter	Race
Female coupling	Penetrating oil	Welding tip
Jump starting	Preloaded	

INTRODUCTION

In the Classroom Manual, we described the types and use of basic tools the technicians use to repair cars. This chapter focuses on some of the most common tool tasks that the technicians perform. We describe how to connect and use an impact wrench, how to adjust and use a torque wrench, how to use a gear puller to remove parts, how to remove a broken stud or screw, and how to remove a damaged nut. Specific safety measures are also discussed in this chapter.

TOOL AND EQUIPMENT INSPECTION AND MAINTENANCE

Basic Tools

Basic technician's tool set: Eye protection

Before beginning work with any power tool or equipment, it is best to perform a general inspection of the devices and their power connection. Although you may be comfortable picking up one of your own wrenches or screwdrivers and starting work, using a tool or piece of equipment powered by electricity, air, or hydraulics presents a chance to do more damage than a set of skinned knuckles. Plugging the tool into its power source is not part of the pre-use inspection. If there is damage, it may become obvious in a harmful manner. A general inspection at the item in question takes less than a minute and could prevent damage or injury.

Power Tools

Generally, power tools in an automotive shop use either air or electricity for power. Look at the hose or cord for any damage or cuts. A ruptured air hose can cause almost as much injury as a shorted 110-volt electrical cord (**Figure 3-1**)—it just takes longer. Inspect the hose for any swelling along its length with emphasis near the connections. The end connections of the hose or cord should be firmly fastened without damage. On an electric cord, ensure the third prong is present and tight in the connector.

The tool itself should be clean and clear of any burrs or cover cracks, and the connector should be tight and fully functional. On an electric-powered tool, check the area where the cord enters the tool body. This is a common site for fraying of the protective insulation and the wires. The air fittings must be tightened into the cavity.

Though most shops do not use tools driven by hydraulics, technicians working on certain types of equipment may be exposed to tools powered by up to 10,000 pounds per inch (psi) hydraulic pressure. Do not work or disconnect these types of tools until you are sure the pressure is at zero psi and you know how the tool operates.

Classroom Manual
pages 54 and 55

Figure 3-1 Look for bulging hose around the air hose connector and frayed or exposed wiring around electrical connections.

AUTHOR'S NOTE Many electric utility trucks use drills that operate on 10,000 psi. During a typical leak repair on a drill of this type, the technician did not check to ensure the pressure was off. A small leak was found at the hose/connector crimp. There appeared to be no pressure behind the leak; however, when the technician started to disconnect the tool, he twisted the hose "just right" and opened the leak up to about a 1-inch stream of highly pressurized fluid. The fluid entered his skin at the bottom of his wrist just behind the heel of the hand. Before he could move, several ounces of hot fluid were injected under the skin to a point above the elbow. The lower arm was swollen to more than double its natural size. He was evacuated to the emergency room where the fluid was sucked out. Luckily, he suffered no lasting damage and was back at work in 2 days. Never trust what appears to be a minor problem when dealing with hydraulic, electrical, and pneumatic systems.

Equipment

Most automotive equipment is powered by either electricity or hydraulics. In most cases, electricity is used to power the hydraulic pump for hydraulic pressure to drive the cylinder. The most common type are the vehicle lifts (**Figure 3-2**). The system is a simple one

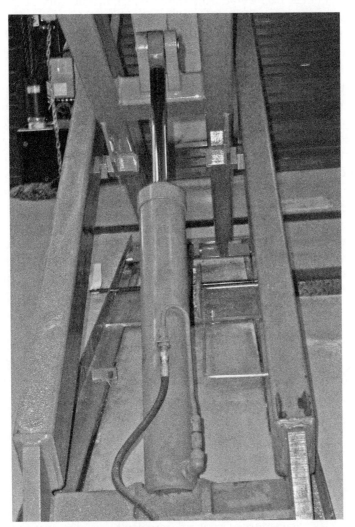

Figure 3-2 This is a lift cylinder for a drive-on scissor lift. The hoses can be damaged by careless work or maintenance, as well as old age. Always inspect for leaks before using lift.

Figure 3-3 This lift arm lock was apparently damaged during careless operation. It must be replaced before further use of the lift.

that can injure people or damage vehicles. Inspect the visible hoses and around the area where the lift cylinders are located. Any fluid present on the floor or the lift mechanism covers indicates a leak. In most cases, the lift cylinders are not completely visible, so the leaked fluid may be the only indication of a problem. The electrical cords for a lift are usually encased in a conduit and cannot be inspected, but the conduit itself can be checked.

Walk around the lift and inspect the lift arms, pads, and pin connections. Of particular interest are the arm locks. When the arms are moved into lift position and moved a few inches up, the mechanical locks drop in place to prevent the arms from swinging or kicking out when the vehicle weight is applied (**Figure 3-3**). Trouble may be caused if a technician removes or disables the locks to make it easier to position the pads. If the lift has not been used for a period of time or has just been repaired, operate the lift until the mechanical locks on the lift cylinder drop in place. Never work under a vehicle if the lift locks are not engaged.

Hoists have to be periodically maintained. Check the owner's manual on the hoist for the specific requirements. Most of the time, a hoist with large pistons needs to be greased once a month or more frequently. Sometimes, pulleys, wheels, and cables that move need to be lubricated with a light oil. Hoists should be inspected by the technician every time he or she uses it but calling in a professional hoist repair company to do an annual inspection of the shop's hoist is usually done.

Classroom Manual
pages 66 and 67

LIFTING AND SUPPORTING A VEHICLE

To work safely and quickly, it is sometimes required to raise the vehicle to a good working height. If brakes or tires are being repaired, a good working height has the wheels at about chest level. Engine and transmission oil changes can be done more easily with the vehicle high enough for the technician to walk under it.

Double-Post Lifts

Special Tools

Vehicle double-post above-the-ground lift

Service manual showing lift points

Double- and single-post lifts work similarly, with the primary difference being the position of the lift cylinder(s). Procedures discussed here can be used to some extent with single-post lifts. Before lifting any vehicle with any type of lift or jack, consult the driver's manual or service manual for the location of the vehicle's lift or jack points (**Figure 3-4**).

Do not lift or support on track bar

☐ Drive-on hoist   ▨ Frame-contact hoist   ▨ Twin-post hoist

A   Unibody vehicle

▨ Drive-on hoist   ▨ Twin-post suspension hoist   ■ Frame-engaging hoist

B   A vehicle with a full frame

Figure 3-4 The highlighted points are typical lift and jack points.

⚡ WARNING **Never guess at the vehicle's lifting points. What appears to be solid metal may bend or be crushed when the vehicle's weight is transferred to it. Damage to the vehicle and possible injury could occur.**

Move the vehicle so it is centered between the posts. The front doors should be roughly even with the posts. Most lifts of this type have shorter lift arms at the front. This places most of the vehicle's weight closer to the lifting cylinder and provides better balance. Most double-post lifts have a **spring-loaded switch** button that operates the motor and the pump. Pressing and holding the button down will power the motor. Any time the switch is released for any reason, the spring will automatically push the switch to OFF.

After identifying the lift points, rotate the lift arms under the vehicle and position the pads directly under the lift points. On longer cars and trucks, it may be better to raise the lift arms until the pads are just below the vehicle. This makes it easier to properly place the pads.

With the pads aligned, lift the arms until there is contact between the pad and vehicle. Check the contact points to ensure there is good proper alignment between the pads and lift points. Adjust as necessary.

As the vehicle is being raised, note that both ends of the vehicle are moving upward evenly. Pickup trucks are hard to keep even because of their length and the position of the lift points. If the vehicle is not even from end to end and side to side, lower it back to the ground and move the vehicle or arms until both ends rise at the same time. Continue to raise the vehicle until all wheels are off the floor. Attempt to rock the vehicle by hand. If positioned properly on the lift, it should not rock at all. In fact, it will probably have a more solid side-to-side movement than when sitting on its wheels.

When the vehicle is at the desired work height, stop the upward movement. Lower the vehicle until the safety locks on the lift engage. The vehicle can now be repaired.

To lower the vehicle, first raise it high enough to clear the locks. There is usually a lock on each post. Move both the locks to the unlock position. Some lifts require the technician to hold an unlocking lever in place while the lowering valve is operated. A small

This type of **spring-loaded switch** is sometimes referred to as a "dead-man switch."

lever attached to another spring-loaded switch usually operates the lowering valve. Lower the vehicle until the pads break contact. Continue lowering the lift arms until they are at the bottom of their travel. Swing the arms from under the vehicle.

> **AUTHOR'S NOTE** There are two types of two-post, above-ground lifts: asymmetrical and symmetrical. The primary visual difference is that the four lift arms on a symmetrical lift are of the same length, whereas the front arms on an asymmetrical lift are noticeably shorter than the rear arms. Each type of lift can be used with almost any vehicle, but each works best with a particular type of vehicle. The shop may have all lifts of the same type or a variety. Honda dealerships have a special-purpose lift designed for the express service bays. The vehicle can be driven onto the lift head. The lift head is designed to make contact along the vehicle's undercarriage, and the ramps drop down during lifting so that the wheel assemblies can be removed. This lift saves time and reduces possible damage to the vehicle.

Floor Jacks and Jack Stands

Some work can be done underneath the vehicle using a floor jack and jack stands. It may be more difficult, but for some jobs it may be quicker than using a lift.

Before jacking up the vehicle, decide where the jack's lift pad will be placed and how much of the vehicle is being lifted. Shift the transmission to PARK (forward gear on manual transmissions) and set the parking brake. Position the wheel blocks behind and in front of at least one wheel left on the floor. The surface should be level enough for the jack wheels to roll.

Unlock the jack handle and align the jack's lift pad under the jacking point of the vehicle. Rotate the valve knob on the end of the handle clockwise to close the valve. Move the handle up and down to raise the jack pad until it contacts the vehicle's lift point. Check the contact and, if it is correct, continue to jack up the vehicle. The jack should roll further under the vehicle as it is raised. This is to keep the pad directly under the jacking point. If the jack cannot roll, it will either pull the vehicle toward the jack handle or slide the pad from its contact point.

When the vehicle is at the desired height, slide the jack stands under the vehicle. They should be placed at a point that will support the weight of the vehicle. This is usually under the frame, under the lower end of the suspension, or under the axle housing. The stands should be placed opposite to each other and at the same height to keep the vehicle level. With the stands in place, slowly rotate the jack's valve knob counterclockwise. The knob will probably stick a little because of the hydraulic pressure against the valve. As the knob begins to turn, the jack will begin to lower. Proceed slowly until the vehicle rests securely on the jack stands. Remove the jack, stand the handle straight up, and lock it in place. Do not leave the jack in a walking area, and *never* leave the handle lying down.

To remove the stands and lower the vehicle, carefully place the jack in position to make the same contact as when raising the vehicle. Unlock the handle and close the valve. As before, once the pad contacts the vehicle, check it before raising the vehicle. Raise the vehicle just enough to remove the stands. With the stands removed, open the jack valve slowly and lower the vehicle to the ground. Store the jack properly. Ensure that the parking brake is set before removing and storing the wheel blocks. See **Photo Sequence 2**.

CONNECTING AND USING AN IMPACT WRENCH

The air-operated impact wrench is frequently used by all technicians. **Photo Sequence 3** shows a typical procedure for connecting and using an impact wrench to remove **lug nuts**. The impact wrench is often misused by technicians who do not understand that its main

---

Special Tools

Floor jack
Wheel blocks
Service manual showing jack point
Jack stands

⚠ **Caution**

Never perform any work under a vehicle supported only by a jack. Severe injury or death could result if the vehicle slips from the jack or the jack moves.

⚠ **Caution**

Never perform any work other than a tire change using a tire-changing jack. This jack is designed to raise one corner of the vehicle to change a tire. It is not made for other repairs. Serious injury could result if the jack and vehicle should move.

⚠ **Caution**

Never work under a vehicle supported only by a jack of any type. Injuries could occur if the jack slips.

Lug nuts are the hex-sided fasteners used to hold the wheel rim on the studs of the brake rotor or drum.

PHOTO SEQUENCE 2
Typical Procedure for Jacking and Supporting a Car

P2-1 Drive the car on to a solid, level surface.

P2-2 Set the parking brake.

P2-3 Put the transmission in PARK.

P2-4 Block the front and rear of one of the wheels that is not being lifted.

P2-5 Position the jack pad on a major strength point under the car. Check the service manual as needed.

P2-6 Turn the jack handle valve to the lifting position.

P2-7 Raise the car carefully.

P2-8 Adjust and position jack safety stands to properly support the car.

P2-9 Slowly turn the jack handle to the lowering position and lower the car down on the jack safety stands.

P2-10 Shake the car to test its stability on the jack safety stands.

PHOTO SEQUENCE 3
Typical Procedure for Connecting and Using an Impact Wrench to Remove Lug Nuts

P3-1 Raise the car on a hoist or jack and jack safety stands.

P3-2 Remove the wheel cover from the wheel.

P3-3 Connect the air impact wrench to the air hose quick-disconnect coupling.

P3-4 Install the correct size impact socket on the wrench drive.

P3-5 Switch the impact wrench to the reverse (counterclockwise) drive direction.

P3-6 Place the wrench and socket over the lug nut and pull the trigger to remove the lug nut.

P3-7 Remove each of the other lug nuts and remove the wheel.

P3-8 Install the wheel back on the car and start each lug nut by hand.

P3-9 Switch the impact wrench to the forward (clockwise) direction.

P3-10 Carefully run the nuts up but do not tighten.

P3-11 Look up the tightening torque and tightening sequence for the lug nuts in a shop service manual.

P3-12 Set the torque wrench to the specified torque.

P3-13 Tighten each lug nut in the correct order to the specified torque.

P3-14 Replace the wheel cover and lower the car.

P3-15 A torque stick may be used instead of a torque wrench to torque lug nuts only.

function is disassembly. The typical air-operated impact wrench is not torque controlled. This means the actual torque applied is not known when this tool is used. Using an air-operated impact wrench to assemble components can result in overtightened or undertightened fasteners.

A common use of the air-operated impact wrench is to take off wheel lug nuts to remove a wheel from a car. First raise the wheel to be removed safely off the floor. Follow the safe procedure described earlier in this chapter to raise the vehicle on a hoist or to raise and support the car on a floor jack and jack safety stands.

Many cars use a wheel cover that must be removed to access the lug nuts. Other wheels use the lug nuts to hold on the wheel cover. If the car has wheel covers that must be removed first, use a prying tool to gently pry the wheel cover off the wheel rim as shown (**Figure 3-5**). Some cars have locks that prevent wheel theft. And some vehicles use plastic screw caps to hold the wheel cover tight on the lug nuts. Be careful not to damage these. Use a ratchet and socket to remove them, not the air impact wrench. The lock mechanism typically fits over one of the lug nut studs. Use the key to remove the lock, then pry off the cover. Be careful not to bend or distort the wheel cover or it will not fit back on properly.

The lug nuts hold the wheel onto studs. Passenger cars typically use either four or five lug nuts to hold on the wheel. Passenger trucks typically use four to eight lug nuts. These studs are attached to the brake rotor, axle flange, or hub. There are several designs of lug nuts (**Figure 3-6**). Lug nuts that hold on steel wheels are often different than those that

Classroom Manual
page 55

Figure 3-5 Removing the wheel cover.

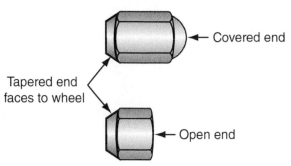

Covered end

Tapered end faces to wheel

Open end

Figure 3-6 Parts of a typical lug nut.

hold on aluminum wheels. One side of the lug nuts are chamfered. The chamfered side always goes on toward the wheel rim. The chamfer fits in a matching chamfer on the wheel and centers the wheel over the studs as it is tightened.

> **SERVICE TIP** You should make match marks on the wheel rim and one of the studs with chalk or grease pencil. Always replace the wheel in the same position in case it has been balanced on the car.

The lug nuts are hex-shaped or six-sided to fit standard wrench sizes. The wrench sizes are different for different cars. Select a six-point impact socket (**Figure 3-7**) that snugly fits the hex sides of the lug nut.

Sometimes the lug nut is covered with a chrome-look cap. At times, this cap will shear and come off in the socket. If this happens, do not use the same socket in an attempt to loosen the nut. It will be slightly too big and may round the nut corners. Use a slightly smaller (e.g., 11/16-inch in place of a 3/4-inch) socket to remove the nut. Removing a lug nut with rounded corners sometimes becomes an almost impossible task without damaging the wheel. Make sure you install a new lug nut when you find a damaged one. You may not be able to properly torque the worn lug nut to the correct specification. Leaving it loose is not an option.

> **Quick-disconnect coupling** Used to provide a fast and easy connection and disconnection to air hoses for air-operated tools.

> A **female coupling** houses the locks that hold the male coupling in place after hookup. This coupling also has the release mechanism.

> A **male coupling** slides into a mated female unit similar to the manner in which a ratchet's drive stud fits into a socket. The male is fitted with grooves and ridges for locking.

⚠ **WARNING** Using a loose-fitting impact socket on a wheel lug nut can cause the corners of the hex to be rounded off and not properly fit a wrench.

At this time, set up and connect the air-operated impact wrench (**Figure 3-8**). The wrench has a socket drive on the end. Install the impact socket onto the wrench drive.

The wrench will need to be connected to the shop air source to get its power. The shop air hose is connected to the impact wrench with a part called a **quick-disconnect coupling**. This coupling allows air lines to be attached to tools quickly without shutting off any air source. The coupling has two basic parts, as shown in **Figure 3-9**. The part attached to the air hose is called the **female coupling**. The part attached to the wrench is called the **male coupling**. To attach the hose to the gun, pull back on the female

Figure 3-7 A six-point impact socket is used with an impact wrench.

Figure 3-8 Parts and controls on a typical air impact wrench.

Socket drive

Quick-disconnect coupling

Reverse switch

Trigger

Air hose

Figure 3-9 Parts and operation of a quick-disconnect air line coupling.

coupling slide and insert it over the male coupling. Release the slide, and it will lock the two parts together in an airtight fit. The coupling may be disconnected in the same way. Just pull back the slide, and the two parts of the coupling will come apart.

With the line connected, the wrench is ready for use. Depress the trigger. The wrench drive will spin the socket at a high speed. Notice the direction the socket spins. To remove most lug nuts, set the reverse switch on the wrench so the socket is driven counterclockwise. Try the reversing switch to see how it works, then set the wrench in the correct direction to remove the lug nuts.

⚡ **WARNING Some cars made in the 1940s through the 1960s used left-handed threads on the lug nuts on the left side of the car. The designers did this so that if the nuts were loose, they would be automatically tightened by the rotation direction of the wheel. Look for this on old cars and do not mix up the lug nuts.**

Insert the impact socket over one of the lug nuts. Pull the trigger and spin off the lug nut (**Figure 3-10**). Repeat this procedure on each of the other lug nuts. Support the wheel as the last lug nut is removed to prevent the wheel from falling off the studs. Carefully lift the wheel off the studs. Use the proper lifting techniques described in Chapter 2 of the Shop Manual.

To reinstall the wheel, first place it in position on the studs. Always start the lug nuts by hand (**Figure 3-11**). Make sure they thread onto the stud several turns before using any wrench. Using a wrench to start the lug nuts could damage the threads.

⚠ **Caution**

Always use impact sockets with an impact wrench. Standard sockets can fracture and explode when used on an impact wrench. Always wear eye protection when using an impact wrench.

⚠ **Caution**

Do not let the impact wrench spin off the lug nut. The nut will fly out of the socket and can damage the vehicle's finish or your face, body, or that of a bystander. Slow the wrench's rotation speed by easing up on the trigger once the nut has broken loose to prevent possible damage or injury.

Figure 3-10 Removing lug nuts with an air impact wrench.

Figure 3-11 Always start lug nuts by hand.

Figure 3-12 Wheel torque sockets can be used with an impact wrench to torque lug nuts.

Special Tools

Impact wrench
Wheel torque socket
Vehicle lift or jack

Use the air impact wrench to spin the nuts in contact with the wheel, but do not use a typical impact wrench to tighten the nuts. Reverse the direction of the impact wrench with the reverse switch. If the impact wrench has a pressure switch, set it to the lowest setting. Carefully spin each nut on the stud until it just begins to contact the wheel. Tightening must be done with a torque wrench, as explained in the next section.

SERVICE TIP Install each lug nut just until it pulls the rim to the hub or axle. This will prevent the rim from "cocking", preventing a flat fit against the hub or axle. Once all of the rim seems to be centered and in place, the lug nuts can be spun into place and then torqued.

If wheel torque sockets are available, select the correct one for the torque desired. Place the torque socket in the impact wrench in place of the regular impact socket used to remove the lug nuts (**Figure 3-12**). Do not hold the bar with your hand. The vibration is part of the calculated torque. Use the impact wrench to torque the lug nuts afterward. If the torque sockets are not available, set the air impact wrench on its lowest setting and then hand torque them with a torque wrench afterward.

ADJUSTING AND USING A TORQUE WRENCH

Installing and tightening the lug nuts is a good way to learn how to use a torque wrench. Lug nuts should *always* be torqued with a torque wrench. Using the correct torque setting ensures that the lug nuts are tightened the right amount and evenly. If the lug nuts are tightened unevenly, the drum or rotor can warp.

The first step is to look up the torque specifications for the lug nuts on the car being serviced. An example of a wheel lug nut torque specification table is shown in **Figure 3-13**. This particular car has a recommended wheel nut torque of 140 N • m or 100 ft.-lb. The symbol N • m is for the metric system measurement newton-meter. The symbol ft.-lb. is for the conventional (English) system measurement foot-pound.

Figure 3-13 Many torque sticks come in a set and are color coded to the designed torque speeding up the process of selecting the correct torque stick.

Figure 3-14 Parts of a break-over hinge-type torque wrench.

The measurement can be made with either a metric or a conventional reading torque wrench.

Most charts for wheel nuts will have a column labeled "foot-pounds." However, some of the torque measurements may have a small number or mark that indicates this particular torque is different from the others in this column. The number or mark refers to a note at the bottom of the torque chart. This note may state that this torque specification is in "inch-pounds" instead of foot-pounds. Many fasteners have been broken by not paying attention to the details.

The most common automotive torque wrench in use today is the break-over or hinge-type (**Figure 3-14**). They are available in 1/4-inch, 3/8-inch, 1/2-inch, and larger size drives. The smaller sizes usually have inch-pound scales, whereas the larger use foot-pounds. Most 3/8-inch drives can be purchased scaled in either inch-pounds or foot-pounds. Scale settings commonly range from 20 to 200 inch-pounds or 50 to 250 foot-pounds. A 1/2-inch drive torque wrench is commonly used to torque lug nuts on light vehicles.

Set the wrench up by adjusting the torque wrench setting to the specified torque. This is done by turning the end of the handle until the zero-index mark is aligned with the proper torque. If the torque is to be 100 ft.-lb., then align the zero mark with 100 ft.-lb. (**Figure 3-15**). The wrench is now set to click or break at 100 ft.-lb. of torque.

Special Tools

Dial readout-type wrench

Break-over hinge-type Torque wrench

Figure 3-15 Break-over hinge handle setting of 100 ft.-lb.

After setting the wrench to the required torque, attach the proper size socket and tighten the lug nuts in a star pattern. It is best to tighten each nut a few turns until all are almost tightened to specifications. This pulls the wheel in square with the axle or wheel hub. Once the nut reaches the wrench torque setting, the wrench will "click," indicating that the torque is correct. One thing to remember when applying low torque to smaller fasteners is that the click may not be very audible, but you should be able to feel the hinge "break over."

Electronic torque wrenches are also available and much more accurate. Switch the wrench on and use the buttons to select the desired torque setting. Make sure the correct measurement—such as foot-pounds—is selected. Torque the fastener until a buzz or vibration is heard or felt. This means the desired torque is achieved. Note the reading on the wrench's display. The torque shown is the actual torque applied. For instance, the wrench may be set for 100 ft.-lb., but after torqueing the display indicates that 102 ft.-lb. were applied. Normally, this overtorque is not a problem unless the fastener has a very low torque setting. Most torque specifications allow for some tolerances.

Dial readout torque wrenches are not commonly used in automotive repair shops except for certain types of repairs (**Figure 3-16**). One type of repair involves measuring the turning torque of a gear or shaft. When a technician rebuilds rear-wheel-drive (RWD) differential gears, the bearings must be **preloaded**. Preloading means a certain amount of force is applied against both sides of the bearing to ensure fit within the bearing **race** for quiet, long-life operation.

To do this, the bearing retaining nut is tightened until all back-and-forth, side-to-side movement is removed. Do not overtighten or you will have to start over and possibly install a new part. A dial torque wrench is attached in some manner to the shaft or gear and is used to turn that shaft or gear (**Figure 3-17**). The amount of force required to turn the shaft or gear is shown on the dial readout. If too low, the nut is tightened (a turn or two) and another measurement is taken. Continue repeating these steps until the specified turning torque is reached. The dial readout torque wrench is one of the few tools readily available to perform this type of measurement.

Preloaded Axial or thrust load applied to a bearing that removes excess play.

The term **race** refers to the machined area(s) of the bearing, shaft, or gear where the bearing's rollers or balls contact and roll on.

Figure 3-16 A dial torque wrench is not used often in most automotive repair shops and is only used for special jobs.

Figure 3-17 A dial torque wrench is commonly used to check bearing preload by measuring the amount of force needed to turn the shaft.

USING A GEAR AND BEARING PULLER

As noted in the Classroom Manual, pullers come in many different sizes and shapes, and many can be used for only certain types of pulling operations. The puller in **Figure 3-18** is designed to remove the front crankshaft pulley or harmonic balancer on certain models of Chrysler vehicles. It will not work well on other vehicles. **Figure 3-19** shows a puller set commonly used to remove steering wheels or the harmonic balancer except for certain Chrysler vehicles. However, both types work in a similar manner.

Special Tool
Gear and bearing pullers

Figure 3-18 This is a special-purpose tool used to remove the harmonic balancer from certain Chrysler engines.

Figure 3-19 A typical puller set used to remove steering wheels and harmonic balancers. It can be used in other applications as well.

Removing a Harmonic Balancer

For this exercise, the puller set in **Figure 3-19** will be used. This is the most commonly used puller in an independent shop. The harmonic balancer was selected for this exercise because it is usually easy to reach, especially if the engine is out of the vehicle for training purposes. Other gears and pulleys can be removed in a manner similar to that discussed here. Remove the large fastener securing the harmonic balancer. Inspect the harmonic balancer and select the two or three bolts (depending on the manufacturer) in the set that will screw into the outer threaded holes. Next select a center cone that fits into the large center threaded hole. This cone will help protect the interior threads during the pulling operation. Remove the large pointed threaded bolt from the set and screw it into the center of the puller body. Rest the point of the large puller bolt into the cone and install one of the side bolts selected earlier. Screw this bolt partially into one of the outer holes before installing the other one or two bolts in a similar manner. Screw these bolts in until they exit the back side of the balancer. They may be visible at that point, but they should become much easier to turn. A wrench may be required to screw them all the way. Each of the two (or three) bolts should be of equal length, and the puller body should be squared with the face of the balancer. Tighten the large center puller bolt—a wrench will be needed—and the balancer should begin to slide off the crankshaft. Continue tightening until the balancer is clear. Remove and store the puller in its case.

Classroom Manual
page 63

PHOTO SEQUENCE 4
Typical Procedure for Lifting a Vehicle on a Hoist

P4-1 Refer to the service information to determine the proper lift points for the vehicle you are lifting.

P4-2 Center the vehicle over the hoist, considering the vehicle's center of gravity and balance point.

P4-3 Locate the hoist pads under the lift points. Adjust the pads so that the vehicle will be lifted level.

PHOTO SEQUENCE 4 (CONTINUED)

P4-4 Lift the vehicle a couple of inches from the floor. Shake the vehicle while observing it for signs of any movement. If the vehicle is not secure on the hoist or unusual noises are heard while lifting, lower it to the floor and reset the pads.

P4-5 Once the vehicle is at the desired height, lock the hoist. Do not get under the vehicle until the hoist locks have been set.

P4-6 To lower the vehicle, release the locks, and put the control valve into the "lower" position. Once the vehicle is returned to the floor, push the contact pads to a location that is out of the path of the tires.

Removing a Gear

Pulling a gear from a shaft uses a technique similar to that used for the harmonic balancer remover. There are some steps that must be done correctly, including selecting the correct puller, or the gear or shaft may be damaged. A three-jawed puller is used to pull many gears (**Figure 3-20**). Select a puller or reposition the jaws so the pulling ends fit completely around the gear. The jaws should not pull on the gear teeth but rather on the gear body.

Special Tools
Vise grips
Small cape chisel
Set of screw extractors
3/8-inch portable drill

⚠ Caution
If the puller jaws do not fit the part, or the cross arm will not allow proper placement of the jaws, select another size puller. Using an incorrect size puller or not adjusting the puller correctly can allow the jaws to slip off during pulling and cause injury. Always wear eye protection when using a puller. Parts can break and cause eye injury.

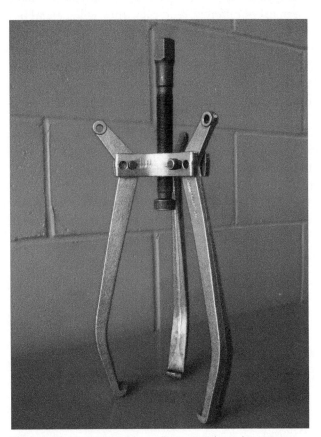

Figure 3-20 A typical three-jaw puller commonly used to remove bearings or gears from a shaft.

Center the long pulling bolt at the center of the end of the shaft and tighten it until the puller holds itself in place. It may be necessary to anchor the shaft and gear to prevent it from turning. Do not clamp the shaft at any bearing journals. Bearing journals are the shiny machined surfaces that can be damaged with a clamping device. Use a wrench to turn the puller center bolt until the gear slips off the shaft.

REMOVING A BROKEN STUD OR SCREW

Regardless of how carefully a technician works, sooner or later the misfortune of breaking a screw or stud occurs. The broken stud or screw will have to be removed before the service work can be completed. Broken screw removal is a very valuable skill. We suggest trying some of the techniques described here on some scrap parts before doing it on a customer's car.

Screws fail for many reasons. Sometimes they are defective when manufactured. Often they are overtightened by a technician not using a torque wrench. Sometimes they are driven too deeply into a threaded hole and bottom out. Regardless of the cause, a broken screw has to be removed and a new fastener used in its place.

There are a couple of removal techniques that should be used before any special tools are tried. Sometimes the screw or stud breaks with a part of the fastener sticking above the surface. This often happens if a screw is bottomed out in the hole. First, try a pair of vise grips. Adjust them to the proper size and grip the broken fastener as shown in **Figure 3-21**. Then turn the fastener in the correct direction for removal.

> **⚙ SERVICE TIP** Sometimes difficult-to-remove broken fasteners can be removed by carefully heating or cooling the area. Because the fastener and the threaded hole are made of different materials, they will expand or contract at different rates. This can cause the fastener to loosen up. Heating can be done with an oxyacetylene torch. This technique can only be used where there is no fire danger or around components made of soft metals like aluminum. Heat the area around the fastener. Try to keep as much heat as possible from the fastener itself. This will speed the process of separating the fastener threads from the component.

If the fastener is broken off even with or below the surface, the removal job will be more difficult. Sometimes the fastener breaks even though it is not extremely tight. A fast removal technique uses a small cape chisel. Place the chisel on the broken surface of the fastener. Use a hammer to tap it in a direction for removal as shown in **Figure 3-22**. If the fastener is loose enough, it may begin to rotate. Tap the fastener around until it sticks above the surface. Then grab and continue turning it with vise grips.

⚠ Caution

Always wear eye protection when drilling and using a screw extractor. Parts could fracture and injure your eyes.

Vise grips

Figure 3-21 Using vise grips to remove a broken fastener.

Cape chisel

Fastener broken
off at surface

Figure 3-22 Using a chisel to back out a broken
fastener.

Figure 3-23 Typical extractors that fit over a nut like a socket and are driven by an impact
wrench or ratchet. Extractors may be purchased individually or in sets.

Flutes

Figure 3-24 The extractor at the top uses a reverse (left hand) thread type screw to grip the
fasteners. Both types can be used in most situations. One key point is using heat to loosen a
fastener. If heat is used, DO NOT use an extractor because heat will warp it and ruin the extractor.

If these techniques fail to work, a screw extractor must be used. There are many
different types of screw extractors. They often come in sets and are often supplied with
the necessary size of drill, as shown in **Figure 3-23**.

Two common types of screw extractors are shown in **Figure 3-24**. One has *reverse
threads* and the other has a set of *flutes*. In either case, the screw extractor is made to
fit in a hole drilled in the fastener. Insert the screw extractor and turn it with a tap
wrench. The reverse threads or flutes grip the fastener and allow it to be backed out of
the threads.

When using a screw extractor, always follow the instructions supplied with the set.
Also, make sure the correct size drill and screw extractor are selected for the job. The
correct size to use is typically shown on the set's instructions.

The first step is to center punch the exact center of the broken fastener as shown in
Figure 3-25. Then install the correct size drill bit in a 3/8-inch-size portable drill motor.
Drill through the center of the fastener to a depth recommended in the screw extractor
instructions, as shown in **Figure 3-26**. Be careful not to drill too deeply and damage the
automotive part, or to drill off to an angle.

With the hole drilled properly, install the extractor. Be sure to choose the correct size
extractor for the hole drilled. Install the extractor as shown in **Figure 3-27**. Some extractors must be tapped into the hole with a hammer.

Use a tap wrench to turn the extractor in a direction to back the fastener out of the
hole as shown in **Figure 3-28**. Some screw extractor sets have drivers that slide over
the extractor and fit a standard automotive wrench.

Special Tools

Cold chisel
Vise grips
Penetrating oil
Nut splitter
Antiseize compound

Figure 3-25 Center punching the center of the broken fastener.

Figure 3-26 Drilling a hole in the broken fastener.

Figure 3-27 Installing the screw extractor in the drilled hole.

Screw extractor

Broken bolt with hole drilled in the middle

Figure 3-28 Using a tap wrench and extractor to remove the broken fastener.

> **SERVICE TIP** Stuck fasteners can often be removed by heating them. You can use an oxyacetylene torch to warm the nut, which causes it to expand and break its bond with the stud or bolt. You can only use this technique on nuts where it is possible to heat without the danger of fire or around components that are not made of soft materials like aluminum.

REMOVING A DAMAGED NUT

A good technician knows how to remove damaged nuts. Cars that have accumulated many years and high mileage that have been repaired many times often have damaged fasteners. For example, hex nuts may have rounded corners due to the use of an incorrect wrench size (**Figure 3-29**), or a nut may have rusted to a bolt or stud like the one shown in **Figure 3-30**.

Figure 3-29 An example of a nut with rounded corners.

Figure 3-30 An example of a nut that has rusted on a stud.

A nut that is rounded at the corners often will not allow the correct size wrench to fit tightly enough to remove the nut. These usually result when a technician uses the wrong size wrench or when a nut is stuck or "seized" to the bolt or stud.

First try to use a file and remove the displaced metal. Then try the largest adjustable, open-end wrench that will correctly fit on the nut. Be careful to push on the wrench with the hand positioned properly in case it slips off the nut. If this procedure does not work, try large vise grips. Adjust these to fit tightly on the nut. Try to turn the nut with the vise grips. Be careful to protect the hand in case the grips slip off the nut. If neither of these steps work, the nut will have to be split as described later.

Another problem is with nuts that are stuck on a bolt or stud. A nut may be stuck because it was overtightened, and its threads are jammed or distorted into the threads of the stud or bolt. Corrosion and rust build up between the two fasteners and lock them together. This is called **condition seizing**. In either case, the nut may be impossible to loosen, and pulling on it with too much force can cause the bolt or stud to break. This makes it even more difficult to remove the broken stud or bolt.

condition seizing A bolt or nut stuck due to the conditions causing the metal to rust.

If a stuck nut is encountered, first apply a **penetrating oil** as shown in **Figure 3-31**. Penetrating oil is made to seep into the threads between stuck fasteners and remove the corrosion. Follow the directions on the oil container. Typically, the penetrating oil must be applied several times, and a long period of time is needed for it to work.

penetrating oil A very low-viscosity oil that can be used to free rusted mechanical parts.

When penetrating oil fails to remove the nut or when the nut is too damaged to be removed, the last resort is to split the nut. Use a sharp flat chisel to make a complete cut between the nut and the stud or bolt as shown in **Figure 3-32**. Try not to contact the

Figure 3-31 Penetrating oil can free a rusted nut.

Figure 3-32 A nut splitter.

Figure 3-33 A nut splitter.

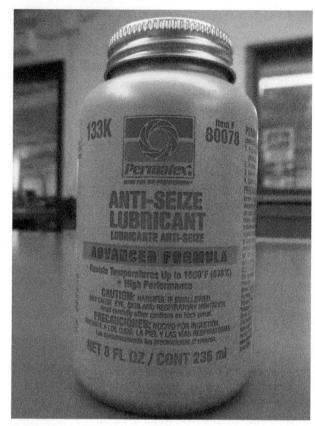

Figure 3-34 Antiseize compound prevents nuts from sticking to bolts and studs.

 Caution

Oxygen and acetylene cylinders must be secured in an upright position. Never allow a cylinder to fall over. If a cylinder were to fall over and break off its valving, it could become a rocket. When cylinders are being stored, they must have a steel protecting cap over the valve area and the regulators must be closed.

Classroom Manual pages 68–69

threads of the bolt or stud. Usually one cut will free the nut. Sometimes you will have to cut it on opposite sides to get it off.

A tool called a **nut splitter** is available to split a nut (**Figure 3-33**). It is positioned over the stuck nut. A forcing screw driven by a wrench causes a cutter to penetrate and split the nut.

Fasteners used in areas of the car subject to heat and corrosion often seize over time. An example is the fasteners used under the car on the exhaust system. Seizing is also a problem when steel fasteners are used on aluminum parts. A good prevention technique is to coat these fasteners with an antiseize compound when they are assembled. This will make your job much easier if you must remove the parts later. An example of an antiseize compound is shown in **Figure 3-34**.

AUTHOR'S NOTE When a little extra time is taken to do a professional job, like using a torque wrench or coating fasteners with antiseize compound, the repair job will last longer. A technician's reputation as a craftsperson will pay off in the long run when customers want her or him to work on their cars.

SETTING UP AND USING AN OXYACETYLENE TORCH FOR HEATING

The oxyacetylene welding and cutting outfit is used in the shop for welding, cutting, or heating. **Photo Sequence 5** shows a typical procedure for setting up and lighting an oxyacetylene torch for heating. Welding and cutting are skills many technicians develop

PHOTO SEQUENCE 5
Typical Procedure for Setting Up and Lighting an Oxyacetylene Torch for Heating

P5-1 Connect the green or black oxygen hose to the torch handle connection marked "OX" by turning it clockwise (right-hand threads).

P5-2 Connect the red acetylene hose to the torch handle connection marked "AC" by turning it counterclockwise (left-hand threads).

P5-3 Connect the correct size torch tip to the end of the torch handle.

P5-4 Close both torch handle valves by turning them clockwise.

P5-5 Turn the main cylinder valve on the oxygen cylinder valve on by turning it counterclockwise.

P5-6 Turn the main cylinder valve on the acetylene cylinder valve on by turning it counterclockwise.

P5-7 Turn the oxygen regulator valve clockwise until the working gauge reads 10 psi.

P5-8 Turn the acetylene regulator valve clockwise until the working pressure gauge reads 5 psi.

P5-9 Turn the torch handle acetylene valve on slightly.

P5-10 Use a striker to ignite the acetylene gas at the tip of the torch.

P5-11 Open the oxygen valve at the torch handle and add oxygen to achieve a neutral flame.

Figure 3-35 Turning off the main cylinder valve.

Special Tools

Welding goggles
Welding gloves
Oxyacetylene welding
 and cutting outfit
Striker

Caution

The oxyacetylene torch develops dangerous ultraviolet light rays. Always wear shaded welding goggles when using this equipment. The flame at the end of the torch is very hot. Wear gloves and be careful not to direct the flame at another person or anything that could ignite.

A **welding tip** is the part attached to an oxyacetylene torch handle where the gases are mixed to create the flame.

to do many advanced repair procedures. A good first step to mastering these skills is to learn how to set up the system for heating. Heating parts to aid in their removal is an important use of the acetylene torch.

 WARNING The installation of welding regulators and hoses on the oxygen and acetylene cylinders should be done by a technician with special training. We describe the setup of the system from the torch handle only.

Before beginning any setup of the oxyacetylene outfit, make sure the gas flow from both the oxygen and acetylene cylinders is off. Look at the hose pressure and cylinder pressure gauge on top of each cylinder (**Figure 3-35**). Both gauges should be at zero. If not, turn the cylinder valve on top of the cylinder clockwise to close. Repeat this procedure for the other cylinder.

If the valve was open and there is a pressure reading on either gauge, bleed the gas out of the system before going any further. Make sure the cylinder valve is closed. Open the oxygen valve at the torch handle. Turn the oxygen regulator handle clockwise (in). This will allow oxygen in the line to flow through the hose and out of the torch handle. Both gauges will drop to zero. Turn the regulator handle counterclockwise (out). Repeat the procedure with the acetylene cylinder.

Now you are ready to install the torch handle (**Figure 3-36**) onto the hoses. The handle used for heating is similar to that used for welding. It has two connections: one is for the oxygen hose and the other is for the acetylene hose. The torch handle connections are marked "AC" for acetylene and "OX" for oxygen. To ensure that the correct hose is connected to the proper connection, the hoses are color-coded (red for gas, green or black for oxygen) and have left- and right-handed threads. The oxygen hose is green or black, and its connection has standard right-hand threads. The acetylene hose is red, and its connection has left-hand threads.

Connect the red hose to the AC connection on the torch handle, observing the left-hand threads. Connect the green or black hose to the "OX" connection on the torch handle. Use a wrench to gently tighten the hose connections as shown in **Figure 3-37**.

The torch handle can be equipped with a number of different size **welding tips**. Welding tips are stamped with a number. The larger the number, the larger the tip and the more heat that can be developed. Common sizes are 1 to 4 as shown in **Figure 3-38**. Most heating is done with a small tip such as a number 1. Select a size 1 welding tip and screw it onto the end of the torch handle as shown in **Figure 3-39**.

Figure 3-36 Parts of oxyacetylene torch handle.

Figure 3-37 Installing hoses to the torch handle.

Figure 3-38 The larger the welding tip size, the more heat it can develop.

Figure 3-39 Installing the welding tip on the torch handle.

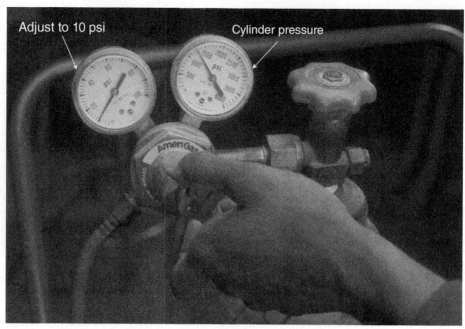

Figure 3-40 Adjusting the working pressure with the regulator handle.

Now adjust the gas pressure for heating. There are two valves on the torch handle. The one next to the oxygen hose controls the flow of oxygen to the tip. The one next to the acetylene hose controls the flow of acetylene to the tip. Close both these valves by turning them clockwise.

Turn the main valve on top of each cylinder counterclockwise to open the valve. The needle on the cylinder pressure gauge will rise to show the pressure in the cylinder. Turn the handle on the oxygen regulator clockwise (in) until the needle registers 10 psi (**Figure 3-40**). Turn the handle on the acetylene regulator the same way until the needle points to 5 psi. This will be the working pressure for heating.

You are now ready to ignite the torch. Put on your welding goggles and welding gloves. The torch is ignited with a tool called a **striker**. The striker (**Figure 3-41**) has a *striker*

The **striker** has a flint that creates a spark to ignite an oxyacetylene flame.

Figure 3-41 The striker is used to create a spark to ignite the torch flame.

Figure 3-42 Igniting the torch flame with a striker.

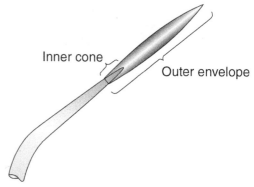

Inner cone

Outer envelope

Figure 3-43 A neutral flame has a small, sharp inner cone.

bar and a *flint* that cause a spark when rubbed together. The spark is used to ignite the torch flame. There is a cup on the striker that protects the operator from burns during start-up.

Hold the torch handle in your hand with the tip pointing downward, away from your body and away from the welding cylinders. Hold the striker by the handle and position it near the tip. Turn the acetylene valve on the torch handle just slightly open. Rub the striker flint over the striker bar to create a spark (**Figure 3-42**). The spark will ignite the acetylene gas coming out of the tip. Open the acetylene valve slowly until the sooty smoke starts to disappear. Then slowly open the oxygen valve on the torch handle.

As you open the oxygen valve, you will see the color of the flame change at the tip. The yellow flame of the acetylene will change to blue as you add oxygen. Continue to slowly add oxygen until you see a small, sharp blue inner cone in the middle of the flame envelope as shown in **Figure 3-43**. This is called a **neutral flame** and is the type of flame wanted for heating.

After heating, it is important to know how to shut the equipment down. First, turn off the acetylene valve on the torch handle. This will extinguish the flame. Then turn off the oxygen valve on the torch handle. Turn the main cylinder valve clockwise on top of both cylinders. Open the valves on the torch handle to bleed the system. Turn the oxygen and acetylene regulator handles counterclockwise until they become loose. Close both valves on the torch handle. Move the cylinders to their storage area.

> ⚠ **Caution**
>
> Always use a striker to ignite a torch. Never use matches or a cigarette lighter because these would put your hand next to the flame. Never allow the flame to point at any part of the welding outfit. When you heat something, always mark it "hot" with chalk so a coworker will not get burned picking it up.

> A **neutral flame** is an oxyacetylene flame with a small, sharp inner cone used for heating and welding.

JUMP STARTING AND CHARGING A DEAD BATTERY

To **charge** a battery means to pass an electric current through the battery in an opposite direction than during discharge. If the battery needs to be recharged, the safest method is to remove it from the vehicle. If the battery is to be charged while still installed in the vehicle, it is important to remove the negative battery cable to prevent damage to the computer.

Depending on the time available, the battery can be charged at either a slow or fast rate. A slow charge rate is 3–15 amperes. A fast charge is more than that. Slow charging has an advantage in that it will not overheat a battery during charging. If a battery overheats during charging, it will become fully charged due to high resistance in the cells.

The **charge rate** is the speed at which a battery can be safely charged. Several factors will depict the maximum charging rate. Refer to **Figure 3-44** for a reference chart for charging a battery. If you are recharging a gel cell battery, you will have to use a special charger. The high voltage battery on a Hybrid Electric Vehicle can only be recharged by special chargers sold by the manufacturer and are available at the dealership.

> **Charge** A battery that has 12.6v of potential.

> **Charge rate** is the rate of current that a battery can be charged at.

Open Circuit Voltage	Battery Specific Gravity	State of Charge	Charging time of Full Charge at 80° F (267° C)					
			at 60 amps	at 50 amps	at 40 amps	at 30 amps	at 20 amps	at 10 amps
12.6	1.265	100%	Full Charge					
12.4	1.225	75%	15 min.	20 min.	27 min.	35 min.	48 min.	90 min.
12.2	1.190	50%	35 min.	45 min.	55 min.	75 min.	95 min.	180 min.
12.0	1.155	25%	50 min.	65 min.	85 min.	115 min.	145 min.	280 min.
11.8	1.120	0%	65 min.	85 min.	110 min.	150 min.	150 min.	370 min.

Figure 3-44 Charging rate chart showing charge times based on battery voltage and charging ampere.

Jump starting a battery uses another battery source to quickly charge a dead battery.

Jump starting a battery will likely occur frequently in your first couple of years as a technician. In colder climates it is more frequent but may still occur in warmer climates, especially in hot temperatures. Consistent hot temperatures can shorten the life of a battery. But it is in colder temperatures that the power of a battery is really put to the test.

The most important safety factor when jump starting a battery is to check to make sure that the battery is not frozen or low on fluid. You can open the caps on the top of the battery to visually check this. If a battery is sealed, you will have to measure the temperature of the battery, but it is always best to just remove it from the vehicle and bring it in to warm up and thaw out instead of jump starting it. A frozen battery, or one with very low fluid, can explode when jump starting or charging it.

You should refer to **Figure 3-45** and read the following steps to safely jump start most vehicles:

1. Make sure the two vehicles are not touching.
2. Place the transmission in park on both vehicles (or set the emergency brake and put in neutral if it is a manual transmission).

Figure 3-45 Proper jumper cable locations and sequence. Note the size of the battery terminals.

3. Turn off the ignition and all accessories on both vehicles.
4. Attach one end of the positive jumper cable to the disabled (Weak) battery's positive terminal.
5. Connect the other end of the positive jumper cable to the booster (Good) battery's positive terminal.
6. Attach one end of the negative jumper cable to the booster (Good) battery's negative terminal.
7. Attach the last end of the negative jumper cable to an engine ground on the disabled vehicle.
8. Attempt to start the disabled vehicle. If it doesn't start right away, wait a few minutes and run the booster vehicle's engine to increase the charge rate.
9. Once the vehicle starts, disconnect the ground-connected negative jumper cable from its engine block.
10. Disconnect the negative jumper cable from the booster battery, then from the other battery.

UNDERSTANDING AND USING SERVICE INFORMATION

Most service data is easily obtained using paper or computerized service manuals. However, there are times when the data is not located in an expected area of the manual. Torque specifications for a certain component, like cylinder head bolts, may be listed in the specification table in one manual, whereas another manufacturer places it in the section concerning engine assembly or in the cylinder head installation section.

The best method is to determine the correct term for the part and the system in which it is located. Most manufacturers use the same terms for the majority of automotive parts and components. They may, however, list a term as part of a subsection. For instance, data on the cooling system may be in a separate section on cooling or in a subsection under engine performance or diagnosis. Experience in the use of manufacturer service manuals is the best method to determine how each is set up.

Data banks are basically a computerized version of a paper manual. The two largest automotive repair data systems, Mitchell and All Data, use manufacturer data for the information, but both have some differences in the way their data bank is set up.

One of the best methods to pinpoint service data is to use the vehicle identification number (VIN) and identify tags or markings on various major components to determine exactly the type of component used. For instance, in a single manufacturer product line, there may be different versions of basically the same automatic transmission. Externally, the transmission may appear to be exactly the same; however, internally, different gear ratios or a different shift strategy will possibly require different valves or gear arrangement. Somewhere on the external casing will be a tag or etched markings that clarify exactly which transmission version is installed. This information is critical for a proper repair of that transmission and identifying the correct replacement parts.

For the purpose of this discussion, let's use a customer noise complaint to determine the initial steps for repairing the problem. The customer says his or her vehicle has a bumping or jerk as the steering wheel is turned to either far right or far left, and it sometimes happens during a slight turn over a speed bump. This fault is probably in the steering and suspension system. Assume that the vehicle is a 2004 GMC 1,500 pickup. During the test drive and initial diagnosis procedure, it was determined that the left lower ball joint is excessively worn. Using this information, the VIN, and the service manuals, computerized or paper, locate the repair procedures for replacing the ball joint. If there are computerized and paper manuals available, use both and compare your findings between the two because there may be a mistake in moving the paper version to a data bank. Because the author does not know what is available at the reader's site, collect the service information and take it to the instructor or an experienced technician for review. Once the repair procedures are gathered, go to the labor guide and determine the labor hours for this job. Don't forget to add the diagnostic time. If available, consult the parts guide and determine the cost of replacement parts.

Cautions A warning or advise against doing something.

Was there anything in the procedures, warnings, or **cautions** that could cause injury or damage? Are special tools or equipment required? What is the labor time and cost of the parts? Based on this research, what is the total cost of this repair based on an $85.00-per-hour labor rate? Does the instructor or advisor agree with the repair estimate?

It is suggested that other vehicles and other complaints, real or simulated, be used to collect information and develop a repair estimate. If available, use a blank service repair order to record each vehicle's information. One thing that will become obvious during a technician's career is that information is one of the best avenues to gaining a great salary. This information may come from the customer, another technician, service manuals, the internet, or the technician's personal experience. Either way, the route to a quick, first time fix is much easier with the correct information.

CASE STUDY

Older technicians have been heard to say, "I have an educated arm; I don't need a torque wrench." They think they can feel the correct tightness for the fasteners they are tightening and do not need a torque wrench.

A group of technicians got a chance to prove this at a recent automotive tools trade show. A torque wrench manufacturer had an engine on a stand with the cylinder head bolts ready for tightening. Technicians were offered a chance to tighten a head bolt to the specified 85 ft.-lb. without a torque wrench. If they could get to within 10 ft.-lb. of the specification, they would win a prize. Technician after technician took a turn. After each tightening, the torque on the head bolt was measured. The trade show lasted three days and not one prize was given. So much for the educated arm theory!

ASE-STYLE REVIEW QUESTIONS

1. The use of an air impact wrench is being discussed.

   Technician A says the air impact wrench works well for tightening wheel lug nuts.

   Technician B says the air impact wrench should be used primarily for disassembly work.

   Who is correct?

   A. A only

   B. B only

   C. Both A and B

   D. Neither A nor B

2. The use of an air impact wrench is being discussed.

   Technician A says standard 12-point sockets can be used on the impact wrench.

   Technician B says 6-point impact sockets must be used on an impact wrench.

   Who is correct?

   A. A only

   B. B only

   C. Both A and B

   D. Neither A nor B

3. The use of an air impact wrench is being discussed.

   Technician A says to check the direction of the reverse switch before using the wrench.

   Technician B says eye protection must be worn when using an air impact wrench.

   Who is correct?

   A. A only   C. Both A and B

   B. B only   D. Neither A nor B

4. The use of a torque wrench to tighten lug nuts is being discussed.

   Technician A says torque specifications are found in the shop service manual.

   Technician B says the rule is to get the lug nuts as tight as possible.

   Who is correct?

   A. A only   C. Both A and B

   B. B only   D. Neither A nor B

5. The use of a torque wrench is being discussed.

   Technician A says wheel lug nut specifications are in newton-meters.

   Technician B says the specifications are in foot-pounds.

   Who is correct?

   A. A only   C. Both A and B

   B. B only   D. Neither A nor B

6. The use of a torque wrench is being discussed.

   Technician A says fasteners must be tightened to the correct torque.

   Technician B says fasteners must be tightened in the correct sequence.

   Who is correct?

   A. A only   C. Both A and B

   B. B only   D. Neither A nor B

7. The use of a gear puller is being discussed.

   Technician A says gear pullers can be used to pull parts of a shaft.

   Technician B says gear pullers can be used to pull parts out of a hole.

   Who is correct?

   A. A only   C. Both A and B

   B. B only   D. Neither A nor B

8. The removal of a broken stud is being discussed.

   Technician A says first try to remove the broken stud with a screw extractor.

   Technician B says first try to remove the broken stud with a chisel.

   Who is correct?

   A. A only   C. Both A and B

   B. B only   D. Neither A nor B

9. The removal of a rounded-off nut is being discussed.

   Technician A says it can be removed with a flat chisel.

   Technician B says it can be removed with a nut splitter.

   Who is correct?

   A. A only   C. Both A and B

   B. B only   D. Neither A nor B

10. A nut is rusted onto a stud.

    Technician A says soak it in antiseize compound.

    Technician B says soak it with penetrating oil.

    Who is correct?

    A. A only   C. Both A and B

    B. B only   D. Neither A nor B

ASE CHALLENGE QUESTIONS

1. *Technician A* says the lug nut should be tightened in a circular sequence because the wheel's lug nut area is rounded.

   Technician B says tightening the lug nuts in a star sequence will cause uneven stress on the rim.

   Who is correct?

   A. A only   C. Both A and B

   B. B only   D. Neither A nor B

2. *Technician A* says to use chrome sockets when tightening chrome lug nuts because the chrome indicates a match of fastener to wrench.

   Technician B says if the chrome covering comes off the lug nut, then a smaller size socket may be needed. Who is correct?

   A. A only   C. Both A and B

   B. B only   D. Neither A nor B

3. Using gear pullers are being discussed.

   Technician A says the puller jaws should not pull on the outer circumference of the gears.

   Technician B says the center bolt of the puller should be aligned with or on the bearing journal.

   Who is correct?

   A. A only

   B. B only

   C. Both A and B

   D. Neither A nor B

4. *Technician A* says to remove any broken screw drill completely through the screw so the extractor can get a full-length grip.

   Technician B says a chisel may be used to remove a broken screw.

   Who is correct?

   A. A only

   B. B only

   C. Both A and B

   D. Neither A nor B

5. *Technician A* says the use of an antiseize compound may reduce the chance of a fastener sticking or corroding to its mating threads.

   Technician B says a nut splitter can be used to remove a stud from its thread bore.

   Who is correct?

   A. A only

   B. B only

   C. Both A and B

   D. Neither A nor B

TABLE 3-1   Guidelines for Service
Raise a vehicle with a lift.

Steps	Classroom Manual	Shop Manual
1. Position the vehicle between the posts.	67,68	60–61
2. Position the lift arms and pads under the vehicle.	67,68	47–50
3. Raise the vehicle until its wheels clear the floor. Check for stability.	67,68	60–61
4. Lift the vehicle to work height and set the locks.	67,68	60–61

TABLE 3-2   Guidelines for Service
Lift a vehicle with a jack.

Steps	Classroom Manual	Shop Manual
1. Position the vehicle, place in park, and block wheels.	66	50–51
2. Place jack lift pad under the vehicle's lift point.	66	49
3. Unlock the handle and knob.	66	50–51
4. Operate the handle to raise the jack.	66	50–51
5. Place jack stands under the vehicle.	66	50–51
6. Lower the vehicle to the jack stand.	66	50–51

TABLE 3-3   Guidelines for Service

Connect and use an air impact wrench to remove the lug nuts on a wheel.

Steps	Classroom Manual	Shop Manual
1. Raise car on a hoist or floor jack and support on safety stands.	66	52
2. Remove wheel cover and mark lug to wheel position.	55	52
3. Connect air impact wrench to air source.	44,45	52
4. Set impact wrench to reverse and install correct size socket.	55,56	53
5. Remove lug nuts.	55,56	53
6. Reinstall wheel and start lug nuts by hand.	55,56	50,51
7. Reverse switch on impact wrench and spin lug nuts into position to be torqued.	—	—
Safety		
Use lift locks.		
Wear eye protection.		
Use impact socket.		

TABLE 3-4   Guidelines for Service

Look up the torque specifications for wheel lug nuts and torque wheel lug nuts to the correct torque.

Steps	Classroom Manual	Shop Manual
1. Look up torque setting for lug nuts and tightening sequence in service or owner's manual.	44,45	52
2. Connect correct size socket to torque wrench.	55	52
3. Tighten the lug nuts in stages to correct torque.	47,45	52
4. Tighten lug nuts in the correct sequence.	55,56	53
5. Replace wheel covers.	55,56	53
6. Lower car from jack or hoist.	55,56	50,51
Safety		
Wear eye protection.		

TABLE 3-5   Guidelines for Service

Select, install, and use a gear puller to correctly remove a pulley from a shaft.

Steps	Classroom Manual	Shop Manual
1. Remove the retaining nut from the pulley.	63	61,62
2. Select the correct size and type of gear puller.	63	61,62
3. Adjust the cross arm and jaws to install the puller on the pulley.	63	61,62
4. Drive the forcing screw in to force the pulley off the shaft.	63	61,62
Safety		
Wear eye protection.		

TABLE 3-6 Guidelines for Service

Use a screw extractor to remove a broken stud or screw.

Steps	Classroom Manual	Shop Manual
1. Select correct size screw extractor and drill for broken fastener.	58	63
2. Drill correct size hole in center of broken fastener.	58	63
3. Install screw extractor in drilled hole.	63	64
4. Install correct wrench or driver on screw extractor and rotate it to remove broken fastener.	63	64
Safety		
Wear eye protection.		

TABLE 3-7 Guidelines for Service

Use a flat chisel to split and remove a stuck nut from a stud or bolt.

Steps	Classroom Manual	Shop Manual
1. Soak stuck nut with penetrating fluid.	56,57	63
2. Cut the nut completely along one side with a cold chisel.	56,57	63
3. Remove the nut with a wrench or pliers.	46,47,48	62
Safety		
Wear eye protection.		

Name _____   **Date** _____

RAISING A VEHICLE

Upon completion and review of this job sheet, you should be able to raise a vehicle using a vehicle lift.

ASE Education Foundation Correlation

This job sheet addresses the following **RST** tasks:

1-3. Identify and use proper placement of floor jacks and jack stands.

1-4. Identify and use proper procedures for safe lift operation.

We Support
ASE | **Education Foundation**

Tools and Materials

Goggles or safety glasses with side shield

Vehicle hoist

Procedure	Task Completed
1. Identify the vehicle's lift points.	☐
Make _____ Model _____ Year _____	
2. Center the vehicle between the two posts. The front door handle should be about even with the posts.	☐
3. Rotate and extend lift arms under the vehicle.	☐
4. Align lift pads with lift points.	☐
5. Raise lift arms until pad contacts points by holding the switch in the ON position.	☐
6. Check contact between pads and lift points.	☐
7. Continue to raise the vehicle until the center of the wheel assembly is at eye level.	☐
8. Lower the vehicle until the locks engage the lift arms.	☐
9. Raise the vehicle to remove weight from the locks.	☐
10. Install the filter and gaskets or O-ring. Torque the filter bolts.	☐
11. Place the locks in the unlocked position.	☐
12. Lower the vehicle to the floor and the arms to their lowest travel.	☐
13. Swing the arms from under the vehicle and move the vehicle from the bay.	☐

Instructor's Response

_____

_____

_____

Name _____   **Date** _____

FLOOR JACK OPERATION

Upon completion of this activity, you will be able to demonstrate the ability to raise a vehicle safely off the ground using a floor jack and jack stands.

ASE Education Foundation Correlation

This job sheet addresses the following **RST** tasks:

1-2. Utilize safe procedures for handling of tools and equipment.

1-3. Identify and use proper placement of floor jacks and jack stands.

We Support
ASE | Education Foundation

Tools and Materials

Goggles or safety glasses with side shields
Floor jack
Jack stands

Procedure   **Task Completed**

1. Describe the vehicle you are lifting.   ☐

   Year _____ Make _____ Model_____ VIN _____

   Using Alldata or Mitchel locate lift points for the vehicle you are working on.

2. Locate floor jack and describe type of jack being used. Rating?

   _____

3. Inspect floor jack and jack stands for proper operation. Note any safety concerns!

   _____

4. Place wheel chocks against front or rear tires on axle not being lifted. Position floor   ☐
   jack at lift point.

5. Locate jack stands and describe type of stands being used. Ratings?

   _____

6. Raise vehicle until wheels or wheel is off the ground. Position jack stand(s) under   ☐
   vehicle on chassis frame.

7. Slowly lower vehicle onto jack stands until fully supported by jack stands. Remove   ☐
   floor jack from under vehicle.

8. Lower vehicle off jack stands: Position floor jack under lift points and raise vehicle off   ☐
   jack stands. Remove jack stands from under vehicle and slowly lower vehicle to the
   ground.

9. Return all equipment used for work sheet back to proper place!   ☐

10. Explain why it is important to locate the correct lift points when using a floor jack.

    _____

    _____

    _____

Instructor's Response

_____

_____

_____

Name _____   Date _____

USING AN IMPACT WRENCH AND A TORQUE WRENCH

Upon completion and review of this job sheet, you should be able to remove and install a tire assembly using an impact wrench and a torque wrench.

ASE Education Foundation Correlation

This job sheet addresses the following **RST** tasks:

1.2 Utilize safe procedures for handling of tools and equipment.

2.5 Demonstrate proper use of precision measuring tools (i.e., micrometer, dial-indicator, dial-caliper).

We Support
ASE | Education Foundation

Tools and Materials

Goggles or safety glasses with side shields
1/2-inch-drive impact wrench
Set of 1/2-inch-drive impact sockets
1/2-inch-drive torque wrench (50–250 ft.-lb.)
Hubcap hammer or long, flat screwdriver
Lift or jack and jack stands

Procedure   **Task Completed**

1. Locate the following vehicle service information.

   Lug nut wrench size _____

   Lug nut torque _____

2. Lift the vehicle to a good working height. Use Job Sheet 2 for instructions.   ☐

3. Remove the hubcap by using a hubcap hammer or a flat tip screwdriver.   ☐

4. Select the correct impact socket and install on impact wrench.   ☐

5. Set the impact wrench to reverse and set the socket over the first lug nut.   ☐

6. Use one hand to support the wrench and squeeze the impact wrench's trigger.   ☐

7. Reduce trigger pressure as soon as the nut breaks loose from the stud.   ☐

8. Repeat steps 5 through 7 to remove other lug nuts. If necessary, use one hand to keep   ☐
   the tire assembly on the hub while the last nut is removed.

9. Lay the wrench aside and grasp the tire assembly at the 4 o'clock and 8 o'clock   ☐
   positions. (12 is straight up.)

10. Keeping the back straight and bending the knees, remove the tire assembly from the   ☐
    hub and place on floor. Installation of the tire assembly is the reverse.

11. Grasp the tire at the 4 o'clock and 8 o'clock positions.   ☐

12. Keep the back straight and use the knees to lift the tire assembly to the hub.   ☐

13. Align the wheel with the studs and slide the wheel assembly onto the studs.   ☐

14. Use one hand to keep the assembly on the studs while the first lug nut is started.   ☐

Task Completed

15. Start each lug nut by hand. ☐

16. Inspect the impact wrench's controls. If there is an adjustment, set the wrench to its lowest forward setting. If there is no adjustment, screw the nuts on as far as they go by hand. ☐

17. Use the impact wrench to screw the nuts into contact with the wheel. ☐

18. Lower the vehicle until the wheels just touch the floor. A second option is to have someone apply the brakes while the nuts are being torqued. ☐

19. Set the torque wrench to the specified torque and transfer the socket from the impact wrench to the torque wrench. ☐

20. With the wheels held, use the torque wrench to tighten the first lug nut. Once the torque is achieved, move to the next nut. ☐

21. The second nut should be across the wheel from the first one. ☐

22. Torque each lug nut following a cross pattern. ☐

23. Install the hubcap. ☐

24. Lower the vehicle and clean the area. ☐

25. Complete the repair order. ☐

Instructor's Response

_____

_____

_____

Name _____   Date _____

REMOVING A HARMONIC BALANCER

Upon completion and review of this job sheet, you should be able to use a harmonic balancer puller to remove a harmonic balancer from a crankshaft.

ASE Education Foundation Correlation

This job sheet addresses the following **AST/MAST** tasks:

1.C.1. Remove, inspect, and/or replace crankshaft vibration damper (harmonic balancer). **(P-1)**

We Support

ASE | Education Foundation

Note: The harmonic balancer was selected to illustrate the procedures for using a bolt-type puller. Other equipment could be used in lieu of the balancer.

Tools and Materials

Goggles or safety glasses with side shields
A crankshaft with harmonic balancer installed in or out of vehicle
Vise if crankshaft is out of vehicle
Harmonic balancer puller set
Hand-tool set

Describe the vehicle being worked on:

Year _____   Make _____   Model _____

VIN_____   Engine type and size _____

Procedure   **Task Completed**

1. If crankshaft is out of vehicle, place it in a vise or other holding device.   ☐

2. Locate the harmonic balancer puller set.   ☐

3. Select the two (or three) bolts that thread easily into the outer threaded holes on the   ☐
   balancer.

4. Select a center cone that fits the best into the large threaded hole in the crankshaft.   ☐

5. Screw the large puller bolt into the puller body and position the bolt point into the   ☐
   center cone.

6. Slide one of two (or three) previously selected pulling bolts through the puller body   ☐
   and screw into the balancer hole. Repeat for the other bolt (or two).

7. Tighten each of the two (three) bolts until all penetrate the balancer. Ensure each   ☐
   protrude an equal length through the puller body. Ensure the puller body is parallel to
   the balancer face.

8. Use a wrench to tighten the center puller bolt. It may be necessary to adjust the shaft   ☐
   anchor as the pulling operation continues.

9. Continue to screw in the large bolt until the balancer is loose and can be removed by   ☐
   hand.

10. Capture and store the woodruff key that pins the harmonic balance to the crankshaft.   ☐

11. Remove the puller components from the balancer and store in case.   ☐

12. Record any problems encountered during this pulling operation.

_____

_____

Instructor's Response

_____

_____

_____

Name _____   Date _____

REMOVING PITTMAN ARM FROM STEERING GEAR BOX

Upon completion and review of this job sheet, you should be able to use a two-jaw puller to remove a pitman arm from a steering gear box.

ASE Education Foundation Correlation —————————————————————

This job sheet addresses the following **MLR** task:

4.B.7 Inspect pitman arm, relay (centerlink/intermediate) rod, idler arm, mountings, and steering linkage damper. **(P-1)**

This job sheet addresses the following **AST/MAST** task:

4.B.16 Inspect, remove, and/or replace pitman arm, relay (centerlink/intermediate) rod, idler arm, mountings, and steering linkage damper. **(P-2)**

We Support

ASE | Education Foundation

Note: A pitman arm was selected to illustrate the use of a two-jaw puller. Other equipment could be used in lieu of the pitman arm, but a two- or three-jaw should be used. It is unlikely a three-jaw puller could be used on a pitman, but it would be preferable if a gear or a shaft is being used instead of the pitman arm.

Tools and Materials

Goggles or safety glasses with side shields
Recirculating ball steering gear, in or out of vehicle
Vise if steering gear is out of vehicle
Lift if steering gear is in vehicle
Two-jaw gear and bearing puller
Hand-tool set

Describe the vehicle being worked on:

Year _____ Make _____ Model _____

VIN_____ Engine Size _____

Procedure   Task Completed

1. If steering gear is out of vehicle, place it in a vise or other holding device.   ☐

2. Loosen the large retaining nut holding the pitman arm. Screw it nearly off but do not remove. This will catch the pitman arm and puller when the arm is pulled free of its shaft.   ☐

3. Locate a two-jaw puller.   ☐

4. Unscrew the puller's large center bolt until the jaws slip over the pitman arm. The center bolt should be centered on the end of the steering shaft.   ☐

5. Tighten the center bolt until the puller holds in place.   ☐

6. Inspect the position of the jaws over the edges and back side of the arm. Ensure the contact surfaces of the jaws are flush against the back side of the pitman arm. If not positioned correctly, either reposition or select a different puller.   ☐

7. With puller positioned correctly, use a wrench to screw the center bolt inward. This should begin to pull the pitman arm off.   ☐

8. Continue to tighten the center bolt until the pitman arm and puller drop loose of the steering gear. ☐

9. Remove the puller from the pitman arm if necessary and store in case. Secure and store the pitman arm. ☐

10. Record any problems encountered during the pulling operation.

_____

_____

⚠ **WARNING** Once the arm is loosened, it will drop off. If the retaining nut was completely removed for use of the puller, be prepared to catch the arm and puller. Failure to catch it could result in an injury.

Instructor's Response

_____

_____

_____

CHAPTER 4

USING FASTENERS AND MEASURING EQUIPMENT

Upon completion and review of this chapter, you should understand and be able to:

- Use a feeler gauge for measuring spark plug gap.
- Operate an outside micrometer to measure shaft diameter.
- Measure a brake disc for thickness with an outside micrometer.
- Operate a dial indicator to measure runout and gear backlash.
- Use a dial indicator to measure shaft endplay.
- Measure and read a pressure gauge.
- Measure and read a vacuum gauge.
- Use a pitch gauge to identify U.S. (English) and metric thread sizes.
- Install a torque-to-yield bolt.

- Operate a tap to repair damaged internal threads in an automotive component.
- Remove and replace a stud with a stud remover.
- Remove and replace a stud with a jam and drive nut.
- Strip a wire and install a solderless terminal and solderless butt connector.
- Use a soldering gun or soldering iron and rosin-core solder to solder an electrical terminal to a wire.
- Use a soldering gun or soldering iron and rosin-core solder to splice two wires together.
- Repair connectors and terminals.

Terms To Know

Anaerobic	Ring travel area	Stud remover
Angle indicator	Rotor	Thread-locking compound
Antiseize	Soldering	Thread-sealing compound
Blind hole	Soldering gun	Through hole Tinning
Bore	Soldering iron	Thrust wear
Dial indictor	Splice	Toothed ring
Flux	Straight edge	Toothed wheel
Journal	Stripped threads	
Out-of-round	Stripping	

INTRODUCTION

This chapter will guide technicians through methods of making measurements using tools studied in the Classroom Manual. The tasks discussed will be typical of the types of measurement jobs performed in a shop. Almost every job a technician does involves the use of fasteners. In the Classroom Manual, we described the types and uses of fasteners. This chapter describes several tasks involving fasteners. It describes how to use a pitch gauge to identify threads, how to use a tap to repair internal threads, how to remove and replace a damaged stud, how to install solderless electrical connectors, and how to solder wires and electrical connectors.

Classroom Manual
pages 83,84,85

MEASURING SPARK PLUG GAP

Even with the best electronics, all gasoline-fueled engines use spark plugs to fire the air/fuel mixture. Although many vehicles require spark plug replacement only every 100,000 miles, the new plugs will have to be correctly gapped (**Figure 4-1**). If the gap is too wide, the spark may be weak in strength and short in duration because of the high voltage needed to bridge the gap. With a too-narrow gap, the spark will have a low voltage and not much heat. In both cases, the spark will have difficulty igniting the air/fuel mixture, resulting in a poorly operating engine.

There are or will be scrapped spark plugs in almost every shop or technical school. Select five or six of them and a round wire feeler gauge (**Figure 4-2**). If necessary, clean the plugs before gapping. Assume that the correct gap for the first one is 0.035 inch.

Figure 4-1 The electricity bridges the gap and creates a spark to ignite the air/fuel mixture.

Figure 4-2 A typical round wire feeler gauge used for measuring spark plug gap.

Figure 4-3 The gauge wire should have a slight drag as it is moved between the two electrodes.

AUTHOR'S NOTE When possible, identify the vehicle from which the plugs were removed. The student can use the vehicle service data to locate the correct gap for the plug.

Locate the 0.035-inch wire from within the gauge set. Attempt to slide the wire between the plug's two electrodes (**Figure 4-3**). If the gap is too small or too wide, use the gauge set's small bending tool to move the outer electrode. Check the gap again and correct any further space by using the bending tool on the gauge set. With one plug correct, select different gaps for the other plugs and set them. Most of the current vehicles use spark plug gaps between 0.035 and 0.046 inch. Some older General Motors vehicles use 0.080-inch gaps.

AUTHOR'S NOTE One of the two easiest means to practice the measuring or setting of an air gap is using the wheel speed sensor for an antilock brake system (ABS) or the crankshaft position sensor on some engines.

Measuring Air Gap

For this discussion, we will use an ABS wheel speed sensor on a front-wheel-drive vehicle. Some vehicles have ABS only on the rear wheels and will not have the sensor at the front wheels. Use the service data for the vehicle being used and retrieve the necessary instructions and specifications for the wheel speed sensor air gap. Each speed sensor will have a **toothed wheel** or **toothed ring** that triggers the sensor (**Figure 4-4**). The ring may be on the axle, rotor hub, or on a gear inside the differential. In this instance, we'll assume the vehicle has front-wheel drive with the tooth ring on the brake rotor hub. Ensure the ignition is in the off position and the transaxle is in neutral. Lift the vehicle

Special Tools

Round wire feeler gauge set or tapered plug gauge

Nonmagnetic feeler gauge set

Service information

The **toothed wheel** or **ring** is a metal ring or plate with teeth at a specific distance apart. As each tooth passes the sensor, it will cause the sensor to generate a voltage.

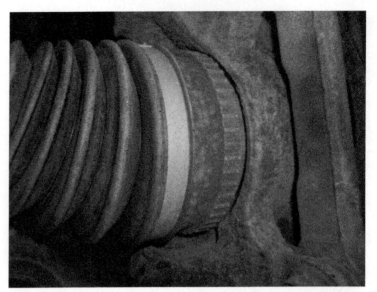

Figure 4-4 This is a version of a toothed ring. This ring is mounted to the drive near the outer CV joint. The raised ridges on the ring create a voltage signal as they pass the sensor.

and remove one of the front wheels to gain access to the speed sensor. Locate the nonmagnetic feeler gauges and select the correct blade based on the air gap specification. Rotate the hub until a tooth on the ring is aligned perfectly with the center of the sensor. While holding the hub in place, slide the nonmagnetic feeler gauge between the tooth and the sensor (**Figure 4-5**). If it does not go, the air gap is too small. If it slides in and out with too little drag, the air gap is too large. Consult the service information to determine the proper method of adjusting the air gap. Adjust as needed and measure to ensure the air gap is correct.

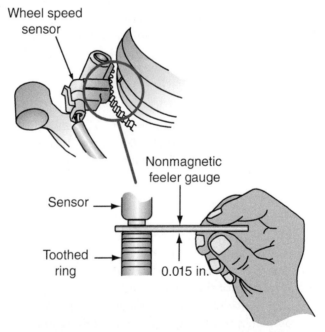

Figure 4-5 Measuring the air gap between the sensor and a tooth or trigger. A nonmagnetic (usually brass) feeler gauge is used to measure the air gap.

MEASURING SHAFT DIAMETER AND WEAR

There will be times when a technician is required to measure various components to determine if they are serviceable or need machining or replacement. Many shops do not rebuild engines in-house but may still have to determine if a replacement is needed. The same applies to transmissions and transaxles. Sometimes the noise made by a bad engine does not mean that the engine is only good for scrap metal. The amount, loudness, or lack of noise does not accurately indicate the amount of damage.

Classroom Manual
pages 86–94

Select a shaft that has one or more precision-machined areas on it (**Figure 4-6**). An output shaft from a scrapped transmission, a crankshaft, or a camshaft works well for this exercise. Select a set of United States Customary (USC) and metric micrometers that will measure the various areas of the shaft.

Special Tools
Outside USC
   micrometers
Shafts

Using a USC outside Micrometer to Measure a Bearing Journal

Before starting the measurement, inspect the shaft and journals. It is a waste of time to measure a journal if it is damaged beyond repair unless it is being done for practice. We will review the classroom information on reading measurements as we proceed through the rest of this chapter. Select the correct USC micrometer for the first area to be measured. The area selected should be machined (Figure 4-6). Measuring the rough or unmachined area is not needed in most instances. If possible, locate some data for the shaft from the instructor or the service manual, specifically data on the correct diameter and the amount of wear that may exist before the shaft has to be replaced.

Move the spindle away from the anvil far enough for the micrometer to be slipped over the shaft (**Figure 4-7**). Rotating the thimble counterclockwise opens the gap between the two measuring faces. With the two measuring faces, anvil and spindle, 180 degrees apart and placed over the shaft, rotate the thimble clockwise until it contacts the shaft. Keep the micrometer in place as the ratchet is turned clockwise until it slips. Lock the spindle and check the alignment of the faces on the shaft. If they are positioned correctly, remove the micrometer so the measurement can be easily read. Unlock the micrometer and reposition it if the faces are not in place correctly.

Begin reading the measurement by noting the size of the total width of the micrometer (**Figure 4-8**). For instance, if this is a 1- to 2-inch micrometer, the first number will be 1 and will be to the left of the decimal point.

The second number and the first to the right of the decimal point will be read from the index line on the sleeve (**Figure 4-9**). This will be the highest visible number along the index line. We will assume for this exercise that the number is 3.

Figure 4-6 The shiny, highlighted areas are the crankshaft bearing journals.

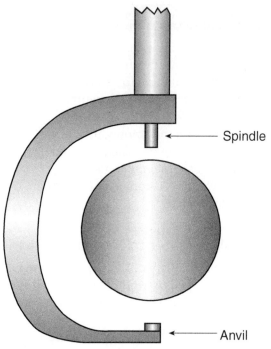

Figure 4-7 Open the micrometer and slide it over the component being measured.

Figure 4-8 The micrometer frame size is the digit to the left of the decimal point.

Figure 4-9 The highest exposed sleeve number is the first digit to the right of the decimal point.

The second number to the right of the decimal comes from the graduations under the index line on the sleeve (**Figure 4-10**). Each mark is worth 0.025 inch. If the second graduation mark is the last one showing, then the number will be 0.050 inch. Using the values from the last three paragraphs, we can say the shaft is at least 1.350 inches thick. The numbers are found as follows:

1 inch from the micrometer frame plus
0.3 inch from the index line plus
0.050 inch from below the index line for a total of 1.350 inches

Figure 4-10 The number of exposed lines past the top number is counted and multiplied by 0.025.

Figure 4-11 The closest thimble graduation to the index line is always selected as the last number of the measurement.

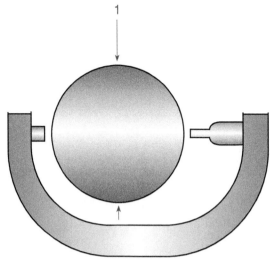

Figure 4-12 The second shaft measurement is at a right angle to the first (arrows).

The last number comes from nearest graduation mark on the thimble that most nearly aligns with the index line on the sleeve. There may be some room for interpreting the reading at this point. If the index line is near the center between two thimble marks, two technicians may come up with two different readings (**Figure 4-11**). This 0.001-inch difference between the two is not of concern on this type of shaft. If the two readings are at the limits of the wear tolerance, the shaft should probably be replaced anyway. The last number derived from the thimble is added to the previous readings. Assume that the number is 0.004 inch; when added to the previous readings, then the shaft is 1.354 inches thick at this **journal**.

A second measurement should be made to determine if the journal and shaft is **out-of-round**. If the journal is not a perfect or nearly perfect circle, the journal and bearing will make contact. The life of the journal, shaft, and bearing will be severely shortened. Other damage may also result from the failure of this bearing.

Move the micrometer to a position 90 degrees to the one where the last measurement was made (**Figure 4-12**). Place the measuring faces in the same position relative to each other. Operate the thimble, ratchet, and lock to move the faces to the journal surface. Remove the micrometer and record the measurement. Compare the two measurements to each other and the specifications for the journal.

If one of the two measurements is smaller than the other, the journal is out-of-round and has a slight oval shape. A difference of about 0.0005 to 0.001 inch may be allowed on some shafts. Consult the service data for the allowable tolerance (**Figure 4-13**). The journals on some shafts can be machined and undersize bearings can be used.

While checking for out-of-round data, also check for the specified minimum diameter of the journal. Even a perfectly round journal that is too thin may result in a failure.

Measuring with a metric micrometer is done in the same way except for the actual compilation of the measurement. Once the micrometer has been positioned and locked, remove it for easy reading. The numbers above the index line are graduated 1 millimeter apart. The last and largest exposed number on the index line will be the digit to the left

A **journal** is a finely machined portion of a shaft fitted through a bearing or housing. This machined surface reduces the friction and heat as the shaft operates.

The term **out-of-round** only applies to a journal or shaft that is supposed to be perfectly round.

Mark 0	54.998–55.003 mm (2.1653–2.1655 in.)
Mark 1	54.993–54.998 mm (2.1651–2.1653 in.)
Mark 2	54.998–54.993 mm (2.1649–2.1651 in.)

Figure 4-13 There is only 0.005 mm (0.0002 inch) of wear allowed on a crankshaft. Mark 1 main journal (54.998 − 54.993 = 0.005).

Figure 4-14 On a metric micrometer, the last exposed number on the index line is the digit to the left of the decimal point.

of the decimal point, opposite of the USC micrometer (**Figure 4-14**). If the journal is more than 25 mm, this first number must be added to the frame size.

Digits to the right of the decimal are a combination of the graduations below the index line and the nearest line on the thimble. See if the 0.5 mm line below the index line is exposed (**Figure 4-15**). Determine which line of the thimble most closely aligns with the index line and add that number to the 0.0 or 0.5 mm line found one step back in Figure 4-15. The digits to the left of the decimal point come from above the index line plus the frame size. All the numbers below the line and the thimble go to the right of the decimal. See **Photo Sequence 6** for the use of an outside micrometer.

Another measuring device is the *dial caliper*. The dial caliper is a machined bar with two measuring jaws, one stationary and one moveable. The bar is graduated in either USC or metric and mounts the stationary jaw. The moveable jaw slides along the bar and mounts a dial indictor. The graduations on the bar are used to measure gross dimensions (whole inches, centimeters) whereas the dial indicator provides more precise dimensions (0.001 inch, 0.1 mm). There is an electronic version that is much easier to use and is inexpensive. See **Photo Sequence 7** for details on using a standard dial indicator.

Thimble pass .5 mm

Add .03 mm to .5 mm

Thimble pass .0 mm

Figure 4-15 Ensure that the below-the-line reading is correct. A 0.5-mm error could be made.

PHOTO SEQUENCE 6
Typical Procedure for Reading a Micrometer

P6-1 The tools required to perform this task are a micrometer set and a clean shop towel.

P6-2 Select the correct micrometer. If the component measurement is less than 1 inch, use the 0- to 1-inch micrometer.

P6-3 Check the calibration of the micrometer.

P6-4 Locate the component between the anvil and spindle of the micrometer and rotate the thimble to slowly close the micrometer around the component. Tighten the thimble until a slight drag is felt when passing the component in and out of the micrometer. If the micrometer is equipped with a ratchet, it can be used to assist in maintaining proper tension.

P6-5 Lock the spindle to prevent the reading from changing.

P6-6 Remove the micrometer from the component.

P6-7 Each number on the sleeve is 0.100 inch, and each gradation represents 0.025 inch. To read a measurement, count the visible lines.

P6-8 The graduations on the thimble define the area between the lines on the sleeve in 0.001 increments. To read this measurement, use the graduation mark that aligns with the horizontal line on the sleeve.

P6-9 Add the reading obtained from the thimble to the reading on the sleeve to get the total measurement.

PHOTO SEQUENCE 7
Typical Procedure for Reading a Dial Caliper

P7-1 Set the dial caliper in your right hand so that the gauge can be easily read and your thumb is on the wheel.

P7-2 Make sure the jaws are clean and close them all the way. Zero the gauge again if needed.

P7-3 Using your thumb to slide the jaws of the caliper, move it into the position desired to measure.

P7-4 Make sure that you are using the beveled edge of the caliper jaws as the contact point.

P7-5 Once you have taken your measurement, lock it in place and remove it from the part. This makes reading of the measurement easier.

P7-6 The dial caliper being used in this example has a range from 0 to 6 inches. Each division between the inches is 1/10th of an inch (0.1 inch).

P7-7 View and record the measurement on the sliding ruler. In this example, the caliper reads 0.4 inch.

P7-8 After reading the inches and first decimal place, you can read the dial. In this example, each mark on the dial measures 1/1000th of an inch (0.001 inch). This will give you the second and third decimal places. In this example, the dial reads 55. That means that 0.055 inch is added to the first number.

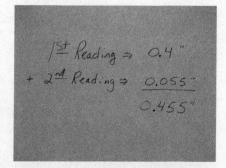

P7-9 The total measurement is the sum of all the recorded numbers.

MEASURING RUNOUT ON A DISC

Runout is the term used to determine if there is a wobble in a wheel or disc (**Figure 4-16**). Generally, a disc or wheel should spin without any side-to-side movement. Movements of this type cause a vibration or shake and if serious enough could cause damage. It can be frustrating to passengers when the vibration is transmitted throughout the vehicle. In most cases, technicians do not specifically measure runout on a brake disc, but it is a good training device.

For this job, we will use a brake system disc mounted on the vehicle, which allows an easy mounting area for the disc and the dial indicator. The dial indicator requires a mounting bracket that can be attached to a solid portion of the engine block. It may be a *clamp type* or a *magnetic type* (**Figures 4-17** and **4-18**). For this setup, a clamp should be used. Sometimes the shape and position of the mounting area may be such that only one type of mount can be used. Regardless of the mount type, ensure that it is securely fastened so the dial indicator can be moved and positioned properly against the component being measured. The dial indicator should be the balanced type, but a continuous type can be used.

Special Tools

Dial indicator set

A vehicle with the disc caliper and wheel assembly removed

Classroom Manual
pages 95–97

Figure 4-16 The side-to-side movement of the brake rotor will cause a brake pedal pulsation. This is known as "runout."

Figure 4-17 A dial indicator with a magnetic mount. The magnetism in this mount can be switched on or off for easier setup and can be fitted to a round component as shown in the figure.

Figure 4-18 The dial indicator can be mounted with an adjustable locking pliers as well.

Set the disc on the hub and install at least half of the lug nuts. The nuts only need to be tight enough to hold the disc firmly in place. Once the disc is positioned, mount the dial indicator bracket to the brake caliper mount. Assemble the various swivels and rods so the links reach over the disc (**Figure 4-19**). Before mounting the dial indicator to the mount, release the indicator's plunger completely. Install the indicator on the mounting rod and shift the linkage until the plunger is resting against the disc about midway from the machined area (**Figure 4-20**).

Move the dial indicator toward the disc until about half of the plunger is retracted into the indicator. At this point, tighten all of the mount's swivel or links to hold the dial indicator in place. Hold the dial indicator with one hand and adjust the scale until the zero mark is directly under the needle (**Figure 4-21**). It may take a few tries to get it correct.

When the dial indicator is firmly mounted, the plunger is partially retracted, and the scale is set to zero, the measurement can be taken. Using a screwdriver between the wheel lugs or by hand, rotate the disc in one direction slowly and evenly. Sharp, sudden movements should be avoided. Instead, try to make a smooth, continuous rotation to prevent moving the dial indicator. As the disc is rotated, read the dial indicator's scale for movement of the needle (**Figure 4-22**).

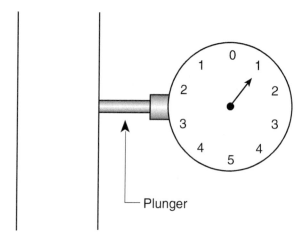

Figure 4-20 With the plunger retracted about halfway, the needle will probably not be at zero.

Figure 4-19 The dial indicator brackets allow the indicator to be positioned for an accurate measurement. Other types of mountings are also available.

Figure 4-21 Rotate the knurled ring to turn the scale so that zero is under the needle. The indicator will have to be held in place.

Figure 4-22 The needle may move to the left and right of zero as the disc is rotated.

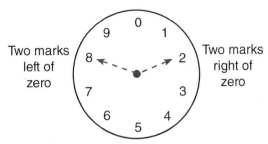

Figure 4-23 A continuous scale dial indicator requires the technician to count the number of graduations the needle moved to the left of zero.

The needle will probably move to the left and right of zero (**Figure 4-23**). Generally, any movement exceeding 0.002 inch indicates the disc needs to be machined or its mountings checked. On a balanced scale, the reading can be read from the numbers directly under the needle. If a continuous dial indicator is used and the disc's rotation caused the needle to move counterclockwise, the marks will have to be counted. Start at zero and count counterclockwise until the mark directly under the needle is reached. Check the size or calibration of the scale. It will be marked on the face as either 1/1,000 or 1/10,000 inch. Each mark on the scale represents one of the calibration units. Usually, it is best to make the measurement two or three times to ensure an accurate reading.

There are several flat, disc-type components, including the brake disc and flywheel, which should be measured for runout. A warped component that is supposed to be flat will cause its mated component to vibrate and shake, thereby resulting in driver discomfort and possible damage.

Special Tools

Brake disc
Brake caliper (micrometer)

USING BRAKE CALIPERS TO MEASURE THE THICKNESS OF BRAKE DISCS

During brake repair, the disc (or **rotor**) should be checked for excessive wear and warpage. The procedure discussed above will give the technician an idea of the flatness of one side of the rotor, but a more common measurement is the thickness and parallelism of the two opposing friction areas. Thickness can be measured with a regular outside micrometer or a special brake disc micrometer (**Figure 4-24**). The primary difference between these micrometers is quick thimble action and larger numbers on the brake micrometer. This

Figure 4-24 An electronic brake (rotor) micrometer.

Classroom Manual
page 94

makes it easier and quicker to use. Warpage or parallelism between the two opposing friction areas can also be checked with a micrometer or a disc parallel gauge.

> ✿✿ **SERVICE TIP** In the last few years, brake rotors have an additional minimum specification: *minimum refinishing thickness.* If the rotor is at that thickness or thinner, then the rotor is replaced. The minimum refinishing limit is usually greater than the minimum or discard thickness. The refinishing specification is found in the service manual and usually not stamped on the rotor.

Brake disc thickness and parallelism can be checked on the car, a table, or a brake lathe. Find the minimum thickness specification for the disc. Many times, it is stamped on the inside or outside of the hub portion of the disc and most are listed as metric measurements (**Figure 4-25**). If the brake disc micrometer is USC, convert the specification to USC.

Place the micrometer's measuring faces on each side of the disc midway to the wear area (**Figure 4-26**). Rotate the thimble until the spindle and anvil faces are against the disc. Read and record the measurement. In almost all cases, the measurement will be less than 1 inch. Repeat the measurement at a minimum of 12 different points around the disc (**Figure 4-27**). Remember to measure at the midpoint of the friction areas at each of the measuring points. Compare your readings to the specified minimum thickness. Any measurement that is less than specified means the disc is not to be reused and should be discarded. Differences over 0.002 inch between any of

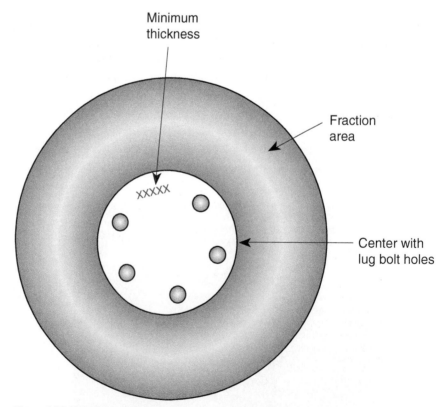

Figure 4-25 The minimum thickness is usually on the disc, but it can also be found in the service manual.

Figure 4-26 The measuring faces should always be at the halfway point of the friction area. The outer edges may be thinner or have a ridge.

Figure 4-27 Check for thickness variation by measuring 8 to 12 points around the rotor.

the 12 points mean the disc's friction areas are not parallel. If the disc is not too thin, it can be machined and reused.

Measuring Shaft Endplay

> **AUTHOR'S NOTE** The easiest component to use for this exercise is either a manual or an automatic transmission/transaxle input (turbine) shaft.

For the purpose of this section, an automatic transmission will be used, and the input shaft endplay will be measured. Determine the type of transmission either from a transmission ID tag or by researching the type of vehicle from which it was removed. Use the service information available and locate the specification for the shaft's maximum endplay. This measurement is one of the very first steps in disassembling any transmission or transaxle. Place the transmission in the holder or solidly on a rigid table. If using a table, it is suggested that blocks of wood or an assistant keep the transaxle from shifting during the measurement. Because most transaxle cases (housings) are made from aluminum, it will be necessary to use a dial indicator with a clamping mount. Before mounting the dial indicator, either push or pull the input shaft until it stops. It does not make any difference whether the shaft is all the way in or all the way out. However, experience seems to suggest that pulling the shaft out makes the final movement easier.

Using the clamp mount over the lip of the transaxle bell housing, position the dial indicator plunger over the center of the outward end of the shaft (**Figure 4-28**). Be careful not to disturb the position of the shaft. Clamp the mount to the housing while keeping the plunger in the general area of the shaft's end. Flex the dial indicator support arm (bracket) until the plunger is directly on the shaft end and approximately halfway in. Tighten all fasteners or clamps to secure the dial indicator in place.

With the dial indicator locked in place, loosen the dial lock and rotate the graduated scale until the zero number is directly under the dial's needle (**Figure 4-29**). Release and recheck the position of the zero to the needle. If correct, lock the scale in place. Have an

Special Tools

Dial indicator with clamp

Assembled manual transaxle

Transaxle holder or table

Classroom Manual
pages 95–96

Figure 4-28 This dial indicator has been mounted on a transaxle prior to measuring shaft endplay.

Unlock

Lock

Rotate scale until zero (0) under needle, then lock in place

Figure 4-29 The scale must be rotated so that the zero mark is aligned with the needle.

assistant hold the transaxle steady while the shaft is pushed (pulled) to its opposite stop. Read the measure from the dial indicator scale. If it is set up properly and the shaft started at its outmost position, the needle should have moved clockwise, and the measurement can be read directly (**Figure 4-30**). If the shaft started at its innermost position, it may be necessary to count the graduations. Compare the measures and determine if the endplay is within specified limits. **Photo Sequence 8** shows the steps in this shaft measurement.

Figure 4-30 After moving the shaft to the opposite position, the needle will move clockwise. The total endplay can now be read.

PHOTO SEQUENCE 8
Measuring a Shaft's Endplay

P8-1 Position the transmission so the input shaft is facing up.

P8-2 Mount a dial indicator to the transmission case so the indicator's plunger can move with the input shaft as the shaft is pulled up and pushed down.

P8-3 Lightly pull up on the shaft, then push it down until it stops. Then zero the indicator.

P8-4 Pull the shaft up until it stops and then read the indicator. This is the total amount of the endplay.

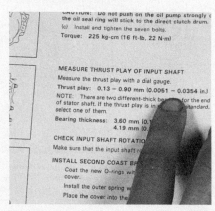

P8-5 Locate the endplay specifications in the service manual.

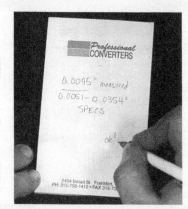

P8-6 Compare your measurements to the specifications.

MEASURING PRESSURE

Pressure on an automobile is normally measured on the fuel, oil, air conditioning, power steering, brake, and automatic transmission systems. With ABS being used on most vehicles, special gauges are used to check brake pressure. Pressure tests on these systems are used to detect malfunctions in fluid delivery. Tapping into the fluid flow in some way is the method to perform any pressure test (**Figure 4-31**). Gauges and taps must be selected depending on the system being tested. Gauges may be either a pressure/vacuum type, measuring only up to 10 psi, or a pressure type that measures up to 200 psi or more. Normally, the pressure or vacuum gauges are used only on carburetor fuel systems. Almost all other systems require gauges that can read high pressures. Only an experienced technician should take pressure measurements, especially on the high-pressure systems. For our discussion, we will measure the pressure of a tire.

To obtain a correct tire pressure measurement, the tire must not have been driven for more than 2 or 3 miles. In other words, it must be "cold." Driving heats the air in the tire, thereby increasing pressure. With the vehicle on the ground or floor, remove the cap from the tire's valve stem.

Place the connection end of the gauge firmly over and down the end of the valve stem (**Figure 4-32**). There should be no noise of escaping air around the gauge. Read the pressure with the gauge held in place. A typical passenger car should have pressure between 32 and 35 psi. If the reading is high, remove the gauge and release some air from the tire. If too low, add air through the valve stem. Do not exceed the tire's maximum allowable pressure.

Figure 4-31 Note that the pressure gauge is connected to the pressure line. This is true in almost every measurement taken on any hydraulic system.

Figure 4-32 Press the gauge firmly against the valve stem.

MEASURING ENGINE VACUUM

Measuring vacuum on an automobile engine can tell the technician a great deal about the internal condition of that engine. The downward movement of the pistons within their cylinders creates engine vacuum. Anything that causes an engine's vacuum to drop will result in lower performance.

⚠ **WARNING An experienced technician or instructor should guide the student or new technician through this section the first time. Disruption to the engine's performance or injury to the technician could occur if skill, attention to detail, and safety procedures are not used.**

A vacuum test is performed using a port tapped directly into the engine's intake manifold (**Figure 4-33**). When possible, a multiple connection, sometimes known as a *vacuum tree*, should be used for the gauge's connection (**Figure 4-34**). If an empty port is not available, the gauge will have to be tied into one of the vacuum lines from the intake manifold. If a vacuum line is disconnected and a gauge is put in its place, the device operated by the vacuum line may cause a malfunction in engine operations and cause errors in the measurements. This can lead an inattentive technician in diagnostic circles.

With the vacuum gauge connected, start the engine and allow it to warm. Observe the vacuum measurements at idle and high idle engine speeds (**Figures 4-35** and **4-36**). Readings between 17 and 20 in. Hg at idle are normal. Note that there will be a drastic drop in vacuum as the engine is accelerated. A steady, below-specified vacuum or a fluctuating needle normally denotes an engine needing repairs. Most scan tools have the ability to display the engine vacuum by measuring it on the engine's manifold absolute pressure (MAP) sensor (**Figure 4-37**).

Special Tools

Vacuum gauge or pressure and vacuum gauge
Vehicle

Classroom Manual
page 97

Vacuum gauge

Special tool

Figure 4-33 This vacuum gauge is connected to the brake booster and can measure vacuum supply or engine vacuum (depending on the setup).

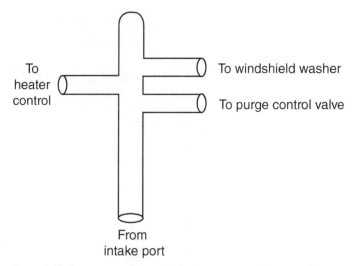

To heater control

To windshield washer

To purge control valve

From intake port

Figure 4-34 A vacuum tee is a small plastic component with connection ports for vacuum hoses. It is usually found on the firewall.

Figure 4-35 This engine has a healthy 18 in. HG at idle.

Figure 4-36 Vacuum gauge readings can indicate an engine's running condition.

Figure 4-37 Reading engine vacuum on a scan tool.

FASTENER TORQUE

One of the most critical procedures that must be followed is applying the correct amount of torque to a fastener, so the fastener can do the job for which it was designed. Overtightening can strip the threads, break the fastener, or damage the components. Under tightening can cause the fastener to loosen and come apart. The bottom line is to torque to specifications.

Torque specifications do not necessarily match the size of the fastener. A physically large fastener may require less torque than expected. The torque is based on the clamping force required and the tensile strength of the fastener. In this section, we will discuss the proper method of collecting torque specifications and installing a bolt into a threaded hole.

> **AUTHOR'S NOTE** Two pieces of scrap metal can be used for this exercise. One piece should have a threaded hole.

> **SERVICE TIP** A widespread practice among senior technicians is to tighten the fastener in stages, particularly when multiple fasteners are used. For instance, with a final torque of 45 ft.-lb., each fastener is tightened to 20 ft.-lb., then to 35ft.-lb., and finally to the specified 45 ft.-lb. This brings the two components together evenly, helping to prevent cocking of one part and a possible bad seal.

Standard Bolts

The first step in the installation of a bolt-type fastener is to clean the internal threads of the hole in which the bolt is to be installed. This can be done with a tap, either a bottom or tapered tap, depending on the hole. Once the tap has reached the bottom of the hole (or through the hole), use compressed air to blow the debris out.

Classroom Manual
page 98

 Special Tools
Face shield
Bottom or tapered tap
Tap wrench
Hinge torque wrench
Angle indicator
Proper-sized socket
OSHA approved air blow gun
Light lubricant

⚠ Caution
A face shield is the appropriate protection. Bits of metal or other material will be blown out of the hole with some force. Injuries to the face and eyes could result.

Item		ft.-lb. (N m)
1st step	in sequence	25 (33.9)
2nd step	in sequence	35 (47.5)
3rd step	in sequence	An additional 90 degrees

Figure 4-38 Many torque specifications require that a specific torque be applied in steps and that an additional amount of rotation be measured in degrees.

Figure 4-39 Typical cylinder head torque sequence.

With the threads cleaned, it is time to collect the torque specification. In a typical torque chart, a multiple step torque sequence may be listed. **Figure 4-38** shows an example of a torque setting chart for a cylinder head. It is also important to note that cylinder head bolts must be tightened in a specific sequence because of the multiple setting. Refer to **Figure 4-39** to see a typical sequence for a cylinder head. For the purpose of this section, assume the following information is provided:

Intake manifold to cylinder head   35 in.-lb. (3.95 N-mm)

Note that this torque is in inch-pounds of force. This indicates a heavy clamping force is not needed, and a 1/4-or a 3/8-inch drive torque wrench would be appropriate. A 1/4-inch drive torque wrench may be easier to set and easier to "feel" when the hinge breaks over. Thirty-five inch-pounds on a 3/8-inch drive torque wrench is a very low setting for a wrench of this size. When choosing a ratchet drive size, a good rule of thumb is to look at the torque range on the torque wrench. For example, if the range is from 25 to 125 ft.-lb., then it is safe to assume that it will be difficult to "feel" the setting when using this torque wrench on a fastener with a torque specification of 25 ft.-lb.

Apply a light coat of lubricant to the bolt threads. The lubricant should be a thin, lightweight penetrating oil such as WD-40. Select the desired wrench and set the scale to 35 in.-lb. Install the correct-size socket onto the wrench and place over the bolt head. Hold the wrench at the end or handle. This will apply the proper amount of force. Move the end of the wrench in a clockwise direction until the wrench "breaks over" or clicks. The fastener is now tightened to the proper torque. Remove and store the tools.

Torque-to-Yield

AUTHOR'S NOTE Old torque-to-yield bolts work well in this exercise without the additional costs of new bolts.

Torque-to-yield bolts are generally tightened in the manner just described except for the final step concerning the stretch of the fastener. For this exercise, follow the steps listed here:

1st step	25 ft.-lb. (2.82 N-mm)
2nd step	35 ft.-lb. (3.95 N-mm)
3rd step	Turn an additional 30 degrees

An **angle indicator** will be needed for step 3 in addition to the socket and torque wrench (**Figure 4-40**). The angle indicator is used to measure the exact degree of wrench rotation. Also, a 3/8-inch drive torque wrench will probably be the best wrench for the torque required.

Select and install the correct socket onto the torque wrench. Set the torque wrench to 25 ft.-lb. Tighten the bolt until the wrench clicks. Remove the wrench from the fastener and set it to 35 ft.-lb. Now tighten the bolt until the wrench clicks at 35 ft.-lb. Remove the socket from the wrench and install the angle indicator onto the wrench. Connect the socket to the drive end of the angle indicator.

The angle indicator has an anchor to hold its scale in place as the wrench is moved. Place the socket over the bolt head and hold in place until the indicator's anchor is positioned and locked (**Figure 4-41**). While still holding the wrench in place, hand rotate the indicator needle (or scale, depending on the model) until it aligns with the zero on the scale. Ensure the anchor is still in place and locked. Hold the wrench end and turn in a clockwise direction until the needle aligns with the 30-degree mark on the scale. The torque-to-yield fastener is now at its specified torque. Remove and store the tools.

Note that the torque wrench was not set to a higher torque when the final 30-degree turn was made. This increased setting is not necessary and the click when the torque is applied acts as a "reminder" of what is intended. If desired, the setting may be adjusted to a higher setting without affecting the final results as long as the fastener is not turned past the 30-degree mark.

Experience is the best method to learn how to set up and use an **angle indicator**. The manual one is awkward to use in tight places.

 Special Tools

U.S. (English) pitch gauge

Metric pitch gauge

Selection of U.S. (English) and metric fasteners

Classroom Manual pages 99–100

Figure 4-40 An electronic angle indicator is used to measure the degree of torque.

Figure 4-41 Note the anchor (upper left) set between the two rocker arms to prevent the scale from moving. The socket is under the scale.

USING A PITCH GAUGE

Replacing damaged or lost fasteners is a common problem when reassembling an automotive component. Whenever a fastener must be replaced, be sure to get a replacement that is the exact quality and thread size as the original. One of the easiest ways to select the correct fastener is to use an original bolt or nut to determine the correct replacement size. For example, if a replacement bolt is needed, take the nut and find the bolt that fits. Make sure the replacement bolt threads into the nut. Be cautious because some metric nuts/bolts may appear to screw onto USC bolts/nuts, and vice versa. However, they will not effectively hold.

There are times when matching bolts and nuts is impossible. In these cases, you may need to measure the thread on a fastener to determine its size. The tool used to do this is the pitch gauge. A pitch gauge (**Figure 4-42**) has a series of blades, each with teeth on one side. Each blade is marked with a thread size. The teeth on the blades are used to match up with threads on a fastener. If they match up, determine the thread size. There are pitch gauges for both U.S. (English) and metric threads.

The best way to learn how to use a pitch gauge is to practice with a bolt or hex head cap screw with a known size. This way you can get to see how the blades of the pitch gauge should match the threads on a fastener.

Begin with a common U.S. (English) thread size like the 1/4 bolt or hex head cap screw shown in **Figure 4-43**. Pull each blade out of the pitch gauge tool and set the teeth on the fastener as shown in **Figure 4-44**. Keep trying the blades until a set of teeth fits perfectly into the threads. When a match is found, thread pitch can be identified. The next step is to simply read the thread pitch off the pitch gauge blade. In our example, the pitch gauge blade reads 16 threads per inch.

Now the technician is ready to try a fastener with an unknown thread size. Choose a metric fastener with an unknown thread size like the one shown in **Figure 4-45**. This time, use a metric pitch gauge. Pull out each blade and set it on the threads as before. When a perfect match is found, determine the thread pitch. The thread pitch of the fastener is 1.5, as shown on the pitch gauge blade in **Figure 4-46**.

Special Tools

U.S. (English) or metric tap and die set

Machinist square

Figure 4-42 A pitch gauge is used to measure the pitch or angle of the threads.

Figure 4-43 A class 8 1/4-inch bolt.

Figure 4-44 The pitch gauge blade that matches the threads reads 16 threads per inch.

Figure 4-45 A metric bolt with an unknown thread size.

Figure 4-46 The thread pitch gauge blade shows that this fastener has a thread pitch of 1.5.

REPAIRING INTERNAL THREADS

Technicians often find fasteners that have been damaged. **Photo Sequence 9** shows a typical procedure for repairing damaged threads with a tap. A common problem is fasteners that have not been started properly by hand. When a fastener is started into a threaded hole crooked, and then a wrench is used to force the fastener into the threaded hole, the mismatched threads are often damaged. This damage (**Figure 4-47**) is often called cross threading or **stripped threads**.

The damaged internal threads must be repaired before the automotive component can be reassembled. The tools made to repair internal threads come in a set called a tap and die set. There are many different types and sizes of tap and die sets. There are sets for both metric and U.S. (English) thread sizes. Most shops have a set that contains the most common sizes used for the work done in the shop.

The tool used to repair or make new internal threads is called a *tap*. The tap (**Figure 4-48**) is a hardened cutting tool. It has a square drive end that is used to rotate it for thread cutting. The shank is the round part of the tap and is where the size of the tap is found. The actual thread cutting is done with sharp cutting threads spaced between flutes where the cut metal can collect. The end of the tap is chamfered or tapered so that it can fit easily into the hole to be tapped.

Before using a tap to repair the threads, make sure to select the correct size and type. This often means you will have to determine the thread size of the screw that fits into the threaded hole you are repairing. This can be done with the pitch gauge, as explained earlier.

If a technician were to measure the thread on the screw that fits into a damaged internal thread as 1/4 28 unified fine threads UNF, then a tap with the size 1/4 28 UNF on the shank would be selected, as shown in **Figure 4-49**.

Stripped threads A fastener failure were the threads have been damaged beyond use.

Classroom Manual
pages 61–62

Figure 4-47 A damaged or stripped internal thread.

Figure 4-48 Parts of a tap.

Figure 4-49 A tap with the size 1/4 28 UNF on the shank.

PHOTO SEQUENCE 9

Typical Procedure for Repairing Damaged Threads with a Tap

P9-1 Determine thread size of fastener that fits into damaged internal threads by matching fastener to pitch gauge.

P9-2 Select the correct size and type of tap for the threads to be repaired.

P9-3 Install the tap into a tap wrench.

P9-4 Start the tap squarely in the threaded hole using a machinist square as a guide.

P9-5 Drive the tap clockwise down the threaded hole the complete length of the threads.

P9-6 Drive the tap back out of the hole by turning it counterclockwise.

P9-7 Clean the metal chips left by the tap out of the hole.

P9-8 Inspect the threads left by the tap to be sure they are acceptable.

P9-9 Test the threads by threading the correct-size fastener into the threaded hole.

A technician must also determine if she or he is repairing a thread that goes all the way through a part, called a **through hole**. The other type of hole does not go all the way through and is called a **blind hole**. Refer to the illustration (**Figure 4-50**) of a blind and a through hole.

There are different types of taps used for blind and through holes. A taper tap is used to repair threads in a through hole. The taper tap has several incomplete threads at the end to help it start in the hole. A bottoming tap is used to repair threads in a blind hole. The bottoming tap has only two incomplete threads at the bottom, so it can tap threads all the way to the bottom of a hole. A taper and a bottoming tap are shown in **Figure 4-51**.

through hole also known as a ("thru-hole") threads that go all the way through the part.

blind hole A hole that is drilled but does not break through the part.

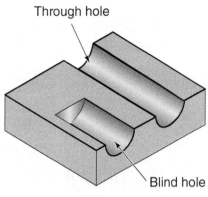

Through hole

Blind hole

Figure 4-50 A through hole and a blind hole.

Figure 4-51 A taper tap is used for through holes and a bottoming tap is used for blind holes.

⚙ **SERVICE TIP** The tap should screw into the hole without much effort. If the tap becomes hard to turn, back it off a few turns to allow the cut metal to drop out of the flutes. If it is still hard to turn, or seems to bottom out, then the tap is either cross threaded or the hole threads may be damaged beyond repair. Do not attempt to turn the tap if you must strain. The tap is brittle and will break in the hole. Removing a broken tap is very time consuming and difficult.

⚠ **Caution**

Always wear eye protection when using a tap. Taps are hardened tools and may shatter if they break.

⚡ WARNING **Practice using a tap on a scrap automotive component before attempting this job on a good part. Using a tap incorrectly can cause further damage to threads.**

After selecting the correct type and size of tap, attach the tap to a tap wrench. There are two common types of wrenches used to drive a tap: A T-handle and a hand-held tap wrench (**Figure 4-52**). The T-handle has an adjustable chuck on the end that is opened and tightened over the square drive of the tap. The hand-held tap wrench also has an adjustable chuck that is opened and closed by rotating one of the handles. The T-handle is usually the best choice for repair work, especially on smaller threads like the 1/4 28 UNF.

Slide the square drive end of the tap into the chuck of the T-handle tap wrench. Tighten the chuck to hold the tap securely in the wrench. Set the tap in the hole to

Handle adjustor

Chuck

Hand-held tap wrench

Adjustable chuck

T-handle tap wrench

Figure 4-52 Two common types of tap wrenches.

Square

Figure 4-53 The tap must be started square in the hole.

be repaired and rotate the tap carefully one or two turns. Make sure the tap enters the hole square and not at an angle. A clever way to check this is to place a machinist square next to the tap as shown in **Figure 4-53**.

Carefully and slowly rotate the tap through the threaded hole. When the bottom is reached, slowly rotate the tap counterclockwise up and out of the thread. Clean up all metal filings that result from cutting out damaged threads. Test the repaired thread with the correct-size screw.

> **SERVICE TIP** Sometimes threads are too damaged to repair. Badly damaged threads can be repaired with the installation of a part called a *helicoil*. A special helicoil tap is used to form a special thread in the hole. The helicoil is a metal coil that is threaded into the hole using a special tool. The coil forms new threads that are of the same size as the damaged ones.

 Special Tools

Helicoil
Helicoil installation tool
Drill and drill bit
Tap and tap wrench
Lubricant

Installing a Helicoil

A **helicoil** is used to replace internal threads. The coil is a steel tread that looks like a miniature coil spring (**Figure 4-54**). The outside of the coil has small barbs that will grip the inside of the hole when a bolt is installed. Several steps must be performed before the coil is installed into the hole.

A **helicoil** is also called a *thread insert*.

The first step is to determine which size coil to use. Use a pitch gauge to determine the thread pitch and then measure the diameter of the bolt's shank. This will provide the information for selecting the helicoil and the tools needed for its installation. For this discussion, assume the bolt is M10 × 1.25 mm × 25 mm. The length of the shank is not needed for this operation.

Find the correct helicoil and the instructions that go with it. The following instructions are general. If the helicoil's instructions are different, follow them.

There is usually a special-sized tap for the helicoil. Ensure that the tap is used or the helicoil may not fit correctly. The required drill bit size is usually etched into the head of the tap. Install the correct drill bit into an air- or electric-powered drill. While drilling the damaged threads out of the hole, ensure the drill bit is exactly aligned to the hole. Once the drill bit reaches the bottom of the hole, remove it and blow out the hole with compressed air.

Figure 4-54 Shown is a set of one-size helicoil. This set does not include the tap and drill bit.

> ⚙ **SERVICE TIP** General Motors and others use specially designed and engineered helicoil sets for their vehicles. Aftermarket helicoils can be used in their place except for critical areas like the cylinder head, engine block, and transmission. Some manufacturers do not recommend the use of helicoils on their vehicles. As always, consult the service manual.

 Caution

Wear a face shield during the drilling and cleaning steps. Injuries to the eyes and face could be caused by flying metal and debris.

Using the tap specified by the helicoil instructions, install the internal threads into the hole. After reaching the bottom of the hole (passing through), remove the tap and blow out the hole again. The hole is now ready to accept the helicoil. Install the helicoil onto its installation tool (**Figure 4-55**). The installation tool is used to screw the helicoil into the hole. Screw the helicoil in until its upper edge is even with the top of the hole. The helicoil

Tool

Helicoil

Figure 4-55 The helicoil is placed (screwed) onto the tool and then screwed into the newly tapped hole.

must not extend above the hole or be lower than the edge. Remove the tool and use either a small screwdriver or punch to break off the installation tab at the bottom of the helicoil. Blow out the hole for the last time. Spread a light lubricant on the bolt threads and screw the bolt into the helicoil. If installed properly, the bolt will screw in with a minimum of force. Remove the bolt and store the tools.

REMOVING AND REPLACING A DAMAGED STUD

A technician will often find damaged external threads on screws, bolts, and studs. These can be repaired with a die much like we used a tap to repair internal threads. However, because the screws, bolts, and studs are easily replaced with new fasteners, we rarely perform this job. All technicians need to master the replacement of studs.

Studs often get damaged from overtightening or from cross threading a nut on the threads (**Figure 4-56**). A damaged stud is replaced by unscrewing it from the part and installing a new one of the same size.

Stud replacement can be difficult because the stud has often been in place for a long period of time. There has been constant heating and cooling, corrosion, and rust buildup between the stud threads and the internal threads of the part. These factors combine to make some studs very difficult to remove.

The first step in stud removal is to use penetrating fluid to remove the corrosion to free the stud from its mating threads. Soak the area of the threads with penetrating fluid, as shown in **Figure 4-57**. Allowing the fluid to soak into the threads overnight will make it easier to remove the stud.

Before removing the old stud, measure the distance it sticks up from the surface. This measurement will be needed later when installing the new stud. Use a 6-inch scale to measure from the part surface to the top of the stud. Write the measurement down so it can be referred to later.

A **stud remover** is used to remove the studs (**Figure 4-58**). It is installed over the stud. The jaws on the stud remover grip the outside of the stud. A wrench fits on the stud

Special Tools

Six-inch scale
Stud remover
Penetrating fluid
Thread sealant
Antiseize compound

Classroom Manual
pages 101–102

stud remover A special tool that can remove a stud with damaged thread by gripping on the damaged stud.

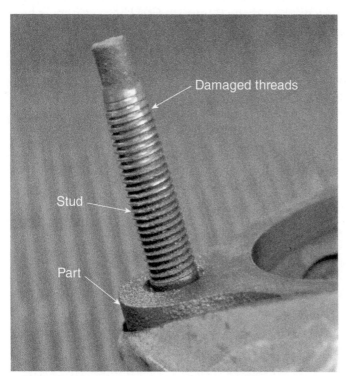

Figure 4-56 A stud with damaged threads must be replaced.

Figure 4-57 Use penetrating fluid around the stud threads before removing them.

Figure 4-58 A stud remover is used to unscrew a stud. Some studs have ends that can be driven with a wrench.

Figure 4-59 Using a stud remover to remove a stud.

remover and allows a technician to rotate the stud in a counterclockwise direction to remove the stud, as shown in **Figure 4-59**.

If a stud remover is not available, a stud can be removed with two nuts. Locate two nuts that are of the correct thread size to thread onto the stud. Start one nut and thread it all the way down to the bottom of the stud. This nut will be the drive nut. Start another nut and thread it down until it contacts the first nut. This is called the jam nut.

Put a wrench on the bottom drive nut and hold it in place. Put another wrench on the jam nut and tighten, or "jam," it against the drive nut. The jam nut will now hold the drive nut in position on the stud (**Figure 4-60**).

Now put an open-end wrench on the bottom drive nut. Turn the nut in a counterclockwise direction. Turning the nut in this direction causes it to want to unscrew up the stud (**Figure 4-61**). The jam nut, however, prevents the drive nut from moving up the stud. Instead the forces cause the stud to unscrew.

When the old stud is out, inspect the internal thread. If it appears rusty or damaged, clean up the thread by running the correct size tap through the threads as previously explained. Compare the new stud with the old one (**Figure 4-62**). The studs should be of the same thread size and the same length.

> ⚙ SERVICE TIP When using **thread-locking** compound or **thread-sealing** compound, make sure you are using the right chemical. They come in similar containers and one will not do the job of the other.

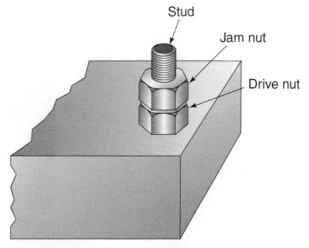

Figure 4-60 A drive nut and jam nut installed on a stud.

Stud

Jam nut

Drive nut

Figure 4-61 Use a wrench on the drive nut to unscrew the stud.

New stud

Old stud

Figure 4-62 The replacement stud must be the same size as the old stud.

CHEMICALS USED WITH FASTENER THREADS

Check the shop service manual for the car being worked on to determine if the threads of the new stud or bolt should be coated. If the stud should be locked in place and not easily removed, the technician may be instructed to use a **thread-locking compound** or **thread-sealing compound** (**Figure 4-63**). Thread-locking compounds are used on studs and other fasteners when vibration or impact might cause them to unscrew. Thread sealants are used when a stud extends into an area where liquids, such as oil or coolant, could get on the fastener. Both compounds are known as **anaerobic** sealants. This means they start the curing process in the absence of air.

thread-locking compound an adhesive applied to threads to prevent a bolt from loosening.

thread-sealing compound an adhesive applied to threads to prevent leaking or corrosion.

Anaerobic means without air. When you drop liquid sealant onto the threads of a bolt, it will remain liquid until you install it. After installing it and removing the air, it starts to harden and become plastic.

Figure 4-63 Threadlocker is a chemical that will assist in keeping a fastener tight. A small amount is placed on the threads of a fastener, usually the bolt.

Figure 4-64 Antiseize is a chemical that will assist in keeping a fastener from seizing in its counterpart during its life. It is most commonly applied to the threads of a spark plug.

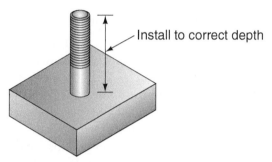

Install to correct depth

Figure 4-65 The new stud must be installed to the correct depth.

Anti-seize A lubricant to prevent mechanical seizing of two components.

Special Tools

Wire stripping and crimping pliers
Solderless connectors

Classroom Manual pages 154

Anti-seize compound (**Figure 4-64**) is used on the stud threads to prevent the stud from reacting with the metal of the internal threads. If this happens, the stud could stick or seize. Antiseize compound prevents this reaction and makes the stud easier to remove next time.

After the new stud is properly coated, it can be installed. Start the stud by hand, making sure it enters the threads squarely. Turn the stud in as far as possible by hand before using any tools. Then use two nuts as described earlier to drive the stud into the part. Use the depth measurement made on the old stud to be sure it is driven to the correct depth (**Figure 4-65**).

INSTALLING SOLDERLESS ELECTRICAL CONNECTORS

Replacing burned or damaged wires is a common electrical service procedure. When installing new wires in an electrical system, a technician will need to be able to install the connecting hardware on the wires. There are two general types of wire connectors: solderless and soldered. This section describes how to use the solderless type. The next section shows how to solder on wire connectors.

⚡ WARNING **Some manufacturers do not recommend the use of solderless butt connectors for making wiring repairs. They recommend soldering the two wires together or complete wire or wiring harness replacement.**

There are many types and styles of solderless connectors, as discussed in the Classroom Manual. The two basic groups are butt connectors used to connect two wires together

Figure 4-66 Parts of a solderless connector.

and terminal connectors used to connect a wire to a terminal. Both types have the same basic parts shown in **Figure 4-66**. The connector is made in two parts. The part that connects to the wire or terminal is made from metal that is a good electrical conductor. A plastic insulator fits over the parts that must not be allowed to touch another conductor. There are crimping tabs on the end of a terminal connector and on each end of a butt connector. The tabs are used to attach the wire. The crimping tabs are formed in a circular shape so that wire can be inserted.

Connectors are classified according to the wire size they fit. A 16-gauge (1.0 mm^2) connector has a crimping tab hole that will fit a 16-gauge (1.0 mm^2) wire. The wire size for solderless connectors is printed on the package.

> **Crimping** is a procedure used to join a wire to a solderless connector.

⚡ WARNING **Always use the correct-size connector for the wire you are install-ing. Wire that is too small or large for the connector will not be retained properly in the connector.**

The basic tool used to install solderless connectors is the **stripping and crimping pliers** shown in **Figure 4-67**. The jaw area of the tool is used to crimp or squeeze the crimping tabs of the connector onto the wire. The area behind the jaws has a set of cutters used to strip away the insulation on the wire.

The first step in installing the connector is to strip the insulation off the end of the wire. The wire stripper has several sets of cutters in the stripping area. Each is labeled with a wire gauge size. When **stripping** a 16-gauge (1.0 mm^2) wire, insert approximately 1/4 inch of the end of wire between the stripper cutters. Squeeze down on the tool handles. This action cuts the insulation but not the wire. Pull the wire out of the stripping area while holding the handles together. The insulation should be removed from the end of the wire as shown in **Figure 4-68**.

> **Stripping and crimping pliers** are used to strip insulation from wire and crimp a solderless terminal on wire.
>
> **Stripping** is the procedure used to remove insulation from an electrical wire.

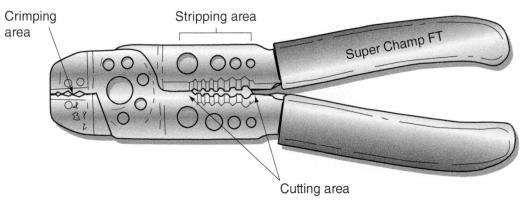

Figure 4-67 A wire crimper is used to strip the wire and crimp on a solderless connection or terminal. Some have the cutting area at the end away from the handle.

Figure 4-68 Stripping the insulation from a wire.

Figure 4-69 The stripped end of the wire is inserted into the crimping taps.

Figure 4-70 How two wires fit in a butt connector.

Insert the stripped end of the wire through the terminal crimping tabs as shown in **Figure 4-69**. When installing a butt connector, install a wire at both ends as shown in **Figure 4-70**. Be sure the stripped wire fits completely into the crimping area.

The crimping jaws on the end of the pliers are used to squeeze the crimping tabs down on the wire. Most stripping and crimping pliers have several sets of jaws that are marked with different wire gauge sizes. Use the one that is correct for your solderless terminal size.

Place the terminal and wire in the tool jaws as shown in **Figure 4-71**. Squeeze on the tool handles to crimp the tabs against the wire. The two parts of the tool jaw will flatten part of the terminal and turn in the crimping tabs on the wire as shown in **Figure 4-72**.

Figure 4-71 Crimping a solderless connector in crimping jaws.

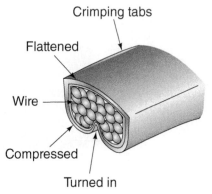

Figure 4-72 The crimping tabs are crimped against the wire.

Test the connection by gripping the wire in one hand and the terminal in the other. Gently try to pull the two apart. A properly installed connector will not have any movement back and forth.

⚙ SERVICE TIP Many manufacturers and most experienced technicians recommend the stripped end of the wire be "tinned" before installing it into a solderless terminal. This seals the strands together and makes a better electrical connection. **Tinning** is done by heating the stripped wire and applying a light coat of solder. When making a solder connection, tinning should also be used. (See Figure 4-79.)

Tinning Application of a thin layer of soft solder to the ends of wires before soldering them.

⚒ Special Tools

Soldering gun or soldering iron
Rosin-core solder
Stripping and crimping pliers

REPAIRING WEATHER PACK CONNECTORS

Hard-shell and weather pack connectors usually provide a means of disassembling the terminal for repair. On some occasions there may be a Technical Service Bulletin that outlines a problem with a connector. On other occasions you may find that wiggling the connector is when the problem occurs.

You can remove these connectors using special tools to release a locking tang inside the connector (**Figure 4-73**). Once disassembled, check the connector for corrosion, worn parts, and a good solder joint. Sometimes you can rebuild the connector, and sometimes you must get a new connector. If you have to purchase a new connector, it will usually come with extra wire already connected to it. You then must solder the wire into the circuit.

⚙ SERVICE TIP A good way to practice the wire stripping and soldering techniques presented in this chapter is to make a set of jumper leads for circuit testing. Cut a 24-inch length of 16-gauge (1.0 mm^2) wire. Strip both ends of the wire. Buy two alligator clips from the local auto parts or electronics store. Solder these on each end of the wire. The jumper is an excellent test tool when used properly.

Push narrow pick between terminal and connector body.

Figure 4-73 Hard-shell and weather-pack connectors can be disassembled and rebuilt.

Classroom Manual
page 154

SOLDERING WIRES AND ELECTRICAL CONNECTORS

soldering A process in which two or more items are joined together by melting solder into the joint.

Another way to join electrical wires or terminals to wires is by **soldering**. **Photo Sequence 10** shows soldering wire and electrical connectors. This is a method of joining wires and connectors using a metal called solder. Solder is a metal combination, or alloy, of tin and lead that melts at a low temperature. It can be melted and then allowed to cool back to a solid to join electrical components. Solder has excellent electrical conductivity.

PHOTO SEQUENCE 10
Typical Procedure for Soldering Electrical Terminals and Joining Wires

P10-1 To solder a terminal to a wire, first place the end of the wire in the correct size cutter on the stripping and crimping pliers.

P10-2 Squeeze the pliers' handles and pull the insulation off the end of the wire.

P10-3 Pull the trigger on the soldering gun to heat the end of the gun.

P10-4 Touch the heated end of the soldering gun to the stripped end of the wire to heat the wire.

P10-5 Melt a small amount of rosin-core solder over the surface of the wire to "tin" the wire.

P10-6 Install the end of the wire into the correct-size electrical terminal.

P10-7 Heat the electrical connector and wire with the soldering gun.

P10-8 Melt solder into the joint between the wire and connector and allow to cool.

P10-9 To join two wires, first use the stripping and crimping pliers to strip the ends of both wires.

P10-10 Slip a piece of heat shrink tubing over one wire and slide it out of the way. Then use the soldering gun to heat each wire end and melt solder on the wire to tin the wire.

P10-11 Wrap the two ends of the wire together to form a tight connection.

P10-12 Melt solder between the two wires to form a connection.

P10-13 Cover the exposed connection with the piece of heat shrink. Use a heat gun to shrink the covering.

To solder parts together, the solder must be heated to its melting point. This is done with either a **soldering gun** or a **soldering iron**. An electric soldering iron is shown in **Figure 4-74**. The unit has a power cord plugged into an electrical outlet.

Electricity flows through a heating element inside the unit and heats the tip hot enough to melt solder. An insulated handle allows you to hold the handle safely. Inexpensive soldering irons do not have on-off switches. Once plugged in, they take a short time to heat up to get the tip hot enough to melt solder.

The soldering gun gets its name from the fact that it looks a little like a gun and has a trigger, as shown in **Figure 4-75**. The power cord of the soldering gun is plugged into an electrical outlet. The tip is heated instantly by a heating element inside the gun whenever the trigger is pulled. When the trigger is released, the tip cools down. Some units have a light that illuminates when the tip is hot enough to melt solder. Both the soldering gun and soldering iron are used the same way to melt solder.

There are many types of solders to do many different joining jobs. Solder used for electrical work is sold in spools (**Figure 4-76**). It is shaped in the form of a wire. Many different diameters are available. A solder wire of 0.062-inch diameter is commonly used for general electrical work. In addition, many different alloys of tin and lead are available. Most electrical work is done with 40/60 alloy. The diameter and alloy are marked on the solder spool.

Soldering must be done with a material called soldering **flux**. Soldering flux is used to remove oxides from the wires and terminals. Oxides are formed when metal is exposed

soldering gun or a **soldering iron:** A tool used to melt solder.

 Caution

The heating tip of the soldering gun and soldering arm gets to a temperature in excess of 300°F. Be careful not to touch the tip or it will cause burns.

flux A substance added to molten metals to clean and bond the surface together better.

Figure 4-74 A typical electrical soldering iron.

Figure 4-75 A typical soldering gun.

Figure 4-76 Spool of 40/60 rosin-core solder.

⚠ **Caution**

Both the flux and the solder cause fumes that are dangerous to breathe for a prolonged period of time. Always solder in a well-ventilated area. Always wear eye protection when soldering.

to air. The flux cleans the wires and terminals. It also helps the solder flow more easily. Although flux is available separately, solder for electrical work typically has the flux inside the core of the solder wire, as shown in **Figure 4-77**.

There are two general types of flux: *acid* and *rosin*. Acid-type flux is used for joining sheet metal. It is too corrosive for electrical work. Electrical soldering is done with a noncorrosive flux called resin. The solder spool will indicate the type of flux in the solder core. Choose solder that is marked "resin core."

⚡ WARNING **Never use acid-core solder on electrical components. It can cause corrosion that can interrupt electrical flow.**

With the correct solder and the soldering gun or iron heated, try soldering a terminal to a wire. Just as in solderless connectors, choose wire and terminals that are of the same gauge size.

Use the stripping and crimping pliers to strip the insulation off approximately 1/4 inch of the end of the wire, as shown in **Figure 4-78**. Tin the wire before it is joined to

Figure 4-77 The center of the solder has a core of rosin flux.

Figure 4-78 Stripping the insulation off the wire.

Figure 4-79 Tinning the end of the wire with melted solder.

the terminal. Tinning improves the final joining of the terminal and wire. Touch the tip of the soldering gun or iron to the wire and allow it to heat a moment. Then touch the solder to the heated wire and allow the solder to flow completely over the wire. Tinning the wire is shown in **Figure 4-79**.

Place the tinned wire into the terminal. Heat the terminal with the soldering iron tip. Give it a moment to heat up, then apply solder into the terminal as shown in **Figure 4-80**. Allow the solder to solidify and cool. Then test the connection by holding onto the terminal and pulling on the wire. A defective solder joint will break under pressure.

Soldering can also be used to join, or **splice**, two or more wires together. The common methods of joining wires are shown in **Figure 4-81**. In all the methods, the wire ends are stripped and tinned, then they are soldered as shown in **Figure 4-82**. The exposed area of the wires and solder will then have to be covered with the correct type of electrical tape or *heat shrink*.

splice Joining two or more electrical wires together using solder.

⚙ **SERVICE TIP** Replacement wires can be covered with a product called *heat shrink*. Heat shrink is a thin-wall plastic tubing. The tubing is available in many sizes to cover a single wire or a number of wires to form a loom. Heat the tubing after the wires are installed through the tubing. The tubing shrinks into a tight fit around the wires. The tubing must be installed onto one section of the wire before the connection is made.

Rosin core solder

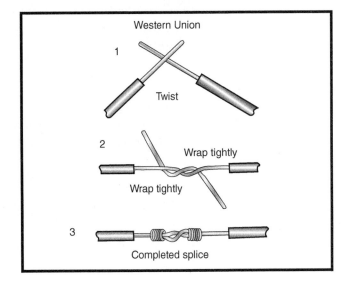

Figure 4-80 Soldering the terminal to the wire.

Figure 4-81 Three common ways of splicing wire.

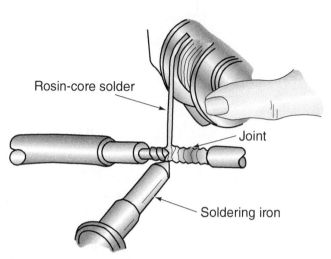

Figure 4-82 Soldering the wire joint.

When you must solder a twisted pair wire, you must make sure to solder it and then not leave much straight wire. Always try to return the wiring to the way the manufacturer designed and produced it.

CASE STUDY

An engine was checked for a poor idle condition. One of the tests was a measurement of the engine's vacuum. A vacuum gauge was connected to an intake port for the adjustment. As soon as the vacuum line was disconnected and the gauge connected, the engine began to run smoothly. A closer inspection found that the hose was connected to a small vacuum motor in the heating system and was disconnected. This allowed a substantial idle leak and leaned the air/fuel mixture. However, it was not enough of a leak to cause a noticeable difference at cruising speeds. The disconnected hose was under the dash, and road noise and the radio muffled the sound of the leak.

CASE STUDY

Using the wrong type of soldering flux can have disastrous results. A friend related a problem several years ago. He was doing a ground-up restoration on an old 356 Porsche. All the old wiring had been pulled out of the car to put in a complete new wiring harness. After pricing commercially available wiring harnesses, he decided to make his own. He bought and cut hundreds of feet of new wire. He wanted to do a good job, so he soldered each terminal and connector in his new loom.

The newly restored car had been on the road a couple of years when he started having electrical problems. Light switches, turn signals, and accessories stopped working for no apparent reason. He began to notice that corrosion was building up on his electrical connections. Further investigation showed that the corrosion was caused by his use of an acid-core solder. He was quite unhappy when he found out that all of the hundreds of connections had to be redone with the new terminals and the correct type of rosin-core solder.

ASE-STYLE REVIEW QUESTIONS

1. Pressure and vacuum measurement is being discussed.

   Technician A says a pressure gauge is used to check some brake systems.

   Technician B says an engine vacuum test may indicate the problems with a poorly performing engine.

   Who is correct?

   A. A only
   B. B only
   C. Both A and B
   D. Neither A nor B

2. *Technician A* says pressure may be used within the steering system.

   Technician B says tire pressure must be checked when the tire is cold to obtain an accurate measurement.

   Who is correct?

   A. A only
   B. B only
   C. Both A and B
   D. Neither A nor B

3. The use of a pitch gauge is being discussed.

   Technician A says the pitch gauge can be used to identify U.S. (English) threads.

   Technician B says the pitch gauge can be used to identify Metric threads.

   Who is correct?

   A. A only
   B. B only
   C. Both A and B
   D. Neither A nor B

4. The use of a tap to repair internal threads is being discussed.

   Technician A says the tap must have the correct thread size on the shank.

   Technician B says one tap fits many different thread sizes.

   Who is correct?

   A. A only
   B. B only
   C. Both A and B
   D. Neither A nor B

5. Measurement systems are being discussed.

   Technician A says a metric micrometer may be used to measure a USC crankshaft journal.

   Technician B says a conversion factor is used to convert measurements between measuring systems.

   Who is correct?

   A. A only
   B. B only
   C. Both A and B
   D. Neither A nor B

6. The removing and replacing of a stud where vibration is a problem is being discussed.

   Technician A says to coat the stud with antiseize compound.

   Technician B says to use a softer grade of bolt so it will grip better.

   Who is correct?

   A. A only
   B. B only
   C. Both A and B
   D. Neither A nor B

7. The installation of a solderless electrical connector is being discussed.

   Technician A says the connector and wire must have the same wire gauge size.

   Technician B says the connector should be crimped to the wire before soldering.

   Who is correct?

   A. A only
   B. B only
   C. Both A and B
   D. Neither A nor B

8. The soldering of a terminal on a wire is being discussed.

   Technician A says to use acid-core solder so the wire is cleaned of contaminants.

   Technician B says to use rosin-core solder so the connection resists corrosion.

   Who is correct?

   A. A only
   B. B only
   C. Both A and B
   D. Neither A nor B

9. Dial indicators are being discussed.

   Technician A says a dial indicator may have a scale numbered from zero and goes up in two different directions.

   Technician B says a brake disc caliper is a modified dial indicator.

   Who is correct?

   A. A only
   B. B only
   C. Both A and B
   D. Neither A nor B

10. Micrometers and dial indicators are being discussed.

    Technician A says a dial indicator can be used without a mount.

    Technician B says the use of a micrometer requires a clamping mount.

    Who is correct?

    A. A only
    B. B only
    C. Both A and B
    D. Neither A nor B

ASE CHALLENGE QUESTIONS

1. A replacement bolt is being discussed. *Technician A* says that thread pitch is the same as thread coarseness or threads per inch. *Technician B* says that bolt head markings specify the grade of a bolt. Who is correct?

   A. A only
   B. B only
   C. Both A and B
   D. Neither A nor B

2. Snap rings hold and fit into grooves in:

   A. pins
   B. gears
   C. shafts
   D. all of the above

3. A bolt is being discussed. *Technician A* says the shank of the bolt is the length of the bolt. *Technician B* says the shank of the bolt is the diameter of the bolt.
   Who is correct?

   A. A only
   B. B only
   C. Both A and B
   D. Neither A nor B

4. *Technician A* says that studs are bolts that are threaded on both ends and do not have a head. *Technician B* says torque-to-yield bolts can be reused if cleaned properly.
   Who is correct?

   A. A only
   B. B only
   C. Both A and B
   D. Neither A nor B

5. Measuring devices are being discussed. *Technician A* says feeler gauges can not be used to measure spark plug gap. *Technician B* says a drum micrometer is considered an outside brake micrometer and can measure drum diameter.
   Who is correct?

   A. A only
   B. B only
   C. Both A and B
   D. Neither A nor B

Name _____   Date _____

CONVERSIONS

Upon completion and review of this job sheet, you should be able to convert one unit of measure to another unit of measure.

ASE Education Foundation Correlation ————————————————————

This job sheet addresses the following **RST** tasks:

6-5.   Contributes ideas and initiative

6-6.   Follows directions

6-11. Uses scientific, technical, engineering and mathematics principles
and reasoning to accomplish assigned tasks

We Support

Education Foundation

Tools and Materials

Calculator

Procedure

In the automotive industry we use metric and standard measuring systems, or occasionally both. It is necessary we know how to convert from one measuring system to the other.

Generally inches, centimeters, and millimeters will be used as units of measure. The largest is an inch, followed by centimeters, and then millimeters.

CONVERSION REFERENCE CHART

10 millimeters = 1 centimeter

2.54 centimeters = 1 inch

25.4 millimeters = 1 inch

When you know the number of:	Multiply/divide	To find:
inches	**Multiply by 25.4**	millimeters
inches	**Multiply by 2.54**	centimeters
centimeters	**Multiply by 10**	millimeters
millimeters	**Divide by 10**	centimeters
millimeters	**Divide by 25.4**	inches
centimeters	**Divide by 2.54**	inches

EXAMPLES:

If you have 2.3″ and want to find millimeters, the formula is: 2.3″ × 25.4 = 58.42 mm

If you have 35 cm and want to find how many inches, the formula is: 35 cm × 2.54 = 13.77″

Do the following exercise to practice using the conversion chart:

Show the formula in the space below. You can use a calculator!

EXAMPLE:

2 inches __2 × 25.4__ = __50.8__ mm

1. 29 inches _____ = _____ cm

2. 30 cm _____ = _____ mm

3. .7 mm _____ = _____ cm

4. 605 mm _____ = _____ inches

5. .05 inches _____ = _____ mm

6. 16 cm _____ = _____ inches

7. 1.024 inches _____ = _____ mm

8. 26 mm _____ = _____ inches

9. 220 cm _____ = _____ mm

10. .850 inches _____ = _____ cm

11. 2.159 cm _____ = _____ mm

12. 2.54 inches _____ = _____ cm

13. 25.4 inches _____ = _____ mm

14. 10 mm _____ = _____ inches

15. .393 inches _____ = _____ cm

Instructor's Response

_____

_____

_____

Name _____   Date _____

DECIMALS

Upon completion of this job sheet, you should have a better understanding of how to read and write decimals properly.

ASE Education Foundation Correlation

This job sheet addresses the following **RST** tasks:

6-5. Contributes ideas and initiative

6-6. Follows directions

6-11. Uses scientific, technical, engineering and mathematics principles and reasoning to accomplish assigned tasks

We Support

 | Education Foundation

Procedure

A quantity or a number less than 1 can be expressed by using fractions or decimals. We use decimals every day when we use money. For example, $2.11 represents two dollars and eleven cents. We use the first two decimal places: tenths and hundredths. Tenths and hundredths are also used when measuring lengths with a ruler.

RULER MEASUREMENTS

1/8″ = .125, 1/4″ = .25, 1/3″ = .33, 1/2″ = .5, 2/3″ = .66, 3/4″ = .75

MONEY

A penny is one-hundredth of a dollar = (.01). A dime is one-tenth of a dollar = (.1)

In the automotive industry, we refer to the size of a diameter by using decimals. The furthest place we use is the <u>ten-thousandths position</u>. Previously, we discussed the first two positions after the decimal: the tenths and hundredths positions. The third and fourth position after the decimal are thousandths and ten-thousandths.

EXAMPLE

2.2 6 3 5

2 is the **WHOLE** number

.2 is the **TENTHS** number

.26 is the **HUNDREDTHS** number

.263 is the **THOUSANDTHS** number

.2635 is the **TEN-THOUSANDTHS** number

NOW TRY TO WRITE SOME DECIMALS ON YOUR OWN:

EXAMPLE

1 part out of 10 parts	0.1 (one tenth)
1 part out of 100 parts	0.01 (one-hundredth)
1 part out of 1,000 parts	0.001 (one-thousandth)
1 part out of 10,000 parts	0.0001 (ten-thousandth)

2 parts out of 10 parts	6 parts out of 1,000 parts
8 parts out of 100 parts	3 parts out of 10 parts
7 parts out of 10,000 parts	41 parts out of 100 parts
214 parts out of 1,000 parts	1 part out of 10,000 parts

WRITE THE FOLLOWING DECIMALS IN WORDS:

EXAMPLE

.5 <u>Five tenths</u>

.05 <u>Five hundredths</u>

.055 <u>Fifty-five thousandths</u>

.0055 <u>Fifty-five ten-thousandths</u>

.0017 _____

.0231 _____

.58 _____

.034 _____

.2 _____

.125 _____

.0009 _____

.05 _____

.0092 _____

.9 _____

.050 _____

WRITE THE FOLLOWING DECIMALS IN WORDS:

EXAMPLE

Seven tenths <u>.7</u>

Seven hundredths <u>.07</u>

Seventy-seven thousandths <u>.077</u>

Seventy-seven ten-thousandths <u>.0077</u>

Thirty thousandths _____

Seven hundredths _____

Three thousand eight ten-thousandths _____

Fifteen ten-thousandths _____

Three ten-thousandths _____

Six tenths _____

Twenty-nine hundredths _____

Eight thousandths _____

One hundred eighty-four ten-thousandths _____

One tenth _____

Nine hundred fifty-five ten-thousandths _____

Twenty-five thousandths _____

Three hundred thousandths _____

One thousandth _____

CONVERTING DECIMALS TO THE LOWEST POSITION:

Sometimes a length dimension may be written in thousandths. When this happens, change the number to hundredths or tenths when **possible.**

EXAMPLE

THOUSANDTHS	HUNDREDTHS	TENTHS
.500	.50	.5
.350	.35	-----
.235	-----	-----

REDUCE NUMBER TO ITS LOWEST POSSIBLE NUMBER:

THOUSANDTHS	HUNDREDTHS	TENTHS
.600	_____	_____
.550	_____	_____
1.400	_____	_____
.800	_____	_____
2.370	_____	_____
.070	_____	_____
.100	_____	_____
.487	_____	_____
.250	_____	_____
.300	_____	_____
2.300	_____	_____
6.550	_____	_____
.366	_____	_____

THOUSANDTHS	HUNDREDTHS	TENTHS
.900	_____	_____
1.100	_____	_____
1.060	_____	_____
5.100	_____	_____
.025	_____	_____
1.170	_____	_____
7.700	_____	_____

ADDING AND SUBTRACTING DECIMALS:

PRACTICE ADDING THE FOLLOWING DECIMALS. **DO NOT** USE A CALCULATOR
AND SHOW YOUR WORK TO THE RIGHT!

PROBLEM   **SHOW YOUR WORK**

.035 + .1295 =

.09 + .0009 =

1.355 + .25 =

.633 + .52 =

3.14 + .956 =

.0077 + .002 =

Instructor's Response

_____

_____

_____

Name _____   Date _____

MEASURING WITH AN OUTSIDE MICROMETER

Upon completion and review of this job sheet, you should be able to measure a brake rotor with an outside micrometer.

> **AUTHOR'S NOTE** For this task, the author uses four different brake rotors. The rotors can be demos or they can be removed from a vehicle. Each rotor has four areas that can be measured. Each machined area is labeled so the answers can be tracked and graded. The ASE Education Foundation task is from the brake repair section.

ASE Education Foundation Correlation

This job sheet addresses the following **MLR** tasks:

5.D.5 Clean and inspect rotor, measure rotor thickness, thickness variation, and lateral runout; determine necessary action. **(P-1)**

5.D.6 Remove and reinstall/replace rotor. **(P-1)**

5.D.7 Refinish rotor on vehicle; measure final rotor thickness and compare with specifications. **(P-1)**

This job sheet addresses the following **AST/MAST** tasks:

5.D.6 Clean and inspect rotor and mounting surface; measure rotor thickness, thickness variation, and lateral runout; determine needed action. **(P-1)**

5.D.7 Remove and reinstall/replace rotor. **(P-1)**

5.D.8 Refinish rotor on vehicle; measure final rotor thickness and compare with specifications. **(P-1)**

We Support
ASE | Education Foundation

Tools and Materials

Goggles or safety glasses with side shields
Brake rotors to be measured
Outside micrometer (at least one should be metric)

Procedure

Task Completed

1. Select the correct size micrometer for the areas to be measured.   ☐

2. Open the micrometer enough to slide over the area to be measured.   ☐

3. Close the measuring faces until they contact with the rotor.   ☐

4. Measure the rotor faces every 90 degrees apart at four different areas.   ☐

5. Rotate the thimble with the ratchet until it slips.   ☐

6. Slide the micrometer back and forth over the rotor to ensure the micrometer is closed and squared correctly on the surface.   ☐

7. Flip the thumb lock on and remove the micrometer from the rotor.   ☐

8. Read the measurement in the following manner:

A. Record the size of the micrometer, 0–1 inch, 1–2 inch, 2–3 inches.

_____

B. Record the last number on the sleeve uncovered by the thimble.

_____

C. Record the number of thousandths of an inch (or millimeters) between the uncovered sleeve number and the edge of the thimble.

_____

D. Read and add the thimble graduation aligned with the index line to the number arrived at in item C.

_____

E. Record the total measurement in the appropriate blank at the end of this job sheet. Note that the number of blanks may not match the number of measurements.

_____

F. Repeat the previous steps for the other measurements.   ☐

Measurements

1. Brake rotor position 1

Area 1 _____ + _____ + _____ = inches/millimeters

size   Sleeve   Thimble

2. Brake rotor position 2

Area 2 _____ + _____ + _____ = _____

3. Brake rotor position 3

Area 3 _____ + _____ + _____ = _____

4. Brake rotor position 4

Area 4 _____ + _____ + _____ = _____

Use the blanks provided for additional rotors to be measured.

_____ _____ _____ _____ _____

_____ _____ _____ _____ _____

_____ _____ _____ _____ _____

_____ _____ _____ _____ _____

Instructor's Response

_____

_____

_____

Name _____   Date _____

MEASURING A SHAFT'S ENDPLAY

Upon completion and review of this job sheet, you should be able to measure the endplay of an automatic transmission turbine (input) shaft.

AUTHOR'S NOTE For this task, the author used a Chrysler 727 automatic transmission because it was readily available, and this measurement is easily done on this transmission. Any other transmission or crankshaft mounted in an engine can be used because the procedure will be the same.

ASE Education Foundation Correlation

This job sheet addresses the following **MAST** task:

2.C.11   Measure transmission/transaxle end play and/or preload; determine needed
                action. **(P-1)**

We Support

ASE | Education Foundation

Tools and Materials

Assigned automatic transaxle or transmission
Dial indicator
Transaxle holder or table
Service manual

Procedure   **Task Completed**

1. Mount the transaxle in a holder or secure to a table.   ☐

2. Identify the transaxle type by using an on-transaxle ID tag or by the vehicle. Record
   the information.

   _____

   _____

3. Locate the transaxle shaft endplay specification and record.

   _____

   _____

4. Secure a dial indicator with a clamp-type mount and mount the indicator to the bell   ☐
   housing of the transaxle.

5. Flex the dial indicator's cable or linkage until the dial is near the end of the shaft.   ☐

6. Push the shaft in until it stops and then pull it out as far as it will go. This procedure   ☐
   may be reversed if desired.

7. Position the dial indicator so the plunger rests against the center of the shaft's end with ☐
   the plunger extended out about halfway of its full travel. Lock indicator in place.

8. If necessary, unlock the indicator's scale and rotate the scale until the zero graduation   ☐
   is directly under the needle. Lock the scale in place and recheck the position of the
   zero and needle alignment.

9. Push (pull) the shaft in (out) until it stops.   ☐

10. Read and record the measurement.

_____

_____

11. Perform steps 5 through 10 again to recheck the measurement. Record the results.

_____

_____

12. Do the two measurements match? _____ If not, remeasure starting at step 4. Record this measurement and determine if it matches the previous reading.

_____

_____

13. Compare the measurement with the specifications and answer the following questions. Is the shaft endplay within limits? If not, what repairs are suggested by the service manual?

_____

Instructor's Response

_____

_____

_____

Name _____   Date _____

BOLT AND NUT IDENTIFICATION

Upon completion of this activity you will be able to identify and describe the different types of bolts and nuts used in the automotive industry. Perform common fastener and thread repair.

ASE Education Foundation Correlation

This job sheet addresses the following **MLR** task:

1.A.6 Perform common fastener and thread repair, to include: remove broken bolt, restore internal and external threads, and repair internal threads with thread insert. **(P-1)**

This job sheet addressees the following **AST/MAST** task:

1.A.7 Perform common fastener and thread repair, to include: remove broken bolt, restore internal and external threads, and repair internal threads with thread insert. **(P-1)**

We Support
ASE | Education Foundation

Tools and Materials

Goggles or safety glasses with side shields

Examples nuts and bolts

Thread pitch gauge

Ruler Inches and metric

Tap and die set

Drill

Drill bit index set

Bolt extraction kit

Procedure

1. Draw a picture of a bolt and label all parts.

    a) Wrench pad

    b) Shank

    c) Threads

    d) Bolt length

    e) Bolt diameter

    f) Bolt pitch

2. Draw a picture of a standard bolt head showing the markings for grade 2, grade 5, and grade 8.

3. Draw a picture of a standard nut showing the markings for grade 2, grade 5, and grade 8.

4. Draw a picture of a metric bolt head showing the markings for grade 4.6, grade 8.8, and grade 10.9.

5. Using a selection of bolts supplied by your instructor, determine the bolt type by circling metric or standard, correct thread pitch, diameter, length, and grade for each.

   A. Metric/Standard: TP _____ Diameter _____
      Length _____ Grade _____

   B. Metric/Standard: TP _____ Diameter _____
      Length _____ Grade _____

   C. Metric/Standard: TP _____ Diameter _____
      Length _____ Grade _____

   D. Metric/Standard: TP _____ Diameter _____
      Length _____ Grade _____

   E. Metric/Standard: TP _____ Diameter _____
      Length _____ Grade _____

   F. Metric/Standard: TP _____ Diameter _____
      Length _____ Grade _____

6. What do the marking on a standard bolt head indicate about the bolt?

   _____

7. Metric bolts are marked with a number, such as 8.8 and 10.9. A higher number indicates a higher what?

   _____

8. Overtightening a bolt can cause the bolt to stretch before it actually breaks. When this happens, the bolt loses what?

   _____

9. Bolts/ fasteners have limits to how tight they can be torqued, and because many components require significant clamping force to remain tight, every bolt/fastener has a what?

   _____

10. What type of tool is used to ensure that a bolt/fastener is tightened correctly?

    _____

11. Perform a bolt extraction. Remove a broken bolt. (See instructor.)   ☐

12. Repair internal threads.   ☐

13. Repair internal threads with a thread insert.   ☐

Instructor's Response

_____

_____

_____

Name _____   Date _____

PERFORM ENGINE VACUUM TEST/USING A VACUUM GAUGE

Upon completion of this activity, you will be able to understand how to properly locate and inspect a PCV system, install a vacuum gauge, and determine necessary action.

ASE Education Foundation Correlation

This job sheet addresses the following **MLR** task:

8.A.2 Perform engine absolute manifold pressure tests (vacuum/boost); document results **(P-1)**

This job sheet addresses the following **AST/MAST** task:

8.A.5 Perform engine absolute manifold pressure tests (vacuum/boost); determine needed action. **(P-1)**

We Support

ASE | Education Foundation

Tools and Materials

Goggles or safety glasses with side shields

Basic hand tools

Engine vacuum gauge

Describe the Vehicle You Are Working on:

Year _____ Make _____ Model _____

VIN # _____ Engine type and size _____

Procedure   Task Completed

1. Using service manuals, locate and record the manufacture's procedures for inspecting the PCV valve and system.

   _____

   _____

   _____

   _____

2. Find and locate the PCV on the vehicle. Describe the location.

   _____

3. With the engine running, carefully remove the PCV valve. Place your thumb over the valve's opening and note the valve's reaction.

   _____

4. Turn off engine and remove the PCV valve from the hose. Install a vacuum gauge in   ☐
   place of the PCV.

5. Start the engine and allow to idle. With engine idling, record vacuum reading.

   _____

6. What's the manufacturer's specifications for engine vacuum at idle?

   _____

   You can use the general specification from the book.

7. Reinstall the valve into the engine and inspect the vacuum hose to the valve. Note the condition of the vacuum hose.

_____

8. Locate the fresh air hose to the engine for the PCV system. Inspect the air hose and note its condition.

_____

9. Explain the purpose of the PCV system.

_____

_____

_____

_____

10. Based on your inspection, what is the condition of the PCV system?

_____

_____

Instructor's Response

_____

_____

_____

CHAPTER 5

ELECTRICAL AND ELECTRONIC SYSTEMS INSPECTION, MAINTENANCE, AND SERVICE

Upon completion and review of this chapter, you should understand and be able to:

- Read and use a wiring diagram.
- Operate a test light in an automotive circuit.
- Operate a digital multimeter (DMM) to test an automotive circuit.
- Utilize a fused jumper wire to test an automotive circuit.
- Inspect circuit protection devices.
- Inspect and test switches and relays.
- Diagnose an open circuit.
- Diagnose a short circuit.
- Diagnose a high resistance circuit.

- Inspect, service, and test the battery.
- Inspect, service, and test the starter and circuit.
- Inspect, service, and test the alternator and circuit.
- Inspect, service, and test the lighting components and circuit.
- Inspect, service, and replace secondary ignition circuit components.
- Perform a power balance test.
- Understand the special procedure for working safely around hybrid electric vehicles.

Terms To Know

Battery terminal voltage drop test
Capacity test
Conductance test
Current draw test
Distributor
Ground reference voltage testing

Hydrostatic lock
Ignition timing
Infinite
Lineman's gloves
Open circuit
Open circuit voltage test
Power balance testing

Self-powered test light
Short circuit
Starter quick test
Test light
Voltage drop

INTRODUCTION

Similar to what was emphasized in the Classroom Manual, understanding the electrical system is just as important as knowing how to test, inspect, replace, and repair problems associated with it. As an entry-level technician, you will be expected to perform some basic electrical checks, inspections, and repairs. This chapter outlines those basic procedures and emphasizes safety.

When diagnosing electrical systems, always try to remember Ohm's law and apply it. **The voltage across a circuit is equal to the current through a circuit times the resistance of a circuit.** These words are highlighted because they will help you understand how to measure electrical circuits.

UNDERSTANDING A WIRING DIAGRAM

Probably the most overlooked and misunderstood part of any electrical diagnosis or test by an entry-level technician is the wiring diagram. The wiring diagram can be complicated and difficult to read at first. One of the best strategies for reading a wiring diagram is trying to remember how an electrical component works. Once you remember how it works, trying to trace the wires will seem easier.

The first step is to find the correct wiring diagram. This usually means that you must look up the right vehicle. This is done by accessing the vehicle identification number (VIN) and correctly entering it into the service manual software or program. Entering it incorrectly or slightly off will result in slightly incorrect wiring diagrams. But even slightly incorrect diagrams will cause headaches. The best example is trying to follow an alternator's wiring harness for a vehicle. If the vehicle came with two different four-cylinder engine options, then incorrectly entering the VIN will result in having the incorrect wire colors, circuit numbers, or harness locations. Here's a good example: A Saturn SL-2 has two different 1.9 L engines. One had a single overhead camshaft and the other had a dual overhead camshaft. The alternators are different, and the wiring harness for each is different. The only difference was in the VIN, so incorrectly entering in the information for the eighth digit will cost you a lot of time and grief.

The next step is to find the information in the service manual that will help you read the diagram. Each manufacturer uses slightly different color codes and symbols for the wiring diagram. Let's look at **Figure 5-1** as an example. In this wiring diagram,

Figure 5-1 Heater system wiring diagram.

the number 181 (by the fuse) stands for the circuit number. The "Br/O" is shorthand for a wire that is brown with an orange stripe. Never take for granted that each manufacturer will use the same notations. The letter "B" for one manufacturer could mean black and for another it could mean blue. The two arrows by the letters "C001" are symbols for female and male spade connectors, whereas other connectors are represented by other symbols. The "C001" stands for the number of the connector. When you look at the fuse, the dots that touch the fuse are connectors, but do you notice how they are inside the dashed lined box? That means that the wires go into the fuse box, and then the connector is on the inside. The dashed line is also interesting. It means there is more than what is drawn. Look at **Figure 5-1**: have you ever opened up a fuse box in a car and found only one fuse? No. That's what the dashed line means. There's more to it. Did you notice that the power side (also known as the B+ or hot side) is on the top of the diagram? This is common. It is just as common as to have the ground on the bottom. With practice, you will be able to easily read wiring diagrams. **Table 5-1** outlines the different electrical symbols that are commonly used in the automotive industry.

USING A TEST LIGHT

There are two types of test lights commonly used in automotive diagnostics and repair: nonpowered and self-powered. A **test light** (**Figure 5-2**) is a tool the technician uses when he or she needs to "look" for electrical power in a circuit. The key here is the word *look*. A nonpowered test light will light up when there isn't full voltage, but it will do so with less brightness because there is less voltage. Sometimes this is hard to see. And, it can be misleading. Some computerized circuits and components will only work with 12 volts. So, if a circuit is measuring 10 volts because of a defect, the test light will illuminate, but the circuit will not have enough power to turn on the component. With some experience, you will be able to tell when the test light is not bright enough, but that's not what a test light is used for. It is there to test for the presence of voltage, not the amount.

The **self-powered** test light (also known as a self-powered continuity tester) has an internal battery that powers the light bulb (**Figure 5-3**). This device will work with a different color light when the test light is in the opposite direction. This type of tester can work on a vehicle with a dead battery.

To use a test light, you have to connect the clip to ground and then move the tester tip around the circuit; wherever it lights up, voltage is present

A **test light** is connected between a positive conductor and ground. It will only show voltage presence, not amount.

A **self-powered** test light is connected across a portion of the circuit. If the lamp lights, then there is continuity between the test points.

Figure 5-2 Typical test light used to probe voltage in a circuit.

Figure 5-3 Typical self-powered continuity tester.

Table 5-1 Common Electrical and Electronic Symbols Used in Wiring Diagrams

COMPONENT	SYMBOL	ALTERNATE
Ammeter		
AND Gate		
Antenna		
Attenuator, Fixed		
Attenuator, Variable		
Battery		
Capacitor, Feedthrough		
Capacitor, Fixed, Nonpolarized		
Capacitor, Fixed, Polarized		
Capacitor, Ganged, Variable		
Capacitor, General		
Capacitor, Variable, Single		
Capacitor, Variable, Split-Stator		
Cathode, Cold		
Cathode, Directly Heated		
Cathode, Indirectly Heated		
Cavity Resonator		
Cell		
Choice Bracket		
Circuit Breaker or PTC device		
Clockspring		
Coaxial Cable		
Coil		
Crystal, Piezoelectric		
Delay Line		
Diode		
Diode, Gunn		
Diode, Light-Emitting		
Diode, Photosensitive		
Diode, Photovoltaic		

Table 5-1 (*Continued*)

Diode, Pin	⟶▸⊣⊢	
Diode, Varactor	⟶▸⊣⊦	
Diode, Zener	⟶▸⊢	
Directional Coupler	⟶⊠⟶	⊕
Dual Filament Lamp	⊗ ⊗	
Exclusive-OR Gate	⟩⟩	
Female Contact	⟶≺	
Ferrite Bead	⟶⊚⟶	▭
Fuse	⟶ᴑᴑ⟶	▭
Fusible link	⎸	
Gauge	⊘	
Ground, Chassis	⏚●	⏚
Ground, Earth	⏚	
Handset	⌢	
Headphone, Double	⌢	
Headphone, Single	⌢	
Heating element	⧢	
Hot Bar	BATT A0	
Inductor, Air-Core	⟞ᴍᴍ⟝	
Inductor, Bifilar	⟞ᴍᴍ⟝	
Inductor, Iron-Core	⟞ᴍᴍ⟝	
Inductor, Tapped	⟞ᴍᴍ⟝	
Inductor, Variable	⟞ᴍᴍ⟝	⟞ᴍᴍ⟶
In-Line Connectors	2 ≷ C123 2 ⤰ C123	
Integrated Circuit	▭	
Inverter	▷○	
Jack, Coaxial	⊙	
Jack, Phone, 2-Conductor	⌐	
Jack, Phone, 2-Conductor Interrupting	⌐	
Jack, Phone, 3-Conductor	⌐	
Jack, Phono	⊙	

Table 5-1 (*Continued*)

Key, Telegraph	
Lamp, Neon	
Male Contact	
Microphone	
Motor, One speed	
Motor, Reversible	
Motor, two Speed	
Multiple connectors	
NAND Gate	
Negative Voltage Connection	
NOR Gate	
Operational Amplifier	
OR Gate	
Outlet, Utility, 117-V	
Outlet, Utility, 234-V	
Oxygen Sensor	
Page Reference	(BW-30-10)
Piezoelectric Cell	
Photocell, Tube	
Plug, Phone, 2-Conductor	
Plug, Phone, 3-Conductor	
Plug, Phono	
Plug, Utility, 117-V	
Plug, Utility, 234-V	
Positive Voltage Connection	
Potentiometer	
Probe, Radio-Frequency	
Rectifier, Semiconductor	
Rectifier, Silicon-Controlled	
Rectifier, Tube-Type	
Relay, DPDT	

Table 5-1 (*Continued*)

Relay, DPST		
Relay, SPDT		
Relay, SPST		
Resistor		
Resonator		
Rheostat, Variable Resistor, Thermistor		
Saturable Reactor		
Shielding		
Signal Generator		
Single Filament Lamp		
Sliding Door Contact		
Solenoid		
Solenoid Valve		
Speaker		
Splice, External		
Splice, Internal		
Splice, Internal (Incompleted)		
Switch, Closed		
Switch, DPDT		
Switch, DPST		
Switch, Ganged		
Switch, Momentary-Contact		
Switch, Open		
Switch, Resistive Multiplex		
Switch, Rotary		
Switch, SPDT		
Switch, SPST		
Terminals		
Test Point		
Thermocouple		

Table 5-1 (*Continued*)

Thyristor		
Tone Generator		
Transformer, Air-Core		
Transformer, Iron-Core		
Transformer, Tapped Primary		
Transformer, Tapped Secondary		
Transistor, Bipolar, npn		
Transistor, Bipolar, pnp		
Transistor, Field-Effect, N-Channel		
Transistor, Field-Effect, P-Channel		
Transistor, Metal-Oxide, Dual-Gate		
Transistor, Metal-Oxide, Single-Gate		
Transistor, Photosensitive		
Transistor, Unijunction		
Tube, Diode		
Tube, Pentode		
Tube, Photomultiplier		
Tube, Tetrode		
Tube, Triode		
Unspecified Component		
Voltmeter		
Wattmeter	W	P
Wire Destination In Another Cell		
Wire Origin & Destination Within Cell		
Wires		
Wires, Connected, Crossing		
Wires, Not Connected, Crossing		

(**Figure 5-4**). It is important to limit the use of a test light to noncomputerized circuits. A test light will draw current from the system to light up the bulb. This may draw more current from a computer circuit than was originally designed, thus burning the circuit. It is good practice to limit the use of a test light to noncomputerized circuits.

Figure 5-4 If voltage is present, the test light will illuminate.

USING A DIGITAL MULTIMETER

The digital multimeter (DMM) is the most effective and most used tool by an automotive technician. It can measure voltage, resistance, amperage, and sometimes more than that. Most voltmeters are sold in digital format (**Figure 5-5**). Analog format DMMs are not used in modern vehicles with computerized equipment. DMMs that are compatible with computerized circuits have internal impedance rating of 10 megaohms or greater. Always check the tool for this specification before working with it on a modern computer-controlled vehicle.

To measure voltage, you need to connect the red and black test leads to the appropriate sockets in the meters, making sure to match the colors and symbols. You can use a voltmeter to measure the voltage across or **voltage drop**. Voltage is always measured "across" a circuit (remember Ohm's law). **Figure 5-6** shows how the voltage drop of a resistor is measured across the circuit. Note that only a limited amount of current will flow through the meter and that the majority of current will flow through the bulb (because the meter has high resistance). **Figure 5-7** shows how to make several voltage checks in the system. Note that the negative test lead never leaves the ground connection of the battery. This is called **ground reference voltage testing**. The loss of voltage due to

> **Voltage drop** is a measurement of the voltage across a portion of the circuit.

> **Ground reference voltage testing** is used to determine the voltage within a circuit.

Figure 5-5 Digital multimeter.

Figure 5-6 Connecting a voltmeter in parallel to the circuit. This will register the voltage drop across the resistor.

Figure 5-7 Checking voltage in a closed circuit.

resistance in the wires, connectors, and loads is called voltage drop. Voltage drop is the amount of electrical energy converted to another form of energy.

For example, to make a lamp light, electrical energy is converted to heat energy; it is the heat that makes the lamp light up. To make voltage checks across the system, you need to determine what is positive and what is negative (or ground). The load in the system is always the splitting device.

Using Figure 5-7 as the example, the motor is the load in the circuit. Everything before the load is the positive side (including the resistor). The meter at point A will measure only slightly less than the battery voltage because there is only a small section of wire between it and the positive terminal, and there are no loads in it. At point B, the voltage reading will measure less because voltage has dropped across the resistor. And at point C, the voltage will be very small because almost all of the voltage was used by the circuit and the load. If you add the voltage drops at points A, B, and C, you should get the voltage of the battery.

Using the DMM to measure resistance is very similar to measuring voltage. The internal battery of the DMM is used for this purpose. To start, you must set the test leads in the correct places (usually the same place as voltage, but each meter is different). Then set the dial or buttons to measure resistance. Touch the two test leads together to calibrate the meter. Next, remove the component from the circuit or de-energize the circuit (**Figure 5-8**). This has to be done because otherwise you will be measuring the resistance

Figure 5-8 Measuring resistance with an ohmmeter. The meter is connected in parallel with the component being tested after power is removed to the circuit.

Figure 5-9 A digital ohmmeter showing an infinite reading.

of the entire circuit and the wiring. Most service manual specifications for electrical component testing are for the component only, not for the component and circuit together.

Connect the meter test leads in parallel with the component or circuit to read the resistance. You can measure the resistance of almost anything that doesn't have voltage applied to it. A reading of O.L. means that the reading is "outside of limits." Another term used for this is **infinite**. This means that the meter cannot read the resistance between the test leads. If you are measuring a circuit, this is an open circuit. If you are measuring a component, there is an open circuit in the component (**Figure 5-9**). If you are measuring a switch, it may just be in the off position. Turning the switch to the on position will show a resistance reading. A common misconception arises when measuring the resistance of a component and the DMM shows a reading of 0.00 (zero) ohms. This does not mean that there is no resistance; it simply means that the reading is so low that it is rounded down. The reading is likely 0.001 ohms or less, and the meter will not display a reading that low. A reading of 0.00 ohms still means the component or circuit has resistance but a very low amount (which in many cases is good).

There are two ways to measure current. Remember Ohm's law? Current must be measured "through" the circuit. One way of making the current go through the circuit is by interrupting the circuit and making all current go through the DMM (**Figure 5-10**). Unlike measuring voltage or resistance (where the DMM is placed in parallel with the circuit or component), you must have the current going through the meter to measure current (amperage).

To keep you and the tool safe, you must use caution and read the tool's amperage fuse rating. Then look at the circuit's fuse. Never place a DMM in a circuit where the fuse rating is greater than the rating of the DMM.

The second way to measure current (and maybe the safer way) is to use an inductive ammeter (also known as an amp clamp). This tool connects with the DMM and clamps around a wire. There is no need to open up the circuit. This comes in handy when a large harness cannot be easily removed for testing or has multiple other wires that need to be connected for testing (**Figure 5-11**).

Infinite means forever or too large. In automobile circuit testing, it means the value being measured is too great for the test device being used.

Figure 5-10 Measuring current flow with an ammeter. The meter must be connected in series with the circuit.

Figure 5-11 An inductive pickup for an ammeter.

USING A FUSED JUMPER WIRE

A very common test tool, and one you can easily make, is the fused jumper wire. It can be used as an extension for testing using a DMM or as a wire replacement. If you suspect that a wire is broken or a switch is not working properly, you can bypass it and verify that the remainder of the circuit is working properly (**Figures 5-12** and **5-13**).

Jumper wires can be used anywhere in the circuit to bypass what would normally be a wire or anything of low resistance. They cannot be placed in the circuit in place of a load. When using the fused jumper wire, always check to see the size of the fuse the circuit you are testing normally uses. Then install that fuse into the jumper wire and test.

INSPECTING POWER DISTRIBUTION AND CIRCUIT PROTECTION DEVICES

There are many kinds of protection devices in automotive systems. Each one is designed to "turn off" the circuit if too much current starts to flow through it. Some of the protection devices will turn back on after they cool down (like a circuit breaker), and others will need to be replaced (fuses and fusible links). In this section, you will learn how to test each type of protection device.

Figure 5-12 Using a fused jumper wire to bypass the switch.

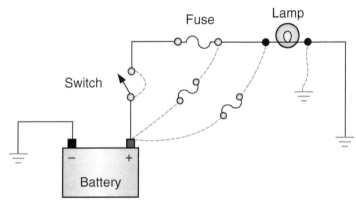

Figure 5-13 Examples of locations a jumper wire can be used to bypass a portion of the circuit, and to test the ground circuit. Remember, if you bypass the circuit fuse, the jumper wire should be fitted with a fuse. Never bypass the load component.

Figure 5-14 A fuse or fusible link can have a hidden fault that cannot be seen by the technician.

To test a fuse, the best method is to visually inspect it for a blown or open link (**Figure 5-14**). This isn't always the easiest thing to see, so testing with a DMM is required. Remove the fuse and test the resistance of it. O.L. means that the fuse is open and requires replacement (**Figure 5-15**). A small resistance reading (usually 0.1 ohms) means it is good. Anything higher than that usually means there is some corrosion. Fuses don't wear out. (This is a very common misconception.) If they are blown, then there is a problem in the circuit.

You can also test the fuse while installed by checking the voltage drop across it (**Figure 5-16**). If there is 12 volts on one side and 0 volts on the other, it is open.

Top view of mini fuse

Figure 5-15 A good fuse will have zero resistance when tested with an ohmmeter.

Figure 5-16 (A) Voltmeter test of a circuit protection device. Battery voltage should be present on both sides. (B) The test light will illuminate on both terminals if the fuse is good.

Non-cycling circuit breaker that is reset by removing from power

Side view (external)

Side view (internal)

Contacts   Coil   Bimetal arm

Manual resetting circuit breaker

Side view (external)

Side view (internal)   Reset button

Spring

30 A

Ampere rating   Contacts   Bimetal arm

Figure 5-18 Operation of a self-resetting circuit breaker.

Ω

Figure 5-17 A fusible link can be tested with an ohmmeter, once it is disconnected from power.

If there is 12 volts on both sides, then it is good and the difference in voltage (voltage drop) is too small to be displayed on the meter. Fusible links are not as common, but to test them you would use an ohmmeter just like you would with a regular fuse (**Figure 5-17**). The only difference is that you have to find the metal connector end of the wire to test it.

To test a breaker, you need to remove it from the vehicle and use the DMM set to ohms. Test the resistance of the breaker. The readings should be similar to a fuse. Some circuit breakers have manual resets on them (**Figure 5-18**).

Testing a relay requires you to use all of the strategies mentioned up to this point. Start by testing the relay without a DMM. Place your hand on the top of the relay and attempt to turn the circuit on. You should be able to feel and hear the contact closing when the coil is energized, assuming it is working properly. If you hear this, you can assume that the relay coil and the control side circuit (low current side) are working properly (power and ground) and that there is a problem with the high current side. If you don't hear or feel anything, remove the relay and test the relay connectors (not the relay itself) for power, ground, and resistance. In **Figure 5-19**, the DMM measures the voltage of the relay when it is installed. Remember that the coil (between C and D) is the load in a circuit, and it controls the switch (between A and B). The switch and load are tested according to the procedures outlined earlier in this chapter.

If everything checks out in the relay connectors, then test the relay. Remember that the relay is a moving part, and moving parts will eventually wear out. In **Figure 5-20**, the DMM measures the resistance of the coil; in **Figure 5-21**, the relay is connected to a battery with fused jumper wires and the DMM measures the resistance of the switch. If the switch measures 0 ohms (like in Figure 5-21), then it is good. If the switch measures O.L. on the DMM, it is bad.

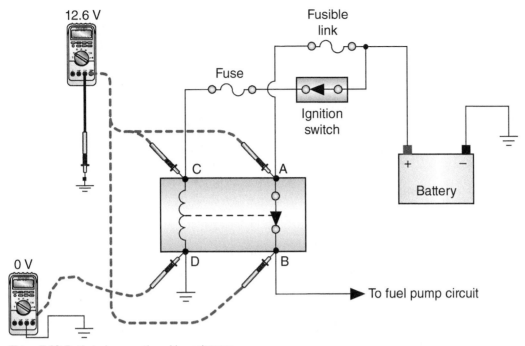

Figure 5-19 Testing relay operation with a voltmeter.

Figure 5-20 Testing the resistance of the relay coil.

Figure 5-21 Bench testing a relay.

DIAGNOSING AN OPEN CIRCUIT

It is possible to test for *opens* using a DMM, test light, or fused jumper wire. An open circuit can be defined in many ways. A broken wire and broken component like a blown light bulb are both examples of an open circuit. When a switch is in the open (or off) position, the circuit is open because current is not flowing through it.

When something doesn't work or doesn't operate as intended, it can be an open circuit. To test for this, it is best to do a visual inspection by checking the fuses, circuit protection, and inspecting the components and wiggling the connectors. If everything checks out, then proceed to turn on the DMM and measure voltage starting at the battery. If the battery has enough voltage, continue testing; if not, connect a jumper box.

Figure 5-22 Locating an open by testing for voltage.

Start by testing at the load that does not work. Test the load itself for resistance. A bad light bulb is an open circuit. **Figure 5-22** shows the DMM checking for positive voltage at the load in the circuit. This is the best technique to use. If voltage is found here (in the case of the picture, it is), then move to test the negative (ground) side of the circuit. This will help you determine which side to start the testing on.

Let's use the same circuit. Look at **Figure 5-23**. It is drawn with an open circuit (broken wire). When the DMM is connected to the same place, it reads 0 volts. This helps us understand that the problem is on the power side. *Note:* Look at both circuits and notice that both of them have the switch closed. In this case, the switch must be turned on to properly test the circuit. **Figures 5-24** and **5-25** show the normal and abnormal (open) operation of the same circuit.

Figure 5-23 An open is present between the point where voltage was measured and where it was not.

Figure 5-24 Properly operating complex parallel circuit.

Figure 5-25 An open in the ground circuit can convert the circuit to a series circuit. The dashed black line represents the resulting path to ground.

DIAGNOSING A SHORT CIRCUIT

A short circuit is when the current has found a lower resistance pathway (a shorter pathway). It can result when two power-side copper wires touch each other (short to power) or when two copper wires (one power and one ground) touch each other (short to ground). It can also result when a copper wire touches a metal ground (short to ground). A short to ground is when the current bypasses the component (because the ground circuit has less resistance). This lower resistance increases the current flow in the circuit and usually blows the fuse or circuit breaker. Diagnosing this can be tricky because it is an open circuit (because of the blown fuse). Remember that fuses don't wear out. If they are blown, there is a problem.

A short to power will cause two power wires to bypass a switch in the circuit. This usually makes a component not turn off or not work correctly in different switch positions.

In **Figure 5-26**, a copper-to-copper short is shown. The short is circled in the drawing. This type of short is a short to power and will cause both lamps to light when switch 1 is closed and nothing to occur when switch 2 closes or opens. In fact, switch 2 could be removed and it wouldn't make any difference to the circuit's operation. Depending on how much current is drawn, fuse B may or may not blow.

A short to ground is outlined in **Figure 5-27**. The short to ground is before the component (load) in the circuit. This will cause the fuse to blow. Checking for a short to

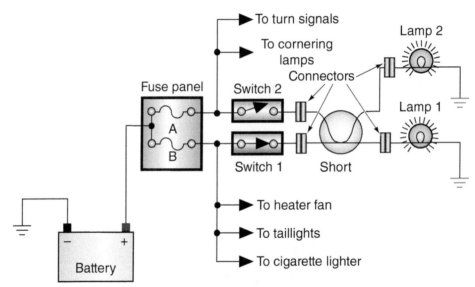

Figure 5-26 A copper-to-copper short (short-to-power) between two circuits.

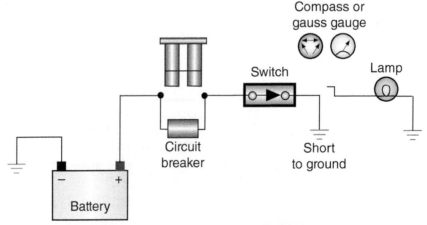

Figure 5-27 The needle of a compass or gauss gauge will fluctuate over the portion of the circuit that has current flowing through it. Once the short to ground has been passed, the needle will stop fluctuating.

Available voltage point A = 12.00 V
Minus available voltage point B = 6.00 V
Voltage drop across lamp 1 = 6.00 V

Available voltage point C = 6.00 V
Minus available voltage point D = 0.00 V
Voltage drop across lamp 2 = 6.00 V

Total voltage drop between points A and D = 12.00 V

Figure 5-28 Using available voltage to calculate voltage drop over a component. This method is used if the wires of the circuit are too long to test with standard test leads.

ground is best done by removing the fuse and physically checking for defects. Replacing the fuse with a circuit breaker and using a compass or gauss gauge will help you find the short (**Figure 5-28**).

DIAGNOSING A HIGH RESISTANCE CIRCUIT

If a circuit performs lesser or not as designed, it could have high resistance. A bulb that is dimmer than usual is an example of a circuit with high resistance. You can use a DMM to check for high resistance, but the most effective method is to measure for voltage drop. This method will provide you with a quick and reliable test result.

The best way to think of voltage drop is to look at **Figure 5-29**. The DMM is placed across the circuit and the values add up to the same voltage as the battery. When testing

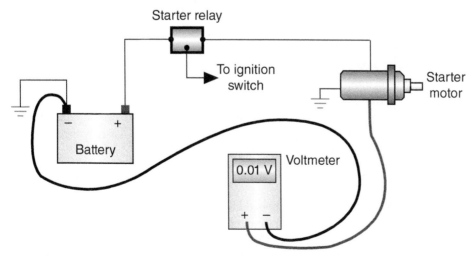

Figure 5-29 Testing the ground side of the starter motor circuit for high resistance by measuring voltage drop. Notice the voltmeter connections.

Figure 5-30 This volt-amp tester (VAT) can test the capacity of the battery, starting, and charging system.

for high resistance, you are usually testing for corrosion and other issues, although a voltage drop test can tell you if you have an open or short circuit.

The most common method of voltage drop testing is depicted in **Figure 5-30**. This is a test for the ground circuit of the starter. When testing for voltage drop, the circuit must be energized and working. This is why it is so effective of a test. So, in the case of Figure 5-30, the starter must be cranking. The steps for performing a voltage drop test can be found in **Photo Sequence 11** (using a voltage drop test to locate high circuit resistance).

INSPECTING, SERVICING, AND TESTING THE BATTERY SAFETY

Understanding batteries is important, but you must first understand how to handle them safely. There are many types of batteries that can be used in vehicles, but the most common type of battery used is the 12-volt lead-acid type. These batteries have metal plates inside that are made of lead and lead dioxide. To create the chemical reaction needed to produce electricity, an electrolyte acid is used.

The acid inside a battery can be unsafe if not handled properly. You must take care not to spill the battery, turn it upside down, or overcharge it. A charging battery will emit hydrogen gas from its vents. This gas is explosive, so keep any sparks away from it. You should also not charge or jump start a battery that is frozen. A battery will freeze when it becomes discharged and cold. Charging a frozen battery could cause it to explode.

When handling a battery, you should also take care not to touch the acid with your skin. Wear the proper safety equipment, such as gloves. Acid will irritate your skin, and it can cause holes in your clothing.

You also need to be aware of the high-voltage batteries in hybrid and electric vehicles. These batteries have a charge level on them that could be lethal. More information about hybrid and electric battery safety is covered in Chapter 5.

Battery Testing

Because of its importance, the battery should be checked whenever the vehicle is brought into the shop for service. A faulty battery can cause engine performance or engine starting problems, or leave a customer stranded. Checking the battery regularly can save diagnostic time and prevent vehicle breakdowns. By performing a battery test series, we can

PHOTO SEQUENCE 11
Using a Voltage Drop Test to Locate High Circuit Resistance

P11-1 Tools required to test for excessive resistance in a starting circuit are fender covers, a digital volt ohm meter (DVOM), and a remote starter switch.

P11-2 Connect the positive lead of the meter to the positive battery post. If possible, do not connect the lead to the cable clamp (terminal).

P11-3 Connect the negative lead of the meter to the main battery terminal on the starter motor.

P11-4 To conduct a voltage drop test, current must flow through the circuit. In this test, the ignition system is disabled, and the engine is cranked using a remote starter switch.

P11-5 With the engine cranking, read the voltmeter. The reading is the amount of voltage drop.

P11-6 If the reading is out of specifications, test at the next connection toward the battery. In this instance, the next test point is the starter side of the relay.

P11-7 Crank the engine and touch the negative test lead to the starter side of the relay. Observe the voltmeter while the engine is cranking.

P11-8 Test in the same manner on the battery side of the relay. This is the voltage drop across the positive circuit from the battery to the relay.

Figure 5-31 An assortment of battery testers.

determine from the state of charge and output capacity of the battery if it is good, in need of recharging, or must be replaced.

There are many different manufacturers of test equipment designed for testing the battery (**Figure 5-31**). Always follow the procedures given by the manufacturer of the equipment you are using.

Battery Inspection

Before performing any electrical tests, the battery should be inspected, along with the cables and terminals. The complete visual inspection of the battery will include the following items:

1. *Battery date code.* This provides information as to the age of the battery.
2. *Condition of battery case.* Check for dirt, grease, and electrolyte condensation. Any of these contaminants can create an electrical path between the terminals and cause the battery to drain. Also check for damaged or missing vent caps and cracks in the case. A cracked or buckled case could be caused by excessive tightening of the hold-down fixture, excessive underhood temperatures, buckled plates from extended overcharged conditions, freezing, or excessive charge rate.
3. *Electrolyte level, color, and odor.* If necessary, add distilled water to fill to 1/2-inch level above the top of the plates. After adding water, charge the battery before any tests are performed. Discoloration of electrolyte and the presence of a rotten egg odor indicate an excessive charge rate, excessive deep cycling, impurities in the electrolyte solution, or an old battery.
4. *Condition of battery cables and terminals.* Check for corrosion, broken clamps, frayed cables, and loose terminals (**Figure 5-32**).
5. *Battery abuse.* This includes the use of bungee cords and 2 × 4s for hold-down fixtures, too small a battery rating for the application, and obvious neglect to periodic maintenance. In addition, inspect the terminals for indications that they have been hit with a hammer and for improper cable removal procedures; also check for proper cable length.
6. *Battery tray and hold-down fixture.* Check for proper tightness. Also check for signs of acid corrosion of the tray and hold-down unit. Replace as needed.
7. If the battery has a built-in hydrometer, check its color indicator (**Figure 5-33**).

Special Tools
Fender covers
Safety glasses

Special Tools
Safety glasses
Voltmeter
Terminal pliers
Terminal puller
Terminal and clamp
  cleaner
Fender covers

⚠ WARNING **Do not attempt to pry the caps off of a maintenance-free battery; this can ruin the case. You cannot check the electrolyte level or condition on maintenance-free batteries.**

Figure 5-32 These battery connections could easily cause a no-start condition.

Figure 5-33 A built-in hydrometer indicates the battery state of charge.

⚡ **WARNING** Always wear safety glasses when working around a battery. Electrolyte can cause severe burns and permanent damage to the eye.

Green indicates the battery has a sufficient charge, black indicates the battery requires charging before testing, clear indicates low electrolyte level, and yellow indicates the battery should be replaced.

When the battery and cables have been completely inspected and any problems corrected, the battery is ready to be tested further. For many of the tests to be accurate, the battery must be fully charged.

⚙ **SERVICE TIP** Grid growth can cause the battery plate to short out the cell. If there is normal electrolyte level in all cells but one, that cell is probably shorted.

BATTERY TESTING SERIES

Battery Conductance Test

A conductance test sends a low-current AC signal into the battery. The return signal is then analyzed.

A newer method of testing a battery is called a **conductance test**. A conductance tester will test the battery's ability to conduct electricity (**Figure 5-34**). This is a good indicator of a battery's health and whether it will be able to provide adequate service when installed in the vehicle. A conductance test is considered to be more accurate than the previous method of testing batteries. During a conductance test, the tester will send a low-current AC signal into the battery. The return signal that is generated is then analyzed. Some conductance battery testers also include a battery charger. These types of testers can recharge a battery if it is not ready to be tested accurately (**Figure 5-35**). During recharging, the tester can measure key items that will indicate if the battery will be strong enough to hold a good charge when it is done.

Battery Capacity Test

A capacity (load) test uses a carbon pile to draw energy out of the battery. The voltage level of the battery should not drop below 9.6 volts during the test.

A **capacity test** (also known as a **load test**) is still a valid battery test, but it has been proven to be less accurate in determining if a battery will provide good service. A carbon pile tester is used to perform this test (**Figure 5-36**). There are many types and brands of testers available, but all of them have a voltmeter and an ammeter. A load test machine works by connecting the high current clamps of the tester to the battery terminals and the inductive pickup lead (amp clamp) to the negative tester cable. The battery must be fully charged before performing the test. For best results, the battery electrolyte should be close to 80°F (26.7°C). The tester then draws half the battery's cold-cranking ampere (CCA) rating from the battery for 15 seconds during engine cranking. On some testers, this is done automatically; on others, there is a control knob for the amperage. If the voltmeter does not drop below 9.6 volts during the 15-second test, then the battery passes the load test.

Battery Terminal Voltage Drop Test

The **battery terminal voltage drop test** checks for poor electrical connections between the battery cables and terminals.

This simple test will establish the condition of the terminal connection. It is a good practice to perform this test any time the battery cables are cleaned or disconnected and reconnected to the terminals. By performing this test, comebacks due to loose or faulty connections can be reduced.

Special Tools
Digital multimeter
Fender covers
Safety glasses

Figure 5-34 This type of tester will test the conductance of the battery. It is quick and can usually be done in less than a minute.

Figure 5-35 This Midtronics tester is becoming an industry standard.

Figure 5-36 This is a VAT machine. It contains a carbon pile in the backside and can perform a capacity (load) test.

Figure 5-37 Disconnecting the coil(s) primary connections to prevent the engine from starting and to protect the coil.

Set your DMM to DC volts. Connect the negative test lead to the cable clamp and connect the positive meter lead to the battery terminal. Disable the ignition system to prevent the vehicle from starting. This may be done by removing the ignition coil primary wires (**Figure 5-37**). (You can also remove the ignition system fuse, powertrain control module [PCM] fuse, or fuel pump fuse.) On vehicles with **distributor**-less ignition, disconnect the coil primary wires (these are the small gauge wires connected to the coils).

Crank the engine and observe the voltmeter reading. If the voltmeter shows more than 5 volts, there is a high resistance at the cable connection. Remove the battery cable using the terminal puller (**Figure 5-38**). Clean the cable ends and battery terminals or *replace* as needed (**Figure 5-39**).

Open Circuit Voltage Test

To obtain accurate test results, the battery must be stabilized. If the battery has just been recharged, perform the capacity test, then wait at least 10 minutes to allow the battery voltage to stabilize. Connect a voltmeter across the battery terminals, observing polarity (**Figure 5-40**). Measure the open circuit voltage, taking the reading to the one-tenth volt.

Distributor To route secondary, or high voltage, current from the ignition coil to the spark plugs in the correct firing order.

⚠ **Caution**

Always refer to the manufacturer's service manual for the correct procedure for disabling the ignition system. Using the wrong procedure may damage ignition components.

Figure 5-38 Use battery terminal pullers to remove the clamp from the battery post. Do not pry the clamp.

Figure 5-39 The terminal cleaning tool is used to clean the clamp and terminal.

Figure 5-40 Open circuit voltage test using a DMM.

Open Circuit Voltage	State of Charge
12.6 or greater	100%
12.4–12.6	70–100%
12.2–12.4	50–70%
12.0–12.2	25–50%
0.0–12.0	0–25%

Figure 5-41 The results of the open circuit voltage test indicate the state of charge.

Figure 5-42 Open circuit voltage test results relate to the specific gravity of the battery cells.

> The **open circuit voltage test** is used to determine the battery's state of charge.

To analyze the **open circuit voltage test** results, consider that a battery at a temperature of 80°F, in good condition, should show at least 12.45 volts (**Figure 5-41**). If the state of charge is 75 percent or more, the battery is considered charged (**Figure 5-42**).

Battery Case Test

A dirty battery can allow voltage to leak across the top of the battery and cause discharging. To test a battery for this condition, place the positive voltmeter lead on a battery post. Place the negative lead at spots across the top of the battery. Repeat the test on the other battery post. Any reading above 0.5 volts indicates that the battery is dirty and allows discharging. Clean the top of the battery with a mixture of baking soda and water or use a battery cleaning solvent. Retest to verify your repair.

> **🔧 Special Tools**
>
> Volt-ampere tester (VAT)
> Electrolyte thermometer
> Safety glasses
> Fender covers

ABSORBED GLASS MAT BATTERY PRECAUTIONS

An absorbed glass mat (AGM) battery is a lead-acid battery that is sealed using special pressure valves. This is truly a maintenance-free battery that cannot and should never be opened. Overcharging is especially harmful to AGM batteries and can shorten their

Figure 5-43 AGM batteries have to be tested using a conductance tester.

service life. You should use a good battery charger that does not allow more than 16 volts, usually the manufacturer of the battery, or the dealership, will have a special charger. A conductance tester is the only type of tester that will work on AGM batteries. You need to select that you are testing an AGM battery when using the tester (**Figure 5-43**).

INSPECTING, SERVICING, AND TESTING THE STARTER AND CIRCUIT

The starter motor must be capable of rotating the engine fast enough so it can start and run under its own power. The starting system is a combination of mechanical and electrical parts working together to start the engine. The starting system includes the following components:

1. Battery
2. Cable and wires
3. Ignition switch
4. Starter solenoid or relay
5. Starter motor
6. Starter drive and flywheel ring gear
7. Starting safety switch

Starting System Troubleshooting

Customer complaints concerning the starting system generally fall into four categories: no crank, slow cranking, starter spins but does not turn the engine, and excessive noise. As with any electrical system complaint, a systematic approach to diagnosing the starting system will make the task easier. First, the battery must be in good condition and fully charged. Perform a complete battery test series to confirm the battery's condition. Many starting system complaints are actually attributable to battery problems. If the starting system tests are performed with a weak battery, the results can be misleading, and the conclusions reached may be erroneous and costly.

Before performing any test on the starting system, begin with a visual inspection of the circuit. Repair or replace any corroded or loose connections, frayed wires, or any other trouble sources. The battery terminals must be clean, and the starter motor must be properly grounded.

The diagnostic chart shows a logical sequence to follow whenever a starting system complaint is made (**Figure 5-44**). The tests to be performed are determined by whether or not the starter will crank the engine.

 Special Tools

Battery charger
Voltmeter
Safety glasses
Fender covers

 Caution

Some engines turn counterclockwise as viewed from the front crankshaft pulley. These engines are rare, but damage may occur if they are turned the other way. Check the service manual for warnings or ask your instructor.

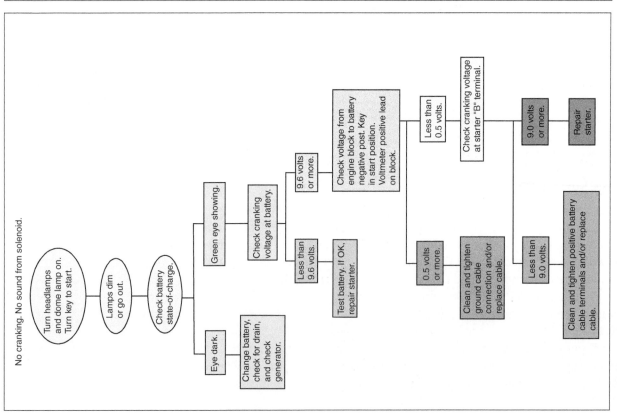

Figure 5-44 (A, B, and C) Diagnostic charts used to determine starting system problems.

Figure 5-44 *(Continued)*

Figure 5-44 (Continued)

Figure 5-45 Excessive wear, loose electrical connections, or excessive voltage drop in any of these areas can cause a slow crank or no-crank condition.

If the customer complains of a no-crank situation, attempt to rotate the engine by the crankshaft pulley nut. Rotate the crankshaft in a clockwise direction through two full rotations, using a large socket wrench. If the engine does not rotate, it may be seized due to operating with lack of oil, **hydrostatic lock**, or broken engine components.

A slow-crank or no-crank complaint can be caused by several potential trouble spots in the circuit (**Figure 5-45**). Excessive voltage drops in these areas will cause the starter motor to operate slower than required to start the engine. The speed with which the starter motor rotates the engine is important to starting the engine. If the speed is too slow, compression is lost, and the air/fuel mixture draw is impeded. Most manufacturers require a speed of approximately 250 rpm during engine cranking.

If the starter spins but the engine does not rotate, the most likely cause is a faulty starter drive. If the starter drive is at fault, the starter motor will have to be removed to install a new starter or drive mechanism. Before faulting the starter drive, also check the starter ring gear teeth for wear or breakage, and for incorrect gear mesh of the ring gear and starter motor pinion gear.

Most noises can be traced to the starter drive mechanism. The starter drive can be replaced as a separate component of the starter.

TESTING THE STARTING SYSTEM

As with the battery testing series, the tests for the starting system are performed on the volt-ampere tester (VAT). Because the starter performance and battery performance are so closely related, it is important that a full battery test series be done before trying to test the starter system. If the battery fails the load test and is fully charged, it must be replaced before doing any other tests.

Hydrostatic lock is the result of attempting to compress a liquid in the cylinder. Because liquid is not compressible, the piston is not able to travel in the cylinder.

Special Tools
Fender covers
Battery cable puller

Starter Quick Testing

If the starter does not turn the engine at all and the engine is in good mechanical condition, the **starter quick test** can be performed to locate the problem area. To perform this test, make sure the transmission is in neutral, and set the parking brake. Turn on the headlights. Next, turn the ignition switch to the START position while observing the headlights.

There are three things that can happen to the headlights during this test:

1. They will go out.
2. They will dim.
3. They will remain at the same brightness level.

If the lights go out completely, the most likely cause is a poor connection at one of the battery terminals. Check the battery cables for tight and clean connections. It will be necessary to remove the cable from the terminal and clean the cable clamp and battery terminals of all corrosion.

If the headlights dim when the ignition switch is turned to the START position, the battery may be discharged. Check the battery condition. If it is good, then there may be a mechanical condition in the engine preventing it from rotating. If the engine rotates when turning it with a socket wrench on the pulley nut, the starter motor may have internal damage. A bent starter armature, worn bearings, thrown armature windings, loose pole shoe screws, or any other worn component in the starter motor that will allow the armature to drag can cause a high current demand.

If the lights stay brightly lit and the starter makes no sound (listen for a deep clicking noise), there is an open in the circuit. The fault is in either the solenoid or the control circuit. To test the solenoid, bypass the solenoid by bridging the BAT and S terminals on the back of the solenoid. The B1 terminal is a large terminal on the solenoid with a heavy gauge wire coming from the battery. Also on the solenoid, the S terminal has a smaller gauge wire coming to it from the ignition switch or starter relay. Use heavy jumper cables to connect the two terminals. If the starter spins, the ignition switch, starter relay, or control circuit wiring is faulty. If the starter does not spin, the starter motor or the solenoid is faulty.

⚠ WARNING **The starter will draw up to 400 amperes. The tool used to jump the terminals must be able to carry this high current, and it must have an insulated handle. A jumper cable that is too light may become extremely hot, resulting in burns to the technician's hand.**

Current Draw Test

If the starter motor cranks the engine, the technician should perform the **current draw test**. The following procedure is used for volts or amperes tester:

1. Connect the large red and black test leads on the battery posts, observing polarity.
2. Zero the ammeter.
3. Connect the ampere inductive probe around the battery ground cable. If more than one ground cable is used, clamp the probe around all of them (**Figure 5-46**).
4. Make sure all loads are turned off (lights, radio, etc.).
5. Disable the ignition system to prevent the vehicle from starting. This may be done by removing the ignition coil secondary wire from the distributor cap and putting it to ground, by removing the ignition system fuse, fuel pump fuse, PCM fuse, or by disconnecting the primary wires from the coils of an electronic ignition (EI) system.

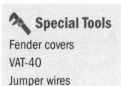

Special Tools

Fender covers
VAT-40
Jumper wires

Figure 5-46 Test lead connections to perform the starter current draw test.

6. Crank the engine and note the voltmeter reading.
7. Read the ammeter scale to determine the amount of current draw.

> **SERVICE TIP** If the engine does not crank and the headlights do not come on, check the fusible link.

> **SERVICE TIP** If the starter windings are thrown, this is an indication of several different problems. The most common is that the driver is keeping the ignition switch in the START position too long after the engine has started. Other causes include the driver opening the throttle plates too wide while starting the engine, resulting in excessive armature speeds when the engine does start. Also the windings can be thrown because of excessive heat buildup in the motor. The motor is designed to operate for very short periods of time. If it is operated for longer than 15 seconds, heat begins to build up at a very fast rate. If the engine does not start after a 15-second crank, the starter motor needs to cool for about 2 minutes before attempting to start the engine again.

 Caution
Always refer to the manufacturer's service manual for the correct procedure for disabling the ignition system. Using an improper procedure may damage ignition components.

After recording the readings from the current draw test, compare them with the manufacturer's specifications. If specifications are not available, as a rule, correctly functioning systems will crank at 9.6 volts or higher. Amperage draw is dependent upon engine size. Most V8 engines have an ampere draw of about 200 amperes, six-cylinder engines about 150 amperes, and four-cylinder engines about 125 amperes.

Higher-than-specified current draw test results indicate an impedance to rotation of the starter motor. This includes worn bushings, a mechanical blockage, internal starter motor damage, and excessively advanced **ignition timing**. Lower-than-normal current draw test results indicate excessive voltage drop in the circuit, a faulty solenoid, or worn brushes.

Ignition timing refers to the timing of spark plug firing in relation to position of the piston.

Figure 5-47 This spark plug is fouled with coolant.

If the current test results are high and the engine does not turn, the starter may not need replacement. Hydrostatic lock or engine seizure may have occurred. To check for this, you should put a socket and ratchet on the crankshaft pulley bolt and attempt to turn the engine over manually by hand. If the engine turns easily, it is most likely a problem with the starter. If the engine does not turn, it is an internal engine problem.

To determine the type of internal problem, start by removing the spark plugs from the engine and attempt to turn the crankshaft again. If it starts to turn easier, look for coolant, fuel, or oil that might be coming out of the spark plug holes (**Figure 5-47**).

These fluids may have gotten into the combustion chamber by a blown head gasket, leaking fuel injector, or a cracked cylinder head or block. If the engine does not turn and no fluids come out of the spark plug hole, the engine may be mechanically seized.

Because the readings obtained from the current draw test were taken at the battery, these readings may not be an exact representation of the actual voltage at the starter motor. Voltage losses due to bad cables, connections, or relays (or solenoids) may diminish the amount of voltage to the starter. Before removing the starter from the vehicle, these should be tested.

STARTER MOTOR REMOVAL

If the tests indicate the starter motor must be removed, the first step is to disconnect the battery from the system.

⚡ **WARNING Remove the negative battery cable. It is a good practice to wrap the cable clamp with tape or enclose it in a rubber hose to prevent accidental contact with the battery terminal. If the positive cable is removed first, the wrench may slip and make contact between the terminal and ground. This action may overheat the wrench and burn the technician's hand.**

The specification for current draw is the maximum allowable, and the specification for cranking voltage is the minimum allowable.

It may be necessary to place the vehicle on a lift to gain access to the starter motor. Before lifting the vehicle, disconnect all wires, fasteners, and so on that can be reached from the top of the engine compartment. Disconnect the wires leading to the solenoid terminals. To prevent confusion, it is a good practice to use pieces of tape to identify the different wires.

⚡ WARNING **Check for proper lift pad-to-frame contact after the vehicle is a few inches above the ground. Shake the vehicle. If there are any unusual noises or movement of the vehicle, lower it and reset the pads. If the lift pads are not properly positioned on the specified vehicle lift points, the vehicle may slip off the lifts, resulting in technician injury and vehicle damage.**

On some vehicles, it may be necessary to disconnect the exhaust system to be able to remove the starter motor. Spray the exhaust system fasteners with a penetrating oil to assist in removal. Loosen the starter mounting bolts and remove all but one. Support the starter motor, remove the remaining bolt, and then remove the starter motor.

⚡ WARNING **The starter motor is heavy. Make sure that it is secured before removing the last bolt. If the starting motor is not properly secured, it may drop suddenly, resulting in leg or foot injury.**

To install the starter motor, reverse the procedure. Be sure all electrical connections are tight. If you are installing a new or remanufactured starter, remove any paint that may prevent a good ground connection. Be careful not to drop the starter. Make sure that it is supported properly. Refer to Photo Sequence 4 for the typical starter removal and reinstallation.

Some General Motors' starters use **shims** between the starter motor and the mounting pad (**Figure 5-48**). To check this clearance, insert a flat blade screwdriver into the access slot on the side of the drive housing. Pry the drive pinion gear into the engaged position. Use a piece of wire that is 0.020 inch in diameter to check the clearance between the gears (**Figure 5-49**).

> The **shims** are used to provide proper pinion-to-ring gear clearance.

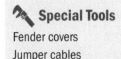

🔧 **Special Tools**
Fender covers
Jumper cables

⚡ **Caution**
Do not operate the starter motor for longer than 15 seconds. Allow the motor to cool between cranking attempts. Operating the starter for more than 15 seconds may overheat and damage starter motor components.

⚙️ SERVICE TIP The major cause of drive housing breakage is too small a clearance between the pinion and ring gears. It is always better to have a little more clearance than too small a clearance.

A 0.015-in. shim will increase the clearance approximately 0.005 in. More than one shim may be required

Figure 5-48 Shimming the starter to obtain the correct pinion-to-ring gear clearance.

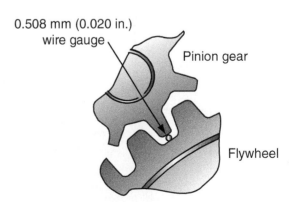

Figure 5-49 Checking the clearance between the pinion gear and ring gear.

If the clearance between the two gears is excessive, the starter will produce a high-pitched whine while the engine is being cranked. If the clearance is too small, the starter will make a high-pitched whine after the engine starts and the ignition switch is returned to the RUN position.

STARTER CIRCUIT TESTS USING A DMM

At several points in the circuit, you will need to test the individual components and wires. To perform a voltage drop test, follow the steps in **Figure 5-50**. Remember to treat the starter circuit just like any other circuit. It can have problems with open circuits, short circuits, and high resistance. Two other common designs to be made aware of when testing the circuit are outlined in **Figure 5-51**.

INSPECTING, SERVICING, AND TESTING THE ALTERNATOR AND CHARGING CIRCUIT

Testing the charging system requires knowledge of the circuit and some of the basic electrical testing tools and procedures previously outlined in this chapter.

The most effective test using a DMM is done by performing both a B+ voltage drop test and a ground side voltage drop test (**Figure 5-52**). Place the DMM at the appropriate places on the system. Then turn the engine on and perform these tests. You should perform these voltage drop tests with the system loaded and unloaded. A loaded test is when you request as much current from the battery as possible, placing the alternator under stress. This means that you should turn on all electrical loads like the headlights, defroster, radio, and fan, and run the engine at 2,500 rpm. *Unloaded* means just what you think it means: the system with almost nothing turned on and at idle. You have to do both tests to check the integrity of the wiring.

Figure 5-50 Voltage drop testing to identify sources of excessive resistance.

Figure 5-51 (A) Test lead connections for a starter mounted solenoid; (B) test lead connections for relay-controlled systems.

Always refer to the manufacturer's specification, but generally you can expect a good system to have less than a 0.7-volt drop on the insulated (power) side and less than a 0.2-volt drop on the ground side.

To perform a current draw output test, you need to connect an amp clamp and a load to the vehicle's battery. Be careful when performing this test not to overload the system. Refer to **Photo Sequence 12** for the steps to perform this test.

B+ voltage drop test

(A)

Ground side voltage drop test

(B)

Figure 5-52 Voltage drop testing of the generator output circuit (A) and the generator ground circuit (B).

PHOTO SEQUENCE 12
Performing the Charging System Output Test

P12-1 Connect the large red and black cables across the battery, observing polarity.

P12-2 Select CHARGING.

P12-3 Zero the ammeter.

P12-4 Connect the inductive pickup around all battery ground cables.

P12-5 With the ignition switch in the RUN position, engine not running, observe the ammeter reading. This reading indicates how much current is required to operate any full-time accessories.

P12-6 Start the engine and hold between 1,500 and 2,000 rpm.

P12-7 Turn the load knob for the carbon pile slowly until the highest possible ammeter reading is obtained. Do not reduce battery voltage below 12 volts.

P12-8 Return the load control knob to the OFF position.

P12-9 The highest reading indicates maximum current output.

SPARK PLUG EVALUATION AND POWER BALANCE TESTING

Spark plug reading and **power balance testing** are two methods you will use to gather information about how the engine is operating. You may use these tests during a routine tune up, when there is a drivability concern, and as part of gathering information about an engine mechanical failure. Spark plugs can help identify if one or more cylinders are acting up or if all cylinders are affected. They can also help guide your diagnosis toward a fuel, ignition, or mechanical issue, for example. Similarly, a power balance test can determine how much each cylinder is contributing and pick out the malfunctioning cylinder(s).

> A **power balance test** evaluates approximately how much each cylinder is contributing to the overall engine power. It is used to help identify a weak cylinder.

SPARK PLUGS

The spark plug must be in good condition to allow the spark to jump the gap. If the electrodes are worn rounded or covered with oil or gas, or if the gap is too large, the spark will not be strong enough to ignite a flame front, if it sparks at all. Spark plugs are generally one of two basic types: the older type plug with nickel alloy steel electrodes and the newer, more common, platinum tip electrode plugs. Most new vehicles use the platinum plugs as original equipment (**Figure 5-53**). The platinum tip plugs have a service interval of 60,000 miles or longer. The older style plugs must generally be replaced every 30,000 miles.

Many times, customers do not bring their cars in for service until there is a drivability concern. Spark plugs have a high failure rate; check them early on in your diagnosis. As a spark plug wears, the gap between the electrodes becomes so wide that it may not allow the spark to jump the gap. This can cause partial or total misfiring on that cylinder. Ignition misfire is usually noticeable at low rpm under heavy acceleration. The engine may buck and shudder when the cylinder(s) misfires. It is often difficult to see wear on a platinum plug; the electrodes are small and do not show much visible degradation. Many technicians will replace rather than inspect and reinstall platinum plugs with over 30,000 miles on them. Given today's labor rates, this is often more cost effective for the consumer. If the engine is running poorly and you have the spark plugs out, it may be wise to perform a compression test, which is described later in this chapter.

Figure 5-53 A platinum tip spark plug.

Spark Plug Removal and Installation

While spark plug removal is a very common task in the automotive repair shop, it is not always a simple one. Many times, you will have to disassemble parts of the engine to access the spark plugs. There are vehicles that require you to remove an engine mount or remove the turbocharger just to access the plugs. Every time you replace a set of spark plugs, you want to be sure you do not damage the plugs or the threads while installing them. Always be sure to replace the whole set of plugs. If one has failed, the others are likely not far behind.

If access to the plugs is not simple, read through the service information to pick up any steps that will make the job easier. This task often requires a universal joint and an assortment of extensions (**Figure 5-54**).

Figure 5-54 This spark plug is readily accessible; not every engine makes it this easy.

Blow away any sand and grit from around the spark plug holes before removing the plugs, to prevent this from falling into the cylinder head. Remove all plugs and keep them in order, so you can evaluate the condition of each plug and correlate it to the correct cylinder. Keep the spark plug wires (if applicable) organized, so you return them to their correct positions. Carefully inspect the ends of the plug wires. They can easily be damaged during removal and may require replacement. Also look closely at the plug wires for signs of electrical leakage or arcing to a ground path. If the wire's insulation is damaged, you can often see a light gray or white residue on the wire where the high-voltage arcs to a ground—on the valve cover, for example—rather than across the spark plug gap.

> **⚙ SERVICE TIP** In one semester, three students had come to me with a misfiring problem. In each case, they had recently replaced spark plugs using a popular, low-cost brand of spark plugs. Installing higher quality or original equipment manufacturer (OEM) spark plugs repaired each of the three late-model vehicles. Use OEM or high-quality spark plugs on your customers' cars to prevent comebacks.

Check the spark plug gap before installing the new set of plugs. Sometimes the gap gets closed if the plugs have been dropped. Note that some platinum spark plugs do not have adjustable gaps. Clean the area around the spark plug hole. If you get grease on the electrodes during installation, you will likely wind up with a misfiring spark plug. Install the new plugs by hand; *do not* use air tools. If you can't reach the plug itself, turn it in a few turns using only the extension.

Finish tightening the plugs to the correct torque specification using a torque wrench. After your work is complete, always give the vehicle a thorough road test to be sure the engine runs perfectly.

Caution
Never use air tools to remove or install spark plugs. It takes a lot longer to repair a spark plug hole than it does to loosen plugs with a ratchet.

Spark Plug Reading

Each spark plug can tell you a story about what is occurring in the cylinder. During a routine service, you can use this information to help determine if the engine is mechanically sound. If one spark plug comes out caked with oil deposits from oil leaking past the rings or valve seals, you will be able to let the customer know that the engine is in need of more serious work. When performing drivability diagnosis, the spark plug can help guide your diagnosis. If one spark plug was wet with fuel, you would first confirm that the ignition system is delivering adequate spark to the plug. And whenever you are trying to narrow down the cause of an engine failure, analyze the spark plugs as part of your pre-teardown investigation. The more information you have going into the job, the more confident you can be that you will find the real cause of the problem.

> **⚙ SERVICE TIP** When the plugs are very difficult to reach, it is helpful to put a piece of vacuum line on the end of the spark plug. This flexible connection allows you to reach the spark plug hole. Turn the plug in a few turns using the end of the vacuum line. If you start to cross thread the plug, the vacuum line will spin on the spark plug and prevent thread damage.

Caution
No technician has a torque wrench built into his elbow, no matter how adamantly he insists he can torque something properly. Spark plug torque specifications are very light on many new engines, and a stiff elbow can easily strip the spark plug threads on an aluminum cylinder head.

Spark plugs that are wearing normally should show a light tan to almost white color on the electrodes with no deposits (**Figure 5-55**). The electrodes should be square, and the gap should be at or near at the specified measurement. You will see plugs that have a red or orange tint on them. This is the result of fuel additives found in certain fuels or in fuel system cleaning additives.

Figure 5-55 This spark plug is wearing nicely; it has no deposits and is a light tan color.

Figure 5-56 Notice the wide gap and rounded electrodes on these worn spark plugs.

A spark plug that has many miles on it but is wearing normally will still be light tan in color. There should be only very light deposits, if any, on the plug (**Figure 5-56**). On nonplatinum plugs, you will be able to see that the corners of the electrodes are rounded. The gap will usually be wider than specified. On platinum plugs, the tip may look brand new even when the plug is not firing well. Often you will have to judge wear by the miles since the last change. If the plug is not firing at all, it will be gas-soaked and smell like fuel.

Fuel-fouled plugs will be wet with gas when you remove them from the engine. They will smell like fuel and may have a varnish on the ceramic insulator from heated fuel. The most common cause of this is inadequate spark, although a leaking fuel injector will also flood the plug. If the valves are burned or the compression rings are badly worn, the engine may lack adequate compression to make the air/fuel mixture combustible. A compression check will either verify or rule out that problem.

> **SERVICE TIP** The classic symptoms of an ignition misfire are bucking and hesitation on hard acceleration. A fuel-related problem, such as a restricted fuel filter, is more likely to cause a steady lack of power as engine speed slowly increases.

When a vehicle is burning oil, the spark plugs will develop tan to dark tan deposits on and around the electrode. As the oil is heated during combustion, it bakes onto the plug, leaving a residue. You will have to rule out any other possible causes of oil consumption, such as a plugged PCV system, badly worn valve seals, or a faulty turbocharger, but this is usually a telltale sign that the engine is mechanically worn. If you find this problem during a routine service, it is important to warn the customer about the failing condition of his engine. If the oil is wet and thick on the plug, look for a cause of heavy oil leakage into the cylinder. On rare occasions, a head gasket will crack between an oil passage and the cylinder, allowing oil into just one cylinder. This can cause an oil-soaked plug. Similarly, a hole blown in a piston can cover a plug with oil, but you will have other indicators of a serious mechanical failure.

Refer to the spark plug chart in **Figure 5-57** to see what the spark plug inspection can tell you about the engine's condition.

Normal Condition

Sooted–carbon-fouled

Oil-fouled

Lead fouling

Pronounced
lead fouling

Formation of ash

Center electrode covered
with melted deposits

Partially melted
center electrode

Partially melted
electrodes

Heavy wear on center
electrode

Heavy wear on
ground electrode

Insulator-nose
fracture

Figure 5-57 A good inspection of used spark plugs can reveal much information about the internal operation of the engine and its subsystems.

Engine coolant temperature sensor

Figure 5-58 A faulty engine coolant temperature sensor can cause the engine to run rich.

> ⚙ **SERVICE TIP** Some spark plugs are advertised to last for more than 100,000 miles. This doesn't guarantee they'll last that long; these platinum plugs have a high failure rate after 50,000 miles and often become welded into the cylinder head by 75,000 miles. A 60,000-mile service interval is a reasonable compromise that is likely to better serve the customer.

Black-colored, soot-covered spark plugs are found when the air-fuel mixture is too rich. This means there is too much fuel and not enough air. This can be caused by something as simple as a plugged air filter or as involved in the powertrain control system as an inaccurate engine coolant temperature (ECT) sensor (**Figure 5-58**). An engine that does not have good compression can also create carbon because combustion will not be complete. The key is to notice the condition. Locate the cause before considering a maintenance service to be complete. Write down your observations if you are evaluating an engine for serious mechanical repairs.

Blistered or overheated spark plugs are a sign of potentially serious trouble in the combustion chamber. The porcelain insulator around the center electrode will actually have blisters in it, or there may be pieces of metal welded onto the electrode. It is possible that the wrong spark plugs were installed in the engine. Otherwise, it is likely that pinging or detonation is occurring. If there is no engine damage yet, it is critical to find the cause of the problem before returning the vehicle to the customer. If you find this as you are diagnosing an engine mechanical failure, it is essential to record this information and find the source of the problem during your repair work. A cooling system passage with deposits in it could be causing one corner of a combustion chamber to get so hot that detonation occurs. If you were to replace the damaged piston and rings but not repair the cause of the problem, the customer would be back soon with a repeat failure. Your thorough evaluation of the engine and analysis of the causes of the symptoms will ensure customer satisfaction.

POWER BALANCE TESTING

On a rough-running or misfiring engine, you can perform a power balance test to identify a cylinder that is not equally contributing power to the engine. To perform the test, you disable one cylinder at a time while noting the change in engine rpm and idling condition. The more the rpm drops or the rougher the engine runs, the more the cylinder is contributing. If a cylinder is producing little or no power, the change in rpm and engine

Figure 5-59 This engine analyzer shorts one cylinder at a time and displays the engine rpm drop.

running will barely be noticeable. You must note the rpm drop quickly as the cylinder is initially disabled; today's PCMs will boost the idle almost immediately to compensate for the drop. You will also be able to feel and see the roughness of the engine as a functional cylinder is disabled. We'll discuss a few different methods of performing a power balance test.

Power Balance Testing Using an Engine Analyzer or Scan Tool

Many engine analyzers can perform a manual or automated power balance or cylinder efficiency test. With the analyzer leads connected to the ignition system as instructed in the user's manual, select the power balance test from the menu (**Figure 5-59**). The tester will automatically disable the ignition to one cylinder at a time for just a few seconds and display the rpm drop. The analyzer disables spark in the firing order; an engine with the firing order 1-3-4-2 would be tested on cylinder number 1 first, then number 3, number 4, and number 2. On some analyzers, the power balance results will be displayed in bar-graph format for easy comparison of cylinder contribution.

On some vehicles, the PCM is programmed to be able to perform a power balance test. The manufacturers and some aftermarket scan tools will be able to access this test. The only hookup required is to the diagnostic link connector (DLC). Then select the power balance test from the menu system. At the prompt, the scan tool will communicate with the PCM to initiate its disabling of cylinders one by one. The rpm drop for each cylinder will be displayed after the test.

Manual Power Balance Testing

Even without an engine analyzer or a capable scan tool, you can perform a power balance test. The procedure will require that you manually disable either fuel or spark to each cylinder. It is also important that you check the vehicle for diagnostic trouble codes (DTCs) after performing this test and clear the codes if any are present. You can access DTCs using a scan tool; the code can help you locate the source of the problem. If the PCM detects a misfire, it is likely to set a DTC. When the malfunction indicator light comes on, a DTC has been set. Using the menu on a scan tool, you can erase any DTCs when your testing is complete.

On vehicles that use spark plug wires to deliver the spark to the plug, you can short the spark to ground on each cylinder in turn to test the power contribution of cylinders.

 Caution

You must be careful not to disable spark for long periods of time and must allow the engine to run on all cylinders for 1 minute between tests. Failure to follow these guidelines can overload the catalytic converter with raw fuel and damage it from overheating.

Figure 5-60 Use insulated pliers to remove a live plug wire. Ground the wire to a metal component immediately to prevent harming the coil.

One way to do this is to use an insulated pair of pliers (**Figure 5-60**). Carefully pull the wire off the plug and hold the wire to the block or head to allow it to find a ground path. It is important that the spark can find an easy path to ground; do not hold the wire more than a half an inch away from a good metal conductor. The coil can be damaged if it uses all its possible power to find a path to ground. Sometimes the spark will track through the insulation of the coil and destroy it. As you remove the spark plug wire, you should notice a clear change in rpm and idle quality if that cylinder is contributing adequate power.

Many newer engines use coil-on-plug ignition systems. Each spark plug has its own coil, and you cannot safely remove the coil and ground it. Instead, locate the connector going into the coil (**Figure 5-61**). While the engine is running, you can briefly remove the connector to prevent the coil from firing. As you unplug the coil, monitor the rpm drop on a tachometer. Repeat this for each cylinder and compare your results. This testing is very likely to trigger the malfunction indicator lamp (MIL) and set a DTC, so be sure to clear the DTCs after your testing.

Another way to perform a manual power balance test is to unplug the connectors to the fuel injectors one by one while listening to and watching the engine rpm. By preventing

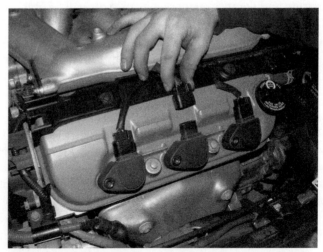

Figure 5-61 During a power balance test, disconnect the low voltage primary wiring to the individual coils and record the rpm drop. Be sure to clear any DTCs after your testing.

fuel from entering the cylinder, you can momentarily disable the cylinder. This allows you to note how much the cylinder is contributing to the overall power of the engine.

Analyzing the Power Balance Test Results

When an engine is running properly, each cylinder should cause very close to the same rpm drop when disabled. Cylinders that are lower by just 50 rpm should be analyzed. Look at the following results:

Cyl. #1	Cyl. #3	Cyl. #4	Cyl. #2
125 rpm	150 rpm	50 rpm	125 rpm

Cylinder number 4 is definitely not contributing equally to the power of the engine. The rpm changed very little when it was disabled, meaning that the engine speed is barely affected by that cylinder. There could be many causes of the problem, but you now know which cylinder to investigate. The problems could relate to fuel delivery; perhaps the fuel injector was not delivering adequate fuel to support combustion. An ignition system problem, something as simple as a spark plug with a very wide gap from worn electrodes, could also cause this. A cylinder with low compression will also produce less power and be suspect during a power balance test. Compression testing and cylinder leakage testing will help you to determine whether the weak cylinder has an engine mechanical problem. They can also provide information about the specific cause.

Special Tool

Lineman's gloves

SPECIAL PROCEDURES FOR WORKING SAFELY AROUND HYBRID ELECTRIC VEHICLES

The service of hybrid electric vehicles (HEVs) is limited to technicians who have received factory training from the manufacturer or are master technicians who have received training and education from a certified program. This text cannot train you to work on an HEV. Working on HEVs is not an entry-level technician's job; however, you must be aware of the safety precautions. As a technician, you should be aware of the safety precautions associated with these vehicles. If you perform towing, you must be fully aware of the safety precautions before approaching a damaged HEV. Your education may well place you at the front of the line for training on these advanced technology vehicles.

Safety must be your first consideration when approaching an HEV (**Figure 5-62**). The voltages on some HEVs may reach 500 volts and regularly store 360 volts. Voltage varies

Figure 5-62 Use caution! Look carefully at this warning label; it says "you will be killed" if you do not use appropriate safety precautions.

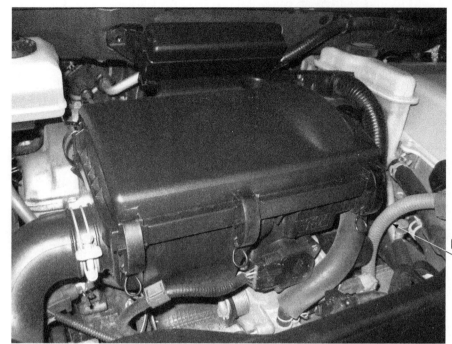

High-voltage
cables

Figure 5-63 The brightly colored big cables hold high voltage; components that use high voltage have warning labels.

with make, but more than 30 volts is potentially fatal to humans; so we are dealing with a potentially lethal mechanism. All high-voltage cables are identifiable by a bright orange coating. Never touch these wires while the electrical system is live. All high-voltage components are labeled as such (**Figure 5-63**). The battery packs are well insulated and protected in an accident. Nickel metal hydride (NiMH) batteries, as used in many of today's HEVs, are sealed in packs designed not to leak even in the event of a crash. If they do leak, the electrolyte can be neutralized with a dilute boric acid solution or vinegar.

> ⚙⚙ SERVICE TIP Something as simple as a blown high-voltage fuse could cause a vehicle not to drive. You'll still need to follow special safety precautions for testing them, though. They are rated between 400 and 600 amps

Understanding proper operation of the vehicle you are working on is essential. Although you should have received factory training, much of this information can be found in the owner's manual. The vehicle may be "ready" and able to move with a tap of the accelerator, even when neither the electric motor nor the gasoline engine is running. The owner's manual will provide you with special operating characteristics of the vehicle and basic safety warnings.

> ⚙⚙ SERVICE TIP It is not uncommon for a customer with a new technology vehicle to come in with service questions. Most consumers do not take the time to read the owner's manual or view the operational CD provided. When you are working on a make of vehicle, be as familiar with it as possible so you can competently answer questions a customer might have about his new vehicle.

SERVICING THE TOYOTA PRIUS

First, familiarize yourself with the layout of the vehicle components (**Figure 5-64**). The figure identifies the location of the following components:

Component	Location	Description
1. 12-volt battery	LR of trunk	Typical lead-acid battery used to operate accessories
2. HV battery pack	Rear seat back support, in trunk	274 NiMH battery pack
3. HV power cables	Undercarriage and engine compartment	Thick, orange-colored cables carrying high voltage (HV) DC and AC current
4. Inverter	Engine compartment	Converts DC to AC for the electric motor and AC to DC to recharge the battery pack
5. Gasoline engine	Engine compartment	Powers vehicle and powers generator to recharge battery
6. Electric motor	Engine compartment	AC permanent magnet (PM) motor in transaxle to power vehicle
7. Electric generator	Engine compartment	AC generator in transaxle used to recharge battery pack

Figure 5-64 Locations of components on the Toyota Prius.

Figure 5-65 The high-voltage service plug is located in the trunk. Use lineman's gloves to switch power off.

Only a certified technician should disable the high-voltage system by disconnecting the negative terminal of the 12-volt auxiliary battery. This is located in the left rear of the trunk. Disconnecting the auxiliary battery shuts down the high-voltage circuit. For additional protection, wait 5 minutes, and then remove the service plug located through an access port in the trunk (**Figure 5-65**). Wear insulated rubber **lineman's gloves** when removing the service plug. Always carefully check the gloves for pin holes. Blow into them like a balloon and ensure they don't leak any air. Even a very small hole could allow a deadly spark to penetrate to you. After removing the service plug, wait at least 5 minutes to allow full discharge of the high-voltage condensers inside the inverter.

Once the high-voltage system has fully discharged, you may diagnose and repair the Prius gas engine as you would any other engine. This is a newly designed engine; be sure to use proper procedures and service specifications.

Lineman's gloves are highly insulated rubber gloves that protect the technician from lethal high voltages.

SERVICING THE HONDA CIVIC HYBRID

The Civic hybrid runs current at 144 volts through brightly colored orange cables. Its Integrated Motor Assist (IMA) System uses a 1.3-liter gasoline engine and a DC brushless motor (**Figure 5-66**). You must shut off the high voltage electrical circuits and isolate the

Figure 5-66 Honda's IMA system "engine" compartment.

High-voltage cables

Figure 5-67 The high-voltage cables are routed through the engine compartment, but you can't miss them due to their bright orange color. Disable the high-voltage system before touching them.

IMA system before servicing its components. Do not touch the high-voltage cables without lineman's gloves (**Figure 5-67**). To isolate the high-voltage circuit, turn the high voltage battery module switch to OFF. Secure the switch in the off position with the locking cover before servicing the IMA system. Wait at least 5 minutes after turning off the battery module switch, and then disconnect the negative cable from the 12-volt battery. Test the voltage between the high-voltage battery cable terminals with a voltmeter. Do not disconnect them until the voltage is less than 30 volts.

HEV Warranties

To reassure the public about the new technology and encourage consumer confidence, most HEV manufacturers offer extensive warranties on the battery packs and the hybrid components. You should always consult the manufacturer's service information before recommending expensive repairs to consumers of these vehicles, even as the vehicles age and you begin to see them in independent repair facilities. A few examples are:

Ford Escape	8 years/80,000 miles on battery pack*
Honda HEV	8 years/80,000 miles on battery pack*
Toyota Prius	8 years/100,000 miles on battery pack* and on the hybrid power plant, including the engine
*Battery packs must be warranted for 10 years/150,000 miles in "green states" (CA, MA, VT, NY, ME)	

CASE STUDY

A friend of mine asked me to jump start her vehicle. While doing so, she explained the battery had been replaced several times along with a new starter but still there was a cranking problem. Later that day, I used a DVOM to check the starting system. A voltage drop test on the battery to starter cable revealed a 6-volt drop. An inspection of the cable found heavy corrosion under the cable covering near the battery connection. Replacement of the cable solved the problem. Always check the cables and connections when dealing with the starting and charging system, particularly on older vehicles.

ASE-STYLE REVIEW QUESTIONS

1. Which of the following describes the function of the starter solenoid?
   A. An electromagnetic device that transfers high current to the starter motor.
   B. An electromagnetic device that moves and engages the starter.
   C. Neither A or B.
   D. Both A and B.

2. The rotor in an alternator:
   A. Is a stationary component.
   B. Controls the speed of the alternator.
   C. Is touching the rectifier.
   D. Is a rotating component.

3. A Hybrid Electric Vehicle uses a DC-to-DC converter to:
   A. Charge the 12-volt battery.
   B. Turn the headlights on.
   C. Power the alternator.
   D. Start the engine.

4. A flywheel is mounted to the:
   A. Crankshaft.
   B. Camshaft.
   C. Cylinder block.
   D. Transmission housing.

5. A fully charged conventional automotive battery has a specific gravity of:
   A. 1.460
   B. 1.265
   C. 1.225
   D. 1.190

6. The battery stores energy in which form?
   A. Chemical.
   B. Electrical.
   C. Mechanical.
   D. Kinetic.

7. Two technicians are discussing relay function.
   Technician A says that a relay can be controlled by a switch.
   Technician B says that a relay can be controlled by a computer.
   Who is correct?
   A. A only
   B. B only
   C. Both A and B
   D. Neither A nor B

8. A recombination battery:
   A. Uses a gel instead of acid.
   B. Contains flat cells and plates.
   C. Can operate only in the upright position.
   D. None of the above.

9. The source of electrical power for everything in the car is the:
   A. Alternator.
   B. Starter.
   C. Computer.
   D. Battery.

10. A computer circuit consists of:
    A. Inputs.
    B. Outputs.
    C. Processing device.
    D. All of the above.

ASE CHALLENGE QUESTIONS

1. Wiring diagrams are being discussed.
   Technician A says you must have correct VIN to locate the proper information.
   Technician B says most wiring diagrams are the same for simple circuits and it is not that important to have the exact diagram.
   Who is correct?
   A. A only
   B. B only
   C. Both A and B
   D. Neither A nor B

2. Two technicians are discussing how to test an electrical circuit.
   Technician A says you can use a 12v test light to test all circuits.
   Technician B says a digital multimeter is the most effective method for testing all electrical circuits.
   Who is correct?
   A. A only
   B. B only
   C. Both A and B
   D. Neither A nor B

3. Two technicians are discussing voltage drops.

   Technician A says always connect the voltmeter in parallel to the circuit.

   Technician B says the voltmeter will show the voltage loss in the circuit you are testing.

   Who is correct?

   A. A only

   B. B only

   C. Both A and B

   D. Neither A nor B

4. *Technician A* says when test a circuit for resistance you should use the ohm setting and calibrate the meter first.

   Technician B says the component or circuit being tested must be operating when testing.

   Who is correct?

   A. A only

   B. B only

   C. Both A and B

   D. Neither A nor B

5. Two technicians are discussing an open circuit.

   Technician A says that a switch that is off is considered an open circuit.

   Technician B says when testing for an open circuit you should start at the load or component that does not work to test for voltage.

   Who is correct?

   A. A only

   B. B only

   C. Both A and B

   D. Neither A nor B

Name _____   Date _____

CHECKING THE BATTERY AND GENERATOR

Upon completion and review of this job sheet, you will be able to test the battery and generator using a digital multimeter (DMM) and make suggestions on possible faults or repairs.

ASE Education Foundation Correlation ————————————————————

This job sheet addresses the following **MLR** task:

6.B.1 Perform battery state-of-charge test; determine necessary action. **(P-1)**

This job sheet addresses the following **AST/MAST** task:

6.B.1 Perform battery state-of-charge test; determine needed action. **(P-1)**

We Support
ASE | Education Foundation

Tools and Materials

Goggles or safety glasses with side shields
Assigned vehicle
DMM
Service manual

⚡ **Caution**

Use extreme care when working around a hot engine. Serious burn injuries could result if contact is made with hot components.

Describe the vehicle and tire being worked on:

Year _____ Make _____ Model _____

VIN _____   Engine type and size _____

Procedure: Use the service manual to collect the following information.

    Recommended battery size by cold-cranking amps: _____
    Generator size based on maximum ampere output: _____

Task Completed

⚡ WARNING Ensure the DMM leads are clear of the cooling fan and drive belts. Keep hands clear of moving engine parts and the cooling fan. Damage to the DMM or personal injury could result.

1. Place the vehicle on level ground or floor and set the parking brake. Shift the transmission to PARK (NEUTRAL on a manual). Ensure the ignition key is in the OFF position. Chock the wheels as soon as you exit the vehicle.   ☐

2. Raise the hood and perform an inspection of the battery and generator drive belts. Record your observations and note any other conditions that may point to problems other than those typically associated with the starting and charging systems.

_____

3. Connect the DMM to the battery for an open circuit voltage test (red on positive post, black on negative post). Record the measurement. Is it within the normal range for a fully charged battery? If not, what tests or repairs do you suggest be performed at this point?

_____

_____

4. Connect the meter to the battery for a voltage drop test on the positive side of the battery (red on terminal, black on post). Record the measurement and then do the same test on the negative side. Is there a voltage drop on either terminal? If so, what possible fault would cause a voltage drop at this point?

_____

_____

5. Use the DMM to perform a voltage leakage test on the battery. Is there a reading of voltage and what is the average of the readings? What do you recommend if there is a measurement of voltage leakage?

_____

6. An assistant can help perform the generator output test but is not required. ☐

7. Connect the DMM to the battery for an open circuit test. ☐

8. Start the engine and allow it to idle. Measure and record the voltage present at the battery. Turn on a heavy load, either the air conditioning or bright headlights. Measure and record the battery voltage. Does there seem to be a problem with the generator at this point? If so, what do you recommend for the next test or a possible repair?

_____

_____

9. Shut down the engine and close the hood. Based on the measurements just completed, what are your recommended actions for the starting and charging systems?

_____

_____

Instructor's Response

_____

_____

_____

Name _____   Date _____

TESTING ELECTRICAL CIRCUITS USING TEST LIGHTS AND A DIGITAL MULTIMETER (DMM)

Upon completion and review of this job sheet, you should be able to test and diagnose circuits using a test light or DMM.

ASE Education Foundation Correlation ———————————————————

This job sheet addresses the following **MLR** tasks:

6.A.3 Use wiring diagrams to trace electrical/electronic circuits. **(P-1)**

6.A.4 Demonstrate proper use of a digital multimeter (DMM) when measuring source voltage, voltage drop (including grounds), current flow, and resistance. **(P-1)**

6.A.6 Use a test light to check operation of electrical circuits. **(P-2)**

This job sheet addresses the following **AST/MAST** tasks:

6.A.3 Demonstrate proper use of a digital multimeter (DMM) when measuring source voltage, voltage drop (including grounds), current flow, and resistance. **(P-1)**

6.A.5 Demonstrate proper use of a test light on an electrical circuit. **(P-1)**

6.A.7 Use wiring diagrams during the diagnosis (troubleshooting) of electrical/electronic circuit problems. **(P-1)**

We Support
ASE | Education Foundation

Tools and Materials

Goggles or safety glasses with side shields
Wiring diagram
Vehicle with a designated test circuit
Test light
Self-powered test light
Digital multimeter (DMM)
Fender cover
Service manual
Safety glasses

Describe the vehicle and tire being worked on:

Year _____ Make _____ Model _____

VIN# _____   Engine type and size _____

Procedure   **Task Completed**

1. Determine the following information from the service manual:

   Using wiring diagrams, trace the headlight and tail light circuits' layout in the vehicle.

   Are the conductors in these circuits in the same electrical conduits as the airbag circuit? If yes, consult your instructor. ☐ Yes ☐ No

Will any of the tests here involve working around airbag components? If yes, consult your instructor. ☐ Yes ☐ No

Is either circuit an electronic circuit? ☐ Yes ☐ No If yes, disregard the use of any test light. Use only the digital multimeter.

⚡ **WARNING Do not work on or near airbag circuits without clearing your work with the instructor.**

⚡ **WARNING Before working on a circuit with a test light, ensure that it is not an electronic circuit or part of one. A test light can damage electronic components.**

⚠ **Caution**

Disable the airbag system by disconnecting the battery's negative cable. Wait for the airbag capacitor to discharge following the manufacturer's specific instructions. Severe injury could result from a deploying airbag.

⚠ **Caution**

Electrical power must be provided for this circuit test. If the airbag system was completely disarmed, disconnect all the airbag system's bright yellow connections before reconnecting the battery. After the test(s) are completed, disconnect the battery negative cable, reconnect all the yellow connections and then reconnect the battery. Consult your instructor before disconnecting or connecting any airbag circuits.

Headlight Circuit

2. Connect the test light across the battery terminals. Did the test lamp light and is it bright? ☐ Yes ☐ No

3. Set the DMM to read DC voltage.   ☐

4. Connect the DMM across the battery terminals. What voltage reading did you get?

_____

5. What is your next action if the battery voltage is less than 11 volts?

_____

6. Use the wiring diagram to locate the easiest point to access a connection in the circuit to be tested. Disconnect the circuit.   ☐

7. Turn the circuit on (power on)   ☐

8. Connect the test clip to a good ground.   ☐

9. Touch the test light's probe to the electrical terminal on the vehicle harness. What happened to the test light and what does it indicate?

_____

10. Remove the test light. Leave the power on.   ☐

11. Connect the DMM's black probe to a good ground. Ensure the DMM is set to read DC   ☐
    voltage.

⚡ **WARNING Do not connect an ohmmeter or a DMM set to the ohm's scale to a powered circuit. The test could be damaged.**

Task Completed

12. Probe the vehicle harness terminal with the DMM's red lead. What is the result and what does it indicate?

_____

The next several steps require that the power is off within the circuit portion being tested. Follow the WARNING instructions.

⚠ WARNING **Do not power up the circuit being tested for continuity. Using a continuity tester to test a powered circuit will damage the tester and possibly damage the circuit.**

13. Ensure the power to the circuit is off. Locate the next circuit connection towards the load and disconnect it. ☐

14. Use the self-power test light for this test. Connect the test light's clip to a terminal at one end of the isolated portion of the circuit. Use the test light's probe to touch the terminal at the other end of the isolated circuit. ☐

15. Observe the test light lamp and record your results and what do the results mean?

_____

16. Set the DMM to read resistance. ☐

17. Connect one lead of the DMM to one end of the isolated circuit and the other lead to the opposite end. Record the results and what do the results mean?

_____

18. Reconnect the all the circuit connections you unplugged. ☐

19. Turn on the circuit's power to ensure the circuit is working properly. Turn off the circuit power and replace any items moved to gain access to the circuit. ☐

⚠ Caution

If the airbag was disabled, consult your instructor and follow the manufacturer's procedures to enable the airbag system. Once the system is enabled, start the engine and check to see if the airbag warning lamp is lit. If so, consult your instructor.

Stop (Brake) Light

⚠ WARNING **Before working on the stop light circuit, determine if the under-dash connection is near the airbag circuit. If so, consult your instructor.**

20. Disconnect the left side brake light inside the trunk. It may be necessary to move the trunk carpet or paneling for access. Ensure that any fasteners are not damaged. ☐

21. Connect the test light clip to a good ground. ☐

22. Have an assistant apply the service brakes and hold the pedal in place. ☐

23. Probe each of the stop light terminals on the vehicle harness. Which terminal is the power conductor for the stop light? Did the test lamp light up? If not, what is the most probable cause?

_____

Task Completed

24. Once the power terminal is determined, set the DMM to read DC voltage.   ☐

25. Connect the DMM's black lead to a good ground and touch the red lead to the power terminal. Ensure the brake pedal is applied. What is your voltage reading?

_____

26. Release the brake pedal and reinstall any carpeting or paneling removed for access to the stop light connection.   ☐

27. Check the operation of the stop light and head light circuit. Ensure any other items removed to gain access to either circuit are installed properly.   ☐

Instructor's Response

_____

_____

_____

Name _____   Date _____

INSPECT, SERVICE, AND TEST THE STARTER AND STARTER CIRCUITS

Upon completion and review of this job sheet, you should be able to test and diagnose starter circuits using a DMM, determine any faults, and replace a starter motor.

ASE Education Foundation Correlation

This job sheet addresses the following **MLR** tasks:

6.A.4   Demonstrate proper use of a digital multimeter (DMM) when measuring source voltage, voltage drop (including grounds), current flow, and resistance. **(P-1)**

6. A.7   Use fused jumper wires to check operation of electrical circuits. **(P-2)**

6.A.10   Repair and/or replace connectors, terminal ends, and wiring of electrical/electronic systems (including solder repair). **(P-1)**

6.A.11   Identify electrical/electronic system components and configuration. **(P-1)**

6.B.1   Perform battery state-of-charge test; determine necessary action. **(P-1)**

6.B.2   Confirm proper battery capacity for vehicle application; perform battery capacity and load test; determine necessary action. **(P-1)**

6.C.2   Perform starter circuit voltage drop tests; determine necessary action. **(P-1)**

6.C.3   Inspect and test starter relays and solenoids; determine necessary action. **(P-2)**

6.C.5   Inspect and test switches, connectors, and wires of starter control circuits; determine necessary action. **(P-2)**

This job sheet addresses the following **MLR/AST/MAST** task:

6.C.4   Remove and install starter in a vehicle. **(P-1)**

This job sheet addresses the following **AST/MAST** tasks:

6.A.3   Demonstrate proper use of a digital multimeter (DMM) when measuring source voltage, voltage drop (including grounds), current flow and resistance. **(P-1)**

6.A.6   Use fused jumper wires to check operation of electrical circuits. **(P-1)**

6.A.10   Inspect, test, repair, and/or replace components, connectors, terminals, harnesses, and wiring in electrical/electronic systems (including solder repairs); determine needed action. **(P-1)**

6.B.1   Perform battery state-of-charge test; determine needed action. **(P-1)**

6.B.2   Confirm proper battery capacity for vehicle application; perform battery capacity and load test; determine needed action. **(P-1)**

6.C.2   Perform starter circuit voltage drop tests; determine needed action. **(P-1)**

6.C.3   Inspect and test starter relays and solenoids; determine needed action. **(P-2)**

6.C.5   Inspect and test switches, connectors, and wires of starter control circuits; determine needed action. **(P-2)**

We Support
ASE | Education Foundation

Tools and Materials

Hand tools
Floor jack with safety stands or vehicle lift
Wiring diagram
Vehicle with a standard starter control circuits

Digital multimeter (DMM)
Fender cover
Service manual
Safety glasses

⚠ **Caution**

Disable the fuel and ignition system before checking the starter circuits. If the engine should start, serious injury to persons or damage to the vehicle or equipment could occur.

Describe the vehicle and tire being worked on:

Year _____ Make _____ Model _____

VIN _____   Engine type and size _____

Procedure	Task Completed

1. Determine the following information from the service manual: ☐

    Correct battery for test vehicle

    Location of the starter motor

    Location of starter relay, if any

    Location of safety devices within the starter control circuit

    Manual or automatic transmission? _____

2. Ensure the wheels are chocked and the fuel and ignition systems are disabled.   ☐

3. Inspect the alternator drive belt(s) for tightness and serviceability.   ☐

4. Is the battery the correct one for this vehicle? If not, can tests be conducted?

    _____

5. Set the DMM to read DC voltage.   ☐

6. Connect the test leads across the battery posts. Does the battery have at least 12 volts? If not, what is your next step?

    _____

7. Place the DMM's black lead to the battery's positive and the red lead to the battery's   ☐
    positive cable terminal. Record the results.

8. Repeat step 7 on the negative post and negative cable terminal. Record the results.

    _____

9. Lift the vehicle to gain access to the input cable on the starter motor.   ☐

10. Connect the DMM's black lead to a good ground.   ☐

⚠ **WARNING** If the starter is working at all, ensure that the hands and tools are clear of any moving parts like a crankshaft-driven belt.

11. Have an assistant turn the ignition to start and probe the starter positive terminal with the red lead. Record the result. What is your next action if the voltage does not equal battery voltage?

_____

_____

12. Assuming there is no battery voltage at the starter, make a jumper wire long enough to reach from the starter relay output to under the vehicle. Connect one end of the jumper wire at the relay's output and the other end to the black lead of the DMM. ☐

13. Have an assistant hold the ignition switch in the start position while probing the starter's positive terminal with the red lead. What is the result? Is the DMM reading zero or near zero voltage? If the reading exceeds more than 0.2 volts, what is your repair recommendation?

_____

_____

14. If the reading in step 13 is below 0.2 volts, lower the vehicle to gain access to the interior and engine compartment. ☐

15. Locate the starter relay and the conductor from the ignition switch. ☐

16. Disconnect the ignition switch conductor. Set the DMM to read DC voltage. Connect the black lead to a ground and probe the conductor terminal with the red lead. ☐

17. Hold the ignition switch in the start position. Did you get a positive voltage reading? If not, set the DMM to ohms. Did you get a resistance reading of zero or near zero? Typically, the starter relay is grounded by the ignition switch. Checking for voltage first will prevent damage to a DMM that is set to the ohm scale and connected to an energized conductor.

_____

_____

18. Based on the tests results you have found so far and after consulting the instructor and service information, what do you recommend as the next test or the repair needed?

_____

Starter Motor Replacement

⚡ WARNING **Before beginning the replace of electrical components, ensure the battery's negative cable is disconnected. Damage to the vehicle or persons could occur.**

19. Disconnect the battery's negative cable and position it where it will not accidently come in contact with the negative terminal. ☐

20. Lift the vehicle to gain access to the starter motor. Remove any heat shields or other items blocking access. ☐

21. Disconnect the starter's positive input cable and any other electrical connections. ☐

22. Use the correct wrenches to remove the upper bolts holding the starter motor to the engine block/transmission housing. Leave the lowest bolt in place while removing the others. ☐

Task Completed

23. Loosen the last bolt and hold the starter motor in place as the bolt is completely removed. On heavier units, it is best to have an assistant support the starter motor as the last bolt is removed. ☐

24. Remove the starter motor. Compare the old motor to the new one. If they are the same, position the new motor into place. ☐

25. While holding the starter motor in place, finger-tighten the lowest bolt sufficient to hold the motor in place. ☐

26. Start each of the other bolts in place before tightening any of them. Once all are in place, tighten each bolt to torque specifications. ☐

27. Connect all electrical connections to the starter motor. Replace any items removed for access. ☐

28. Lower the vehicle and connect the battery negative cable. ☐

29. Check the starter system for operation. ☐

Instructor's Response

_____

_____

_____

Name _____   Date _____

INSPECT, SERVICE, AND TEST THE ALTERNATOR AND ALTERNATOR CIRCUITS

Upon completion and review of this job sheet, you should be able to test and diagnose alternator circuits using a DMM, determine any faults, and replace an alternator.

ASE Education Foundation Correlation

This job sheet addresses the following **MLR** tasks:

6.D.1 Perform charging system output test; determine necessary action. **(P-1)**

6.D.2 Inspect, adjust, or replace generator (alternator) drive belts; check pulleys and tensioners for wear; check pulley and belt alignment. **(P-1)**

6.D.3 Remove, inspect, and re-install generator (alternator). **(P-2)**

6.D.4 Perform charging circuit voltage drop tests; determine necessary action. **(P-2)**

This job sheet addresses the following **AST/MAST** tasks:

6.D.1 Perform charging system output test; determine needed action. **(P-1)**

6.D.3 Inspect, adjust, or replace generator (alternator) drive belts; check pulleys and tensioners for wear; check pulley and belt alignment. **(P-1)**

6.D.4 Remove, inspect, and re-install generator (alternator). **(P-1)**

6.D.5 Perform charging circuit voltage drop tests; determine needed action. **(P-1)**

We Support
ASE | Education Foundation

Tools and Materials

Hand tools
Floor jack with safety stands or vehicle lift
Wiring diagram
Vehicle with a standard alternator control circuits
Digital multimeter (DMM)
Fender cover
Service manual
Safety glasses

⚠ **Caution**

Do not attempt to work on the alternator when the engine is hot. Serious burns could result from touching the hot surfaces of the engine or exhaust. Allow the engine to cool before beginning charging systems diagnostics.

Describe the vehicle and tire being worked on:

Year _____ Make _____ Model _____

VIN _____   Engine type and size _____

Procedures

	Task Completed
1. Determine the following information from the service manual:	☐

Correct battery for test vehicle

Location of the alternator

Location of alternator protection devices if any

Location of safety devices within the alternator's control circuit

2. Ensure the wheels are chocked and transmission is in park (neutral for manual).   ☐

3. Inspect the alternator drive belt(s) for tightness and serviceability.   ☐

Task Completed

4. Is the battery the correct one for this vehicle? If not, can tests be conducted?

_____

5. Inspect the battery and cable for cleanliness and serviceability.   ☐

6. Inspect the alternator electrical connections for looseness and damage.   ☐

7. Set the DMM to read DC voltage.   ☐

8. Connect the black lead to the battery's positive post and the red to the positive cable terminal.   ☐

9. What is the voltage drop between the two leads? The reading should be zero or near zero ohms. If not, there is excessive resistance between the post and the terminal.   ☐

10. Repeat steps 8 and 9 on the negative post and terminal. What is the reading?

_____

11. Connect the DMM across the battery terminals, red lead to positive, black lead to negative.   ☐

12. With the DMM connected across the terminals, have an assistant start the engine and allow it to idle. What is the voltage reading? If the reading is 12 volts or less, what you recommend as the next test or repair? Consult your instructor and the service information for guidance as needed.

_____

_____

13. If the initial voltage is below 12 volts or it dropped when the electrical loads were added, prepare for the next test by shutting down the engine and access the wiring diagram.   ☐

14. On the wiring diagram, find the voltage output for the alternator and then find it physically on the back of the alternator.   ☐

15. Clear area around the alternator output terminal so the read lead of the DMM can be touched safety to the terminal of the output wire with the engine running.   ☐

16. Ensure the DMM is set to DC voltage and connect the black lead to a good ground.   ☐

17. Have the assistant start the engine and allow it to idle.   ☐

18. Touch the red lead of the DMM to the alternator's output and read the voltage. What is the voltage reading? If the alternator's output reading is more than battery voltage, what is the most probable cause of the dropped voltage? What would be your recommendation in that case?

_____

_____

19. If the alternator's output voltage and battery voltage does not show a charge without or with electrical loads, replace the alternator.   ☐

Instructor's Response

_____

_____

_____

Name _____   Date _____

USE WIRING DIAGRAMS TO LOCATE, INSPECT, TEST, AND SERVICE CIRCUIT PROTECTION DEVICES, SWITCHES, AND CONDUCTORS

Upon completion and review of this job sheet, you should be able to use the wiring diagrams to locate, test, and service circuit protection devices, relays, switches, and conductors for serviceability.

ASE Education Foundation Correlation

This job sheet addresses the following **MLR** tasks:

6.A.3 Use wiring diagrams to trace electrical/electronic circuits. **(P-1)**

6.A.9 Inspect and test fusible links, circuit breakers, and fuses; determine necessary action. **(P-1)**

6.C.5 Inspect and test switches, connectors, and wires of starter control circuits; determine necessary action. **(P-2)**

This job sheet addresses the following **AST/MAST** tasks:

6.A.7 Use wiring diagrams during the diagnosis (troubleshooting) of electrical/electronic circuit problems. **(P-1)**

6.A.9 Inspect and test fusible links, circuit breakers, and fuses; determine needed action. **(P-1)**

6.A.10 Inspect, test, repair, and/or replace components, connectors, terminals, harnesses, and wiring in electrical/electronic systems (including solder repairs); determine needed action. **(P-1)**

We Support

ASE | Education Foundation

Tools and Materials

Wiring diagram

Digital multimeter (DMM)

Test light

Service manual

Safety glasses

Describe the vehicle and tire being worked on:

Year _____ Make _____ Model _____

VIN _____ Engine type and size _____

Assortment of good and bad circuit protection devices, switches, relays, and conductors (to be used before testing is done on a vehicle)

NOTE TO INSTRUCTOR: The protection devices with a visible means of detecting the devices' serviceability should have that portion taped or painted over to prevent the student from doing a visible inspection. The short piece of conductor can be wrapped to cover any damage and should have a terminal at each end.

Procedure   **Task Completed**

1. Determine the following information from the service manual:

    Location of the protection devices for the following circuits: headlight, stop light, turn
    signals, air conditioning, blower motor, cooling fan motors, fuel pump, horn. Record
    the location and amperage for each device.

    _____

    Location of the brake light switch, clutch switch (if equipped), power steering pressure
    switch, air-conditioning low-pressure switch. Record their location.

    _____

2. Begin the tests off-vehicle using the instructor's supplied devices.   ☐

3. Set the DMM to ohms scale and test each device for serviceability. Record your results
    for each device measured. Determine if the device is usable.

    Fuse 1 _____   Fuse 2 _____

    Fuse 3 _____   Fuse 4 _____

    Circuit breaker 1 _____   Circuit breaker 2 _____

    Circuit breaker 3 _____   Circuit breaker 4 _____

    Relay 1 _____   Relay 2 _____

    Switch 1 _____   Switch 2 _____

    Fusible 1 _____

    Conductor _____

4. Before testing the on-vehicle devices, ensure that none are attached directly to an
    electronic circuit. If one is in an electronic circuit, which tester should be used?

    _____

5. Locate the fuse panel housing the fuses and circuit breakers.   ☐

6. Remove the fuse housing panel.   ☐

7. Locate and test the following protection devices in-place on the vehicle using a test   ☐
    light.

    Headlight, Air conditioning, Stop light, Blower motor, Horn, Cooling fan motors, Turn
    signals, Fuel pump

8. What is the specified amperage for each?

    Headlight _____   Air conditioning _____

    Stop light _____   Blower motor _____

    Horn _____   Cooling fan motors _____

    Turn signals _____   Fuel pump _____

9. Check to see if the correct protection devices are in the correct slot. If not, replace   ☐
    with the correct device.

10. Connect the test light clip to a good ground. Touch to each test point for each device. Record the results as good or bad. Good is indicated when the lamp lights at both test points on any single device. If a device is bad, which electrical defect is indicated?

Headlight _____   Air conditioning _____

Stop light _____   Blower motor _____

Horn _____   Cooling fan motors _____

Turn signals _____   Fuel pump _____

11. Use the wiring diagram to locate each of the following components: Air conditioning and horn: relays, switches, easy to access conductor terminal and/or splices. ☐

NOTE: The following tests requires the use of a DMM unless the service manual states otherwise.

12. Locate the relay and low-pressure switch for the air conditioning system. ☐

13. Remove the relay from its mounting and determine which terminals are for the interior coil. Many times, the interior schematic is printed on the outside of the relay case. ☐

14. Set the DMM to read ohms. ☐

15. Connect one lead to each coil terminal. What is the resistance reading?

_____

16. Move the DMM leads to the relay's other terminals. One will be battery (B1) power in and the other battery out. Measure the resistance between the two. Is the relay serviceable?

_____

17. Use the relay schematic and/or the wiring diagram to locate the battery (B1) cavity in the fuse housing. Double check to make sure the right cavity is selected. ☐

18. Set the DMM to read DC voltage. Connect the DMM black lead to a good ground. ☐

19. Probe the B cavity and record the reading. Record the reading. Is it within 1 volt of battery voltage? It may be necessary to do an open voltage measurement of the battery to make the comparison.

_____

20. Locate the low-pressure switch for the air conditioning and unplug the electrical connection. If this AC system has a combined high/low pressure switch, do not test. Instead use the stop light switch under the dash. The same technique applies. ☐

21. Set the DMM to read ohms. ☐

22. The terminals on the switch are usually small pins and difficult to get an accurate connection with the DMM leads. It may be necessary to make a short pigtail to plug into the switch with bare conductor ends so a good connection between the DMM and AC switch can be achieved. If using the stop light switch, a pigtail is probably not required. ☐

Task Completed

23. Connect the DMM leads, one to each switch terminal. What is the reading? Is the switch open or closed? According to the service information, what should it be now?

_____

24. Locate and disconnect the electrical connection at the AC compressor.   ☐

25. With the DMM set to read resistance, connect the leads, one to each of the compressor terminals. It may be necessary to use a pigtail again. Record the reading. Is the circuit open or closed?

_____

⚠ **WARNING** **Ensure that the ignition key is OFF, and the blower motor and AC switches are OFF. The DMM can be damaged if a powered circuit is connected to the leads with the DMM in ohms mode.**

26. Move one of the DMM leads to a good ground. Probe the other lead to each terminal of the vehicle's AC compressor harness connection. Which terminal shows a resistance? What this mean?

_____

_____

27. Reconnect the AC electrical connection.   ☐

Instructor's Response

_____

_____

_____

Name _____   Date _____

INSPECT, SERVICE, AND TEST THE BATTERY

Upon completion and review of this job sheet, you should be able to test and diagnose problems related to the battery.

ASE Education Foundation Correlation

This job sheet addresses the following **MLR** tasks:

6.B.1   Perform battery state-of-charge test; determine necessary action. **(P-1)**

6.B.2   Confirm proper battery capacity for vehicle application; perform battery capacity test; determine necessary action. **(P-1)**

6.B.4   Inspect and clean battery; fill battery cells; check battery cables, connectors, clamps, and hold-downs. **(P-1)**

This job sheet addresses the following **MLR/AST/MAST** tasks:

6.B.3   Maintain or restore electronic memory functions. **(P-1)**

6.B.4   Inspect and clean battery; fill battery cells; check battery cables, connectors, clamps, and hold-downs. **(P-1)**

This job sheet addresses the following **AST/MAST** tasks:

6.B.1   Perform battery state-of-charge test; determine needed action. **(P-1)**

6.B.2   Confirm proper battery capacity for vehicle application; perform battery capacity test; determine needed action. **(P-1)**

We Support
ASE | **Education Foundation**

Tools and Materials

Hand tools
VAT-40/60 or other battery load tester
Digital multimeter (DMM)
Service information
Fender cover
Service manual
Safety glasses

Describe the vehicle and tire being worked on:

Year _____ Make _____ Model _____

VIN _____   Engine type and size _____

Procedure   **Task Completed**

1. Determine the following information from the service manual.   ☐

2. What is the correct battery for test vehicle _____?

3. Location of the battery _____

4. Ensure the wheels are chocked and transmission is in park (neutral for manual).   ☐

5. Is the battery the correct one for this vehicle? If not, can tests be conducted?

_____

6. Begin the inspection of the battery and its related components. Record your results. Before proceeding, are there any odors of possible gases resulting from a discharged or overly charged battery? If yes, stop and consult your instructor.

_____

7. Battery cleanliness

_____

8. Battery cable terminals status

_____

9. Visible portions of the battery cables

_____

10. Alternator drive belts

_____

11. General condition of the engine/engine compartment

_____

12. If not a maintenance-free battery, remove caps and check fluid levels.   ☐

13. Make a quick check of the battery's voltage by connecting the DMM across the battery terminal, red to positive, black to negative. Who is the voltage? Is it sufficient to properly test the battery?

_____

14. Is the battery voltage sufficient to continue the testing? If not, stop and consult your instructor. ☐ Yes ☐ No

15. Measure voltage between the battery posts and their corresponding cable terminals. Is there excessive voltage drop between the posts and the terminals? If so, repair as needed. ☐ Yes ☐ No

16. Review the operating instructions for the battery tester. Connect the battery tester to   ☐
the battery as specified.

⚠ WARNING **Do not put a high current load on the battery longer than that specified by the tester's instructors. Serious damage to the battery or tester could occur.**

⚠ WARNING **If this is a very new model vehicle with many electronic systems, consult the service manual for the proper test procedures. It may be necessary to disconnect the battery from the vehicle wiring before conducting a capacity or conductive test.**

17. Follow the battery's tester instruction, apply the test load to the battery and observe the results. ☐ Yes ☐ No

Task Completed

18. What does this capacity test indicate of this battery?

_____

19. Prepare the conductive tester (may require a different tester) to perform a battery   ☐
    conductive test.

20. Connect the conductive tester to the battery as specified by the tester instructions.   ☐

21. Perform the conductive test on the battery following the tester's instructions. Record
    the results.

_____

22. Based on the test results and your inspection, is the battery serviceable enough for
    vehicle operation. What factors would cause you to recommend a new battery?

_____

23. Disconnect and store the tester(s). Prepare the vehicle for the customer.   ☐

Instructor's Response

_____

_____

_____

CHAPTER 6

ACCESSORY LIGHTING AND ELECTRICAL SYSTEMS INSPECTION, MAINTENANCE, AND TESTING

Upon completion and review of this chapter, you should understand and be able to:

- Perform the proper method to disarm the airbag system.
- Inspect and repair the warning light systems.
- Conduct a performance test of dash-mounted instrument circuits.
- Diagnose, inspect, and repair or replace the interior lighting.
- Diagnose and inspect the power windows and door locks.
- Perform a function test of the antitheft system.

Terms To Know

Instrument voltage regulator (IVR)
KOEO
KOER

Lamp check
Memory saver
Panic

Pigtail
Resistor block
Thermal limiter

INTRODUCTION

Most passenger comfort and convenience features will fail like any other component on the vehicle. Most of the failures will be electrical in nature, either a failure of the wiring or the electrical components. Because many of the accessories on today's vehicles require skills gained from many hours of experience and training, this chapter will cover some of the common tests and repair that could be performed by entry-level technicians.

DISARMING AN AIRBAG

Many of the repairs to the interior of the vehicle present a danger to the technician from the vehicle's supplemental restraint system (SRS). Before attempting any repairs in or around the steering column, the SRS should be disconnected from the battery.

⚡ **WARNING Do not use a memory saver when the battery is disconnected for interior service. Even a small 9-volt battery is sufficient to charge the airbag capacitor. Severe injury could result if the airbag deploys while a person is leaning over or is near the airbag mounting site.**

Confirm the customer's complaint that indicates a repair is needed within the passenger compartment. If repairs or testing are deemed necessary in that area, disarm the airbags by first disconnecting the NEGATIVE cable from the battery. Consult the service manual for the required waiting period needed for the capacitor to naturally discharge. Typical waiting time is at least 10 minutes, but some manufacturers recommend or require a longer time. Always consult the service manual before assuming that the capacitor is discharged.

Classroom Manual
page 182

A **memory saver** is a small battery-powered device used to save the vehicle's various memories, such as radio settings.

In many instances, it will be necessary to have electric power to test components located in or near the airbag areas. There are two ways to do this safely:

1. Use an ohmmeter (a digital multimeter or DMM) to measure resistance of the component and compare it to manufacturer specifications after it has been disarmed and removed from the vehicle.
2. Remove the component(s) from the vehicle and test it at a test bench equipped with the proper power supply and tools.

 Special Tools

Service manual
Vehicle's diagnostic routine

⚡ WARNING **Do not reconnect the battery again until all the service work is completed, including reassembly of components and their mounting. A quick touch of the cable to the battery will instantly recharge the capacitor. This would supply the power needed to fire the airbag, and severe damage and injury could result.**

WARNING LIGHTS

Many of the warning lights are simple, electrical circuits. A good understanding of their operation will give the technician a head start on basic repairs. Most tests must be run when the key is on with engine off **(KOEO)** or key on with the engine running **(KOER).** You may have to turn the key off and back on several times to read and understand all the lights. A light that goes on for only a few seconds and then goes off does so because of the program designed to work with it. This is known as a **bulb check**. Bulb check is a program in which the bulbs will illuminate for only a few seconds. This is done so that the technician (or driver) can check if the bulb is working or not. If the oil pressure warning light bulb is burnt out, then the driver would never know when or if there is a problem with the oil pressure.

Bulb check is a program where certain dash indicator lights will illuminate for only a few seconds so that the technician can check their operation.

KOEO means ignition key on, engine off, a test condition required to performed certain tests.

KOER means ignition key on, engine running, another test condition required for certain tests.

⚠ **Caution**

The following tests are for nonelectronic circuits only. Sensor circuits can be determined by observing the number of wires attached to the sensor or sending unit. True sending units will have only one wire, whereas sensors will have two or more wires. Grounding a sensor circuit could damage the computer or the circuit.

KOEO and **KOER** running are two test conditions associated with powertrain control module (PCM) test routines, but they appear in other system test procedures as well.

Older vehicles use a mechanical switch to turn the light on when certain conditions are present. Other vehicles use a sending unit for the warning light and gauge. One basic function occurs on both types: the ground to the warning light is completed.

⚡ WARNING **Do not grab hoses or other components under the hood if an overheated engine is suspected. Even an engine operating at normal temperature can cause serious burns.**

⚡ WARNING **Never remove the radiator cap on a warm or hot engine. Always let the engine cool down. Extremely serious burns and other injuries can occur when removing a hot radiator cap.**

If the red engine light is lit, the problem indicated is either an overheated engine or low oil pressure. On some vehicles, there is a light for oil and one for overheating (**Figure 6-1**). Before starting work on the electrical circuit, do the obvious and check the level and condition of the coolant and oil. For a cooling problem, look for missing or damaged fan belts. The coolant level on a hot engine can be checked in the overflow reservoir. The reservoir may not contain a true representation of the coolant level, but it will be an indication. Note that, on some cooling systems, the reservoir may become pressurized. Notice the condition of the engine by checking for oil or coolant leaks. The hoses should be inspected for swelling. Do not touch hot hoses. In many cases of low oil pressure or overheating, the condition will be found during the inspection.

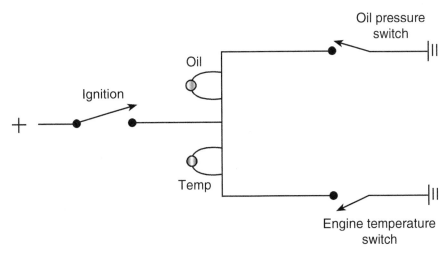

Figure 6-1 Two warning lights require two completely different circuits powered through the ignition switch.

⚡ **WARNING** Do not perform a KOER test if there are any visible or audible signs that the engine is damaged or could possibly suffer damage when running. Doing so could cause more damage.

⚡ **WARNING** Never apply 12-volt direct current (DC) to a sending unit unless directed to do so by manufacturers' test routines. Sending units are almost always on the negative side, may be electronic, and will be damaged by direct 12-volt (DC), positive current.

Assuming that the problem was not apparent under the hood during the visual inspection and the engine operates normally, run the engine just long enough to confirm the customer's complaint and determine if there is more than one warning light for the engine. Before starting the engine, switch the ignition to run and observe the lights. There should be several red lights and possibly an amber light. If all the red warning lights are lit, the circuit to each light is good.

Do not run the engine any longer than it takes to check the lights. If the warning light stays on during engine idle, speed up the engine to see if the light goes out. As the engine speed increases, the oil pump produces more flow and pressure. If the light goes out, the likely problem is in the oil system. In this case, the oil pump and engine could be worn, or sludge is developing in the oil pickup or passageways. Generally, an overheated engine will not cool down enough to turn off the light by simply speeding up the engine. Note the matching gauges, if equipped. When a red light turns on but the gauge reads normal, the problem is usually in the light circuit.

For a vehicle with a shared light for oil pressure and temperature warning, shut down the engine and locate the oil pressure and temperature sending units (**Figure 6-2**). Perform a KOER test by disconnecting the temperature unit. It usually is the easiest unit to reach

⚙ **SERVICE TIP** Before starting the engine, check the repair order or with the customer to determine if there were unusual noises coming from the engine prior to shut down. Noise is usually a sign of internal damage. Further operation of the engine may not give the technician additional information for diagnosis, but it will probably cause even more damage.

A bulb test or proof-out test of the warning light is done every time the ignition key is switched to ON. Other warning lights are tested while the key is in the START position.

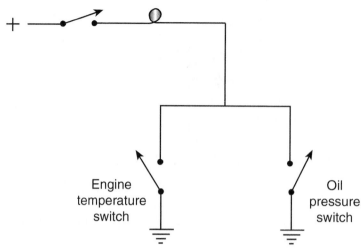

Figure 6-2 One warning light for two systems requires parallel circuits from each sending unit that are powered from the ignition switch.

because it is on or near the thermostat housing in most vehicles. Start the engine. If the light goes off, the problem is either overheating or circuit trouble. A light that stays on may indicate an oil pressure or circuit problem. Repeat the test with the oil pressure sending unit. With the engine running and both circuits unplugged, the red light should go out. If it does not, there is a problem within the circuit. With the light out, connect one of the two sending units. Now, the appearance of a red light indicates a failure in that system. Lack of a red light indicates that the problem is in the disconnected system.

If there are separate lights for temperature and oil, disconnect the sending unit for the affected circuit and test that circuit only (**Figure 6-3**).

On a dual light system, check the electrical circuit from the sending unit to the light by disconnecting the wire at the sending unit. Run a KOEO test. Use a jumper to ground the sending unit's wire to a metal part with ignition switch in ON (**Figure 6-4**). Each time the wire is grounded, the warning light should come on. If the light and circuit function properly, the sending unit is bad. On a temperature warning system, let the engine cool down, if necessary, and replace the sending unit. Do not remove the temperature sending unit on a hot engine. If the lubrication system is being checked, it is possible to manually check the oil pressure before replacing the sending unit. This would also be a direct test of the oil pump output.

Special Tools

Service manual
Jumper wire

Figure 6-3 The oil pressure sending unit is usually different in appearance and operation from the temperature sending unit or switch.

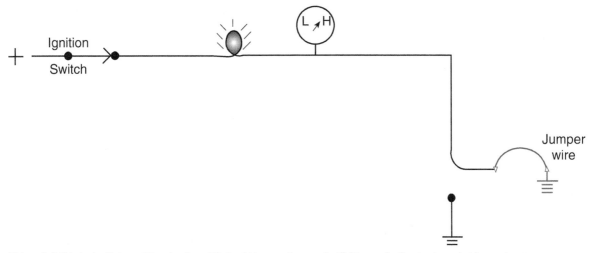

Figure 6-4 This test will ground the circuit, and it should turn on the warning light or make the gauge read at its maximum.

⚡ WARNING **Do not start the engine if the oil level is low or the engine is making unusual noises. Severe damage could result if there is not enough lubrication on the bearings and journals.**

TEST OIL PRESSURE WARNING CIRCUIT

Remove the oil pressure sending unit and install a manual oil pressure gauge (**Figure 6-5**). Start the engine and record the oil pressure. Consult the service manual for exact specifications. Generally, pressures over 20 psi at idle are acceptable. Shut down the engine immediately if oil pressure does not reach 20 psi within a few seconds of engine start-up. If the pressure is at least 20 psi, speed up the engine. There should be an increase in pressure. If the pressure is low or does not increase, the probable cause is a worn oil pump or

Figure 6-5 The gauge for this test is a dash-mounted aftermarket gauge fitted with a long tube or hose for attachment to the engine's block.

Special Tools

Manual oil pressure
gauge

Caution

Electric cooling fans
may turn on at any
time when the
temperature sending
unit is connected.
Observe safety
procedures by
remaining clear of an
electrical cooling fan
during testing.

Caution

Some electric cooling
fans with a
temperature switch
(not to light or
gauge) will turn on
when the switch is
disconnected with
the ignition key on.
Observe safety
procedures by
remaining clear of an
electrical cooling fan
during testing.

a worn engine. However, each of these conditions occurs over a period of time, depending on the condition of the engine. A discussion with the operator can verify this. A broken oil pump will suddenly turn on the light, cause almost immediate engine failure, and the condition will be obvious to the technician.

Test Engine Temperature Warning Circuit

If the temperature light or gauge circuit proves correct during the test, the engine's cooling system must be checked. That includes the fan, water pump, thermostat, and coolant condition and level. A detailed diagnosis routine will be set forth in the service manual. This is particularly true of computer-controlled engines. Many computer-controlled systems use the PCM to turn on the warning lights based on input signals from its sensors. Diagnosing computer systems is best left to experienced technicians, but check the basics such as belts, oil and coolant level, leaks, or anything else that may point to a mechanical failure. If the top of the engine shows signs of recent repairs, check to make sure all connectors are, in fact, properly connected and tight.

Test Alternator Warning Circuit

Low voltage or current in the charging circuit will switch on the alternator warning lights. A loose alternator drive belt may slip and not allow the alternator to produce the required voltage or current. Always check the belt and all battery and alternator connections before testing the charging system or the battery. If the inspection does not indicate a mechanical problem, the battery and charging system can be checked using a VAT-40 or VAT-60 as outlined in Chapter 5. Do not use a VAT-40 on a computer-controlled alternator.

The voltage or current sensor is usually in the regulator and, depending on the system, the alternator may have to be replaced, rebuilt, or an external regulator may need to be replaced (**Figure 6-6**).

> **SERVICE TIP** If the alternator has been replaced, it is possible that the wrong alternator pulley has been installed. An oversized pulley will slow down the speed of the alternator, thereby reducing its output. It will be most notable at idle speeds.

Figure 6-6 The voltage regulator on some model vehicles is internal to the generator.

Other warning light systems can be checked in the same manner. Before condemning the electrical part, always check the mechanical portion of the system. Most of the time, the warning system is doing what it was designed to do. The most common failure in a warning system is the switch or sending unit. Disconnecting the switch or sending unit should turn off the light.

⚡ **WARNING** **Do not check electronic circuits using a test light, grounding a conductor, or jumping a connection unless specifically directed to do so by the service manual. Incorrect test equipment and procedures can damage electronic circuits and components.**

GAUGES

Before proceeding, determine the type of gauges used on the vehicle. If the instrumentation is electronic and operated by a computer module, **DO NOT** use the test procedures suggested here. Most sending units with one wire are not considered to be electronic whereas two or more wires are generally part of an electronic circuit in some manner. Also consider the customer's complaint. If only one gauge is not working, then the problem is in that system. If all gauges are not working, it is probable that the fault lies in some shared component such as a fuse or common conductor. Always consult the service manual to determine the system's operation and the basic diagnostic test procedures. The tests discussed here are very similar to or the same as the ones for warning lights. Some vehicles may use separate circuits for each but share a common sending unit.

Sending units for analog gauges use a material that changes resistance based on the amount of heat or pressure being applied, or they use some variable resistor (**Figure 6-7**). Both types can fail at OFF, ON, or any point in between. If there is a one-wire connection at the sensor, grounding the wire in the KOEO mode will cause a maximum gauge reading. A disconnected wire will allow the gauge to drop to its lowest limit. If the gauge works in this test, then the sending unit is bad. A gauge that does not move when the wire is disconnected or grounded (KOEO) indicates a bad circuit or gauge. Consult the service manual for test procedures for the circuit and gauge.

Variable resistor sending units are used for fuel gauges. A float in the tank moves a wiper arm over a very thin resistor wire (**Figure 6-8**). The position of the arm on the wire determines the amount of resistance between the gauge mechanism and ground. The test for this circuit is like the one for one-wire sending units on the engine. Lift the vehicle and locate the connection for the sending unit. It will be near or on the tank. On older vehicles without in-tank electric fuel pumps, there is usually one wire leading to the tank. On systems with in-tank pumps, there may be several electrical connections. Before disconnecting any of them, consult the service manual. Once the sending unit wire is located and disconnected,

Figure 6-7 A cutaway view of a typical temperature warning light switch.

Figure 6-8 The wiper arm sliding across the resistor wire changes the resistance between the signal wire and ground.

Special Tools
Service manual
Jumper wire

perform a KOEO test. Turn the ignition key on while the wire is grounded. A good circuit will cause the gauge to read full. Removing the wire from ground should cause the gauge to drop to empty. If either condition is not met, then the circuit or gauge is bad. An operating circuit means the fuel tank must be dropped and a new sending unit installed.

> ⚙ **SERVICE TIP** Like in all electric circuits, when nothing works, start with the most common point if none of the circuits work. Common points for gauges and warning lights are fuses, switches, integrated circuits, and the instrument voltage regulator. Also understand that the fuse for the instrument panel light is not the one for the warning lights and gauges.

Here is a word of caution before leaving this section. Most instrument panels use some type of voltage limiter to regulate the amount of voltage being supplied to the various gauges and sometimes to the entire gauge circuits (**Figure 6-9**). It is normally referred to

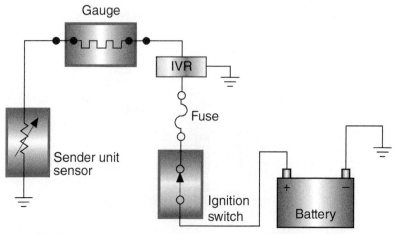

Figure 6-9 Most IVRs can be individually tested and replaced from the instrument panel or any gauge or light circuit.

as an **instrument voltage regulator (IVR)**. Jumping 12 volts (DC) to a gauge or circuit may damage them and cover up the original problem. The tests listed here are simple and cover most initial testing of analog, nonelectronic gauges and circuits. If the test indicates that the circuit or gauge is inoperative, the test procedures to be followed from that point are found in the service manual. Testing of the actual instrument panel components should always be done according to the manufacturer's procedures and performed or supervised by an experienced technician.

An **instrument voltage regulator (IVR)** limits the voltage supplied to the gauges to 12 volt (DC) so the gauges are more accurate.

LIGHTING SYSTEMS

Automotive lighting systems are very simple electrical systems. They have a positive side and a negative side. Almost all systems are series-parallel circuits with the controls and protection devices located in the positive or hot side (**Figure 6-10**). After the last control, the circuit parallels out to two or more lamps. The initial diagnosis is easy. If one lamp on a system does not light but the others do, the circuit, controls, and protection devices are operating properly. The usual problem would be a burned-out lamp. However, there are times when all the lamps on a system go out, indicating a control or

Classroom Manual
pages 167–176

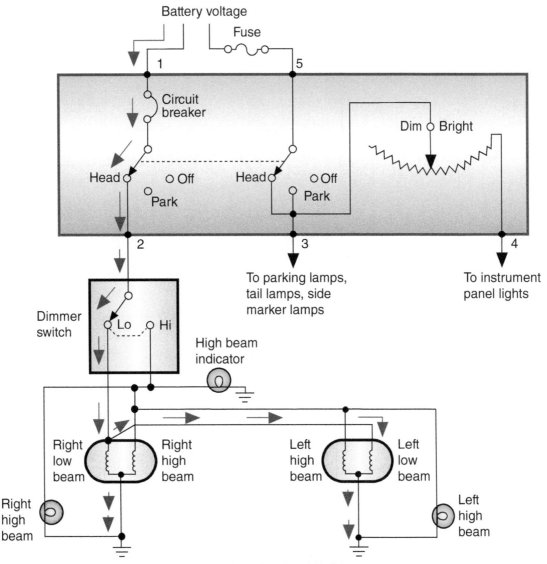

Figure 6-10 Note that the protection devices are on the positive side of this light system.

protection device failure. It is possible, but not probable, that all the lamps on a system burn out at once. Even then the root cause is probably somewhere else in the circuit, such as a protection device not opening when the circuit is overloaded with current or there is a circuit defect. Diagnosing the headlight and marking light circuits and the turn signals will be discussed, covering the most probable causes of complete system failure.

Headlights and Marking Lights

Each of the two systems is turned on by different contacts within the same switch (**Figure 6-11**). This multiposition switch may be found on the dash or steering column and has OFF, PARK (marker), and headlight positions. The switch is normally connected to one or more relays on the grounded side of the circuit. The relays are electrical/mechanical or electronic switches that connect the battery to the light when the switch is on. A fuse protects power to the marking side while a circuit breaker is normally used on the headlight feed. With all of the marking lights (or daytime running lights) and headlights inoperative, the most probable cause is a bad headlight switch or total loss of power from the battery. However, always check the fuse and circuit breaker before proceeding further.

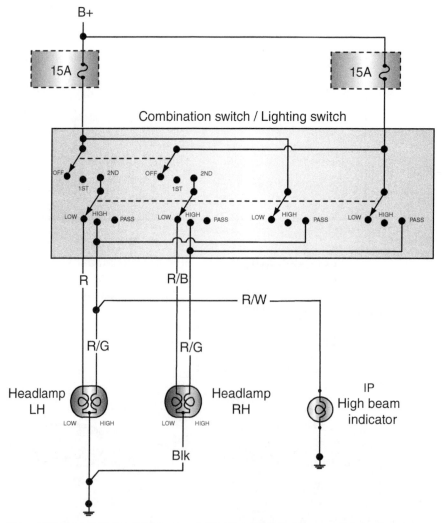

Figure 6-11 The headlight switch has several different contacts that feed parallel circuits.

Special Tools

Digital Multimeter or voltmeter (DMM)
Service manual
Jumper wire

SERVICE TIP A test light could be used in the following tests, but the amount of voltage will not be registered. It is best to use a multimeter when testing for voltage.

Remove the switch or disconnect the switch from the harness, whichever is the easiest. If the headlight switch is mounted on the steering column or near the steering wheel, ensure the **airbag is disarmed**. Locate the two heavy conductors feeding (input power) the switch. They are usually red or orange and will be 10- or 12-gauge wires (**Figure 6-12**). If the wires are much smaller, then the switch is probably controlling the headlight relay and not the headlights directly. For the purpose of the discussion, we will assume there is no relay. The two feed conductors should register within 1 volt of battery voltage. On some vehicles, it is necessary for the ignition switch to be in the RUN position before the marker lights or headlights or both will have electrical power. If necessary, turn the key to RUN (KOEO) and retest the connections for voltage. This confirms the presence

Disabled airbag To remove the power source from the airbag system preventing an unwanted deployment.

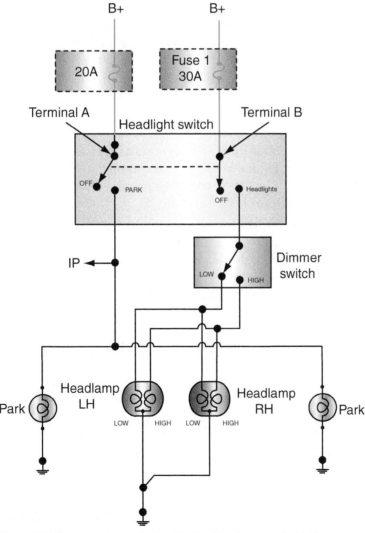

Figure 6-12 The red or orange wire from the fuse(s) is the power feed to the headlight switch. There should be battery voltage (within 1 volt) available at terminals A and B in this illustration.

⚠ **Caution**

Many new vehicles actually use computer modules that turn on lights upon receiving a ground (negative) signal from a specific switch. Do not use a standard test light on this type of circuit. A DMM or LED test light is the only acceptable tool to use when checking an electronic circuit. But remember that an LED test light will show the presence of voltage but not the amount. Damage to the computer and circuit could result from using incorrect test procedures or tools.

or absence of voltage at this connection. If neither conductor shows the presence of voltage or has low voltage and the circuit protection devices are good, then the problem is between the **switch** and battery. Remember that a weak or dead battery will cause problems with the other vehicle circuits. If only one conductor registers a voltage, then a problem exists in the other. In many cases, failure to investigate thoroughly may not reveal the presence of a relay or other electrical device that may be controlling or protecting the dead circuit. Check the circuit layout carefully using the wiring diagram(s) and component locator. If there are none installed, it is obviously a break between the battery and the switch.

In most cases where only one circuit controlled by a gang switch is inoperative, the switch has failed in one area or one set of contacts. However, the first thing to check is the protection device for the failed circuit. Do not just replace the protection device and return the vehicle to the customer. The device failed for a reason. It could be old age, a common reason on older vehicles, but the circuit should always be checked for damage or excessive load. Replacing the device without checking the circuit may result in customer comeback and loss of profit and time.

⚙ **SERVICE TIP** Most switches located in the passenger compartment are switches on the ground side of the circuit. Many times, you will not find a "positive" feed at the switch connection. When studying the wiring diagram, determine the position of the switch in the circuit with regard to the fuse, load, and grounding connection. This will give an indication of whether positive voltage will be present at the switch.

Many headlight systems have a relay that routes a large current flow directly from the battery to the headlamps (**Figure 6-13**). The relay is operated by a set of contacts in the headlight switch. If the relay, normally under the hood, is easier to access, it may be better and quicker to check it before checking the headlight switch. A typical simple relay has four terminals:

- One for battery feed to the relay's coil
- One for battery input feed to the lamps
- One for output feed to the lamps
- One for ground feed from the headlight switch

Figure 6-13 A typical 4- to 5-terminal relay. This is not considered to be an electronic relay.

One of the most common failures in any electrical circuit is a loose ground or a completely missing ground. Before testing the relay, use the wiring diagram to determine which terminal is positive and which is negative. For this exercise, the headlight switch provides the ground for the relay's coil. Once the terminals and conductors have been positively identified, a quick test with a jumper wire can be performed.

Locate the terminal that is connected to the headlight switch. Connect a fused jumper wire from the terminal to a good ground (Figure 6-13). If the system works, the problem is either in the switch or in the conductors between the switch and relay. Because the system(s) worked in this test, it proves the relay coil is working. The next "easier" thing to check will probably be the switch.

Assuming that the system(s) did not work when the jumper wire was installed, more extensive testing must be performed. Remove the relay from its connection to expose the connection's internal terminals. Use a DMM to perform the following steps in sequence:

Step 1. Select DC voltage on the DMM. Connect the DMM's black lead to a good ground and probe the two power terminals (#3 and #2) with the red lead (**Figure 6-15**).

Figure 6-14 Back-probing a jumper wire to the #4 terminal creates a good ground for the relay coil. If the lights function, then the headlight switch is probably bad.

Figure 6-15 Remove the relay for the tests. Use a voltmeter or a DMM set on DC volts to measure power at terminals #2 and #3. It should be within 1 volt of battery voltage.

The meter should register within 1 volt of battery voltage. If not, repair the circuit to the battery. If voltage is present on both terminals, go to **step 2**.

Step 2. Select resistance on the DMM and turn on the headlight switch. Leave the black lead at ground and move the red lead (**Figure 6-16**) to the terminal connected to the headlight switch. There should be continuity between the terminal and the headlight switch with little or no resistance. If not, repair the circuit or switch. If continuity is correct, go to **step 3**.

Special Tools

Jumper wire
Digital Multimeter (DMM) or voltmeter (DVOM)

⚙ SERVICE TIP A common symptom of a loose ground connection is apparently impossible actions with the various circuits on that ground. A customer may complain that one day this works, the next day that one doesn't but this one does, sometimes nothing works, and sometimes everything works perfectly fine. Look for a loose ground common to all the strange-acting circuits or several ground connections in the same area of the vehicle. Failure to recognize the symptoms of a loose ground can be frustrating.

⚙ SERVICE TIP Many relays have a small diagram of their internal wiring printed on the outside of the relay. Use the diagram to determine exactly which terminal connects to the relay coil and internal switch. On relays with more than five terminals, it will be necessary to apply extra diligence to the wiring diagrams before testing.

Step 3. Turn off the headlight switch and move the red lead to the output terminal of the headlights (**Figure 6-17**). The meter should register continuity and a resistance reading dependent on the type and size of the lamps. If an open circuit is indicated or the resistance is very low, then the problem exists either in the circuit or the headlamps. Note that a positive result on each step thus far indicates the relay is probably at fault, but **step 4** will confirm the suggested diagnosis.

Figure 6-16 Test for continuity between terminal #2 and ground with the headlights switch ON. There should be continuity and little resistance.

Figure 6-17 Test for continuity between terminal #1 and ground. There should be continuity and a resistance value dependent on the size of the lamps.

Jumper wire

Relay removed

Figure 6-18 Placing a fused jumper wire between terminals #1 and #3 should activate the headlights. If not, there was an error at the initial diagnosis or during the test procedures.

Step 4. By this time, the problem should be isolated, but one final test can be performed if desired. Locate the headlight output terminal (#1) and the power input terminal (#2). Install a fused jumper wire between the two terminals (**Figure 6-18**). The lights should illuminate. If not, there is a problem between the battery and lights.

The previous tests seem to be more difficult than the ones discussed earlier. However, many times the most difficult part of electrical testing on a vehicle is locating and accessing the conductor or the component being tested. Experience, wiring diagrams, and electrical component locator manuals will show the technician how to conduct the same tests at different and easier-to-access points in the circuit (**Figure 6-19**). The preceding tests on the headlight circuit can be used to test almost any electrical, but not electronic, system on the vehicle by understanding the circuit's operation and tracking the circuit and components back to the voltage source.

> **SERVICE TIP** The power terminals on most four-terminal relays are #30 and #86 with #85 and #87 as the output and ground terminals, respectively. In the illustration, #2 and #3 are the power with #1 and #4 as outputs.

Fuel tank gauge unit

Thermal transmitter

ECM (ECCS control module)

Vehicle speed sensor (VSS)

Figure 6-19 Component locators show wiring routing and components within a circuit. This can help make it easier to find accessible test points.

SERVICE TIP If the fuse in the jumper wire opens (blows), there is a short-to-ground somewhere in the circuit being tested. Some probable locations for short-to-ground in the light circuits are under the battery tray, in and around moving components, at the connections, or at the place where the harness is routed through the bulkhead.

Two tools can assist the technician in testing electrical circuits. One is known as a short finder. It is connected to the circuit and a power source. The short finder indicates the presence of a magnetic field. When electric current flows through a conductor, a magnetic field is established around that conductor. A needle on the face will move from side to side as the finder is moved over a conductor. When the needle stops moving, there is no current in the conductor at that point.

SERVICE TIP A quick check used by many experienced technicians is to locate a working relay exactly like the one under testing. Use the known good relay to replace the "bad" one and switch on the circuit. If it works now, then the relay is bad. However, don't use this method until you are absolutely certain of which relays can be used for this swap test.

A pigtail is a replacement connector with a short length of the wiring for connection to the vehicle wiring.

Special Tools

Jumper wire
Digital multimeter (DMM) or voltmeter (DVOM)

A second tool is designed for testing relays. The tool has a female plug for connection to the suspected relay. A short **pigtail** is plugged into the connection where the relay is usually inserted. Operating the tester and observing the tester's meter and the relay operation can pinpoint the electrical or mechanical problem.

Turn Signals

The turn signal circuits work like most other lighting systems except that they use a fast-acting circuit breaker called a flasher to blink the lamps. Testing of this system is done in a manner similar to other lighting systems.

When the turn signal switch is moved to either the left or right turn position, current flows from a fuse through the flasher and switch to a front and rear lamp. When one lamp burns out, the flasher will do one of three things: blink the other lamp faster, blink the other lamp slower, or turn on the other lamp continuously. Replacing the nonoperative lamp will usually repair the system.

A second problem may be indicated by signals working on one side that do not turn on or do not blink on the other side. Because of the design of the signal switch, one side of the turn signal system may work well, but the other side may not receive any power (**Figure 6-20**). Another condition that could cause only one side to work is the installation of the wrong lamp or excessive or extremely low resistance in the nonworking side. Resistance determines the amount of current needed for operation, and the proper amount of current is needed to make the flasher open and close.

The easiest method of checking a turn signal with one inoperative side is to try the hazard warning system. If all the lights work, the circuits are good, and the turn signal switch is probably bad. Sometimes, replacing the turn signal flasher with a new one will make the system work, but the flasher will not be the cause of the failure. A new flasher will temporarily make the circuit a little better and allow the current to flow. Corrosion or looseness of the switch contacts would be the most likely cause, and the switch must be replaced.

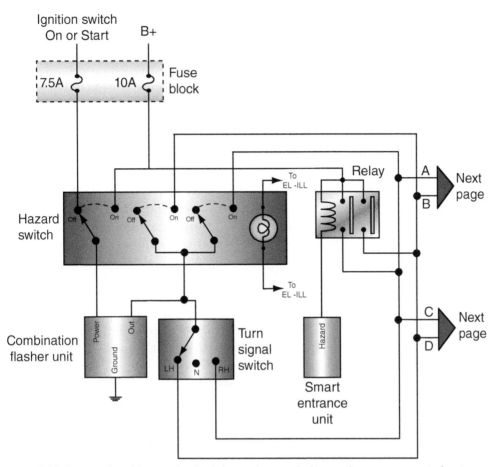

Figure 6-20 The operation of the turn signal switch sometimes results in excessive wear on one set of contacts, causing failure of that circuit.

Figure 6-21 A trailer's electrical connection can be easily damaged if it is left to hang loose, thereby causing shorts and grounds within the circuits.

When the turn signals, marking lights, or brakes work incorrectly, one thing to check is the addition of a trailer electrical connector on the vehicle (**Figure 6-21**). When additional lamps are added to the vehicle's lighting system, a heavy-duty flasher

Classroom Manual
pages 172–174

> **⚙ SERVICE TIP** The most common cause of brake, turn, and marking light failure on a trailer is its grounding wire. If all trailer lights blink or become very bright when the turn signal or brake is used, suspect the trailer's master ground. Usually, it is a white wire connected to the trailer frame near the hitch. The problem may also affect some or all of the towing vehicle lights.

is required. Also, the connection between the trailer harness and the vehicle wiring could be faulty, thereby adding resistance to the circuits. A proper T-connection is shown in **Figure 6-22**.

Newer trucks and some passenger vehicles may have a trailer kit installed by the dealer or other vendor. This kit has the usual mechanical trailer hitch, but the electrical connection may be different. The vehicle side of the trailer connection can be round and have more cavities than the old four-wire connector commonly used (**Figure 6-23**). The additional cavities are most often used for electrical brakes on the trailer or to supply a 12-volt power supply to some light-duty equipment mounted on the trailer. In some cases, this 12-volt supply can be wired to an on-trailer converter so that low-amperage, 120-volt alternating current is available. An adapter is usually included in the trailer kit so that trailers using the four-wire electrical connection can be connected.

Figure 6-22 Vehicle or trailer light connections should use a T-type harness, which will provide good contacts and weatherproofing. It is usually connected into the vehicle's harness just forward of one of rear lights.

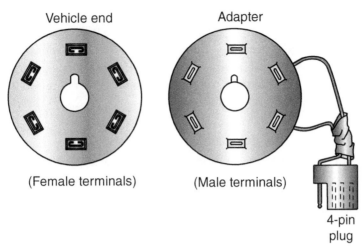

Vehicle end

Adapter

(Female terminals)

(Male terminals)

4-pin
plug

Figure 6-23 Shown is a 7-pin trailer electrical connector (vehicle side) with an adapter that steps it down to the old flat 4-pin connector still being used on most light trailers without electric trailer brakes.

Flasher failure is a common problem and does not mean there is any other problem within the circuit, but repeated failure of the flasher indicates a circuit problem. New electronic flashers may be expensive and require certain procedures and test equipment to properly diagnose them. The procedures in the service manual must be followed to successfully test and isolate the malfunction in electronic circuits.

Like all lighting systems, failure of all turn signal lamps at one time points to a breakdown of the protection device, switch, or some other common component of the system. Most turn signals will not work unless the ignition switch is in the RUN position. Four-way or hazard circuits can be checked in the same manner as the turn signals. The hazard light circuit will work in any ignition switch position. A different switch and a heavy-duty flasher are used for hazard warning lights, but the same conductors and lamps are used after the switch.

POWER WINDOWS, LOCKS, SEATS, AND DEFOGGER SYSTEMS

Service of the power windows, door locks, seats, and defogger consist of checking operation, fuses, and some slight lubrication. Almost any repair to these systems requires disassembly of the inner door panels or seat removal. In this section, preliminary checks will be discussed.

The first step in troubleshooting is to check all operations and determine answers to the following questions:

1. Are all the devices inoperative?
2. Do some of the devices work with one switch but not with the others?
3. Do some or all of the devices work slowly or create noise during operation?
4. Do the noisy devices attempt to move the window, lock, or seat?

Remember from the Classroom Manual that switches for the power windows and door locks are wired parallel to the master switch with a common protection device. If the answer to question 1 is yes, the obvious problem is in the power feed to the master switch. Should the answer to question 1 be no, then the problem is not the power supply.

> **SERVICE TIP** Most defoggers have a timer that will shut down the system after a specific time. If a customer complains of poor defogger action, operate the system to determine if the system operates for the time limit specified by the service manual. Educate the customer concerning the system's operation if the system does operate properly.

An example of the situation in question 2 is the operation of the power windows. If the master switch lowers all the windows, then the power feed, the motors, and the conductors are good. The only thing left to check is the individual switch for each window. The situation is reversed if the individual switches work the windows and the master does not. The same theory of operation applies to the power door locks.

Questions 3 and 4 cover the same general problem. A motor that lowers and raises the window slowly could be worn or the window or regulator is binding. The same is true of power door locks. Worn solenoids or binding linkage may cause the solenoid or motor to hum, hang, or refuse to work at all. A noisy window motor could indicate worn gears, a broken ribbon strip, or loose regulator fasteners. If the motor or solenoid does not attempt to move the glass or lock but continues to operate, the problem is usually stripped gears or ribbon.

Power seats use different motors and switches to move the seat, but a common fuse protects everything. A fuse or other power-source failure will cause the failure of all operational modes. If any of the two or three switches work, the power source and fuse are good. The probable cause would be the switch or motor. Again, if the motor can be heard but the seat component does not move, the problem will probably be found in the cables or screw-nut.

> **SERVICE TIP** When the panic button is activated or the alarm is set off, most systems require the door to be unlocked with the key and the engine started before the alarm will shut down. Some aftermarket systems have a separate disarming procedure that can be difficult to determine if the operator manual for the system is not available.

ANTITHEFT AND KEYLESS ENTRY SYSTEM FUNCTION TESTING

Panic button A button on the key remote used to alert people in a potential emergency.

Before testing the antitheft system, alert everyone in the shop. If possible, move the vehicle outside and away from the shop doors. A remote control that has lock, unlock, **panic,** and trunk buttons will be discussed.

Close and lock all vehicle doors using the remote-control button. Leave one window down in case the door will not unlock. An audible signal or a light should indicate that the system is armed, and the alarms are set. Test the panic mode next. The vehicle's alarm should sound and some of the exterior lights should flash. Disarm the system by pressing the unlock button. Rearm the system and press the button to unlock the trunk. The trunk lock should function, but the alarm should not. Close the trunk and ensure that the system is armed. Next, test the ability of the system to recognize tampering. With the system armed, violently rock the vehicle. The alarms should sound and flash. Disarm the system.

Rearm the system and note if there is a red or green light on the instrument panel. The red light indicates that the vehicle has been tampered with and the green light indicates that the system is armed, and no one has tampered with it since the last time it was armed. Different systems may have different lights with different meanings. Consult the service manual.

CASE STUDY

This is a problem that was national in scope, but it only concerned one brand of vehicle and only for a model year or two. The customers complained that the vent air was not cool when the air conditioning was on in MAX or NORMAL. The heater worked fine. This system used stepper motors on the blend doors. Routine electrical tests did not locate the problem. The doors could be heard moving back and forth as the controls were moved. Studying the service manual's system operation section revealed a possible cause. When the ignition key was moved to ON, the A/C computer ran a test on its circuits. Part of this test was the search for a 35-mA load on each of the motors as they moved the doors through their entire range of movement. A DMM set to amps and connected between the heat blend door motor and the computer showed 35 mA at full cold and at full hot. Supposedly the system was functioning correctly. Further physical investigation found the problem on top of the air box. The arm from the motor was held in place to the door lever with a small plastic clip. The clip had broken and was hanging down. As the motor moved the door toward the cold position, the clip would hook a ridge of the air box. The motor would try to push the door further, but the 35-mA load was reached, and the computer assumed this was the limit of door travel and stopped the motor. The end result was the door never reached its cold position even though the ampere reading was correct. Replacing the clip with a better connection solved the problem.

ASE-STYLE REVIEW QUESTIONS

1. Keyless entry system is being tested.
   Technician A says that all doors and windows should always be closed when test the keyless remote.
   Technician B says the driver's window should be down when testing the keyless remote.
   Who is correct?

   A. A only　　　　　　C. Both A and B
   B. B only　　　　　　D. Neither A nor B

2. Lighting systems are being discussed.
   Technician A says inoperative marking lights may be caused by an open circuit breaker.
   Technician B says testing a lighting system without relays can normally be done with a test light.
   Who is correct?

   A. A only　　　　　　C. Both A and B
   B. B only　　　　　　D. Neither A nor B

3. A customer complains that the headlights turn on and off, generally when traveling on a rough road.
   Technician A says the circuit breaker may be weak or failing.
   Technician B says a loose ground on the relay may be the fault.
   Who is correct?

   A. A only　　　　　　C. Both A and B
   B. B only　　　　　　D. Neither A nor B

4. A power window is stuck about halfway up and will not move up or down. There is a "clunk" each time the switch is operated.
   Technician A says the drive gear is stripped.
   Technician B says the window motor switch is shorting to ground.
   Who is correct?

   A. A only　　　　　　C. Both A and B
   B. B only　　　　　　D. Neither A nor B

5. Instrument gauge diagnosis is being discussed.
   Technician A says a test light can be used to test digital gauge systems.
   Technician B says supplying 12 VDC to a gauge will cause it to read full.
   Who is correct?

   A. A only　　　　　　C. Both A and B
   B. B only　　　　　　D. Neither A nor B

6. Warning light systems are being discussed.
   Technician A says grounding an oil pressure warning light wire at the sending unit will cause the gauge to read low.
   Technician B says grounding the engine temperature sending unit will cause the warning light to come on and the gauge to read hot.
   Who is correct?

   A. A only　　　　　　C. Both A and B
   B. B only　　　　　　D. Neither A nor B

7. A vehicle with low engine oil pressure is being discussed.

   Technician A says using a manual pressure gauge can test the oil pump output.

   Technician B says the low oil pressure reading may be caused by a bad gauge.

   Who is correct?

   A. A only   C. Both A and B

   B. B only   D. Neither A nor B

8. The alternator warning light is on at idle and off at speeds above idle.

   Technician A says the problem may be a misadjusted throttle stop screw.

   Technician B says the battery connections could cause this problem.

   Who is correct?

   A. A only   C. Both A and B

   B. B only   D. Neither A nor B

9. Diagnosing an electrical circuit is being discussed.

   Technician A says you should always check the wiring schematic before any electrical testing.

   Technician B says it's ok to use a jumper wire without a fuse.

   Who is correct?

   A. A only   C. Both A and B

   B. B only   D. Neither A nor B

10. A vehicle power seat has forward and backward movement but not tilt. There is a low, "whirring" sound when the tilt switch is operated.

    Technician A says the motor drive gear may be stripped.

    Technician B says the forward and upward cable may be broken.

    Who is correct?

    A. A only   C. Both A and B

    B. B only   D. Neither A nor B

ASE CHALLENGE QUESTIONS

1. Disarming an air bag is being discussed.

   Technician A says it's OK to use a memory saver when working on an air bag.

   Technician B says it is OK to connect the battery during an air bag service for quick testing only.

   Who is correct?

   A. A only   C. Both A and B

   B. B only   D. Neither A nor B

2. Warning light are being discussed.

   Technician A says a good place to start is by performing a bulb check.

   Technician B says most tests can be performed with the key on engine off or key on engine running.

   Who is correct?

   A. A only   C. Both A and B

   B. B only   D. Neither A nor B

3. An oil pressure light is being discussed.

   Technician A says if the oil warning light is on you should start by checking the oil level in the engine.

   Technician B says the oil light will stay on until the engine is at full operation temperature.

   Who is correct?

   A. A only   C. Both A and B

   B. B only   D. Neither A nor B

4. A four terminal relay is being discussed.

   Technician A says two of the terminals should have battery feed voltage to them.

   Technician B says two of the relay terminals are for control side of the circuit and two are for the load side of the circuit.

   Who is correct?

   A. A only   C. Both A and B

   B. B only   D. Neither A nor B

5. Circuit failure is being discussed.

   Technician A says the most common failure in an electrical circuit is a loose or missing ground.

   Technician B says before testing a relay you should locate a wiring diagram to determine the power and ground for the circuit.

   Who is correct?

   A. A only   C. Both A and B

   B. B only   D. Neither A nor B

Name _____   Date _____

ACCESSORIES PERFORMANCE TESTS

Upon completion and review of this job sheet, you should be able to conduct a performance test on selected accessories.

ASE Education Foundation Correlation

This job sheet addresses the following **AST/MAST** tasks:

6.G.1   Diagnose operation of comfort and convenience accessories and related circuits (such as: power window, power seats, pedal height, power locks, truck locks, remote start, moon roof, sun roof, sun shade, remote keyless entry, voice activation, steering wheel controls, back-up camera, parking assist, cruise control, and auto dimming headlamps); determine needed repairs. **(P-2)**

6.G.2   Diagnose operation of security/anti-theft systems and related circuits (such as: theft deterrent, door locks, remote keyless entry, remote start, and starter/fuel disable); determine needed repairs. **(P-2)**

We Support

ASE | Education Foundation

> **⚠ Caution**
>
> Before beginning a performance test, study the service or operator's manual and the system controls for familiarization and expected results.

Tools and Materials

Goggles or safety glasses with side shields
Service manual

Describe the vehicle and tire being worked on:

Year _____ Make _____ Model _____

VIN _____ Engine size _____

Procedures

Remote-Control Antitheft Systems

Task Completed

1. Confirm that the vehicle has a remote-control antitheft system.   ☐

2. Shut all doors and arm the system from the outside with the remote.

    Did the door locks function? _____

    Did a horn or light function to indicate that the alarm is armed? _____

    If not, check the remote's battery before proceeding.

3. Press the panic mode button. Did the alarms sound? _____

4. Disarm and rearm the system.   ☐

5. Look through the driver's window without touching the vehicle. Is there a light in the dash indicating that the alarm is armed? _____

6. Ensure that the remote is available to disarm the system.   ☐

7. Attempt to open each door. If all are locked, rock the vehicle from side to side. Disarm the system if the alarm sounds. ☐ Yes ☐ No

8. Did the alarm operate correctly? Explain.

    _____

    _____

    _____

Power Windows

1. Block the wheels and apply the parking brake.   ☐

2. Open all doors and switch the ignition to RUN.   ☐

3. Operate each window in turn using the master switch.

   Did each window work smoothly up and down? _____

   Did each window stop smoothly at the top and bottom? _____

4. Operate each window using the individual switches on the door.

   Did each window work smoothly up and down? _____

   Did each window stop smoothly at the top and bottom? _____

5. Set the window lock. Attempt to operate each window, first with the master switch and then with the individual switches.

   Did each window work with the master switch? _____

   Did the right front or either of the rear windows work using their individual switches?

   _____

   Comments

   _____

   _____

Power Seats

1. Sit in the driver's seat with the driver's door open. Switch the ignition to RUN.   ☐

2. Operate the switch to move the entire seat fully forward and fully backward.

   Did the seat move smoothly and quietly? _____

3. With the seat fully back, operate the switch for the backrest. Do not raise or lower the backrest completely.

   Did the backrest perform as expected? _____

4. Set the backrest to a comfortable position and operate the tilt switch.

   Did the seat tilt up and down at the rear? _____

   Comments

   _____

   _____

5. If the other front seat is equipped with power, repeat steps 10 through 18 for that seat. Did the seat perform as expected? _____

Instructor's Response

_____

_____

_____

Name _____   Date _____

DIAGNOSING AN OIL PRESSURE WARNING LIGHT CIRCUIT

Upon completion and review of this job sheet, you should be able to diagnose a nonelectronic oil pressure warning light circuit.

ASE Education Foundation Correlation

This job sheet addresses the following **AST/MAST** tasks:

6.F.1   Inspect and test gauges and gauge sending units for causes of abnormal readings; determine needed action. **(P-2)**

6.F.2   Diagnose (troubleshoot) the causes of incorrect operation of warning devices and other driver information systems; determine needed action. **(P-2)**

We Support

ASE | Education Foundation

Tools and Materials

Goggles or safety glasses with side shields
Hand tools
Manual oil pressure gauge
Jumper wire
Service manual

⚠ **Caution**

Do not use the following test procedures or equipment on electronic circuits.

Damage to the circuit could result.

Describe the vehicle and tire being worked on:

Year _____ Make _____ Model _____

VIN _____ Engine type and size _____

Procedure

Task Completed

1. Locate the service data.   ☐

   Type of instruments _____

   Warning lights only or lights and gauges _____

   Minimum and maximum oil pressures _____

2. Is the oil level correct? _____

3. Are there signs of oil leakage? _____

4. Proceed with this step only after correcting the oil level. Turn the key to KOEO.   ☐

5. Did all the warning lights come on? _____ If not, replace defective lamps and retest.

6. Start the engine and idle.   ☐

7. Is the oil light still on? _____ If yes, slightly speed up the engine.

8. Did the light go out? _____ If no, proceed with the circuit test. If yes, proceed with the manual oil pressure test starting with step 13.

9. Stop the engine, locate the oil pressure sending unit under the hood, and disconnect the wire from the sending unit.   ☐

10. Use jumper wire to ground the sending unit wire.   ☐

Task Completed

11. Set the ignition to RUN (KOEO). Is the light on? _____ If yes, go to the next step. If no, follow the circuit test routine in the service manual.

12. Remove the wire from the ground. Did the light go out? _____ If yes, proceed with the next step. If no, follow the circuit test in the service manual.

13. Remove the oil pressure sending unit.   ☐

14. Install the manual oil pressure gauge.   ☐

15. Start the engine and idle. Record the oil pressure. _____

16. Does the oil pressure meet minimum specifications? _____ If yes, proceed with the next step. If no, stop the engine and install the sending unit. Consult with the customer on all test results and recommended actions.

17. Remove the pressure gauge. Install a new oil pressure sending unit and connect its   ☐
    harness. Check the light operation. Return the vehicle to the customer.

    Dual-Purpose Light-Oil and Temperature Combined

18. Locate the temperature sending unit under the hood.   ☐

19. Disconnect the wire. Set the ignition to RUN (KOEO).   ☐

20. Did the light go out? _____ If yes, check the condition of the engine cooling system and sending unit. If no, check the oil pressure warning circuit following steps 8 through 16.

Instructor's Response

_____

_____

_____

Name _____   Date _____

SOLDERING COPPER WIRES/TERMINAL AND CONNECTOR REPAIR

Upon completion of this activity, you will be able to properly perform wiring harness repairs. You will also be able to understand and properly perform wiring harness, terminals, and connector repairs.

AUTHOR'S NOTE For this task, the author used a GM harness and terminal repair kit because it was readily available. Any other harness and terminal repair kit can be used because the procedure will be the same.

ASE Education Foundation Correlation

This job sheet addresses the following **MLR** tasks:

6.A.4   Demonstrate proper use of a digital multimeter (DMM) when measuring source voltage, voltage drop (including grounds), current flow, and resistance. **(P-1)**

6.A.10   Repair and/or replace connectors, terminal ends, and wiring of electrical/ electronic systems (including solder repair). **(P-1)**

This job sheet addresses the following **AST/MAST** tasks:

6.A.3   Demonstrate proper use of a digital multimeter (DMM) when measuring source voltage, voltage drop (including grounds), current flow and resistance. **(P-1)**

6.A.10   Inspect, test, repair, and/or replace components, connectors, terminals, harnesses, and wiring in electrical/electronic systems (including solder repairs); determine needed action. **(P-1)**

We Support
ASE | Education Foundation

Tools and Materials

Goggles or safety glasses with side shields
Soldering iron
Solder
Crimping tool
DMM
Heat shrink
Electrical tape
Terminals, butt connectors, and bulk wire
Fuses and fuse holders

Procedures

Task Completed

1. Have your instructor choose a wire for you to repair from an old wire harness.   ☐

2. Using a DMM check the resistance of the wire you are going to repair. Record your reading. _____

3. Cut the wire somewhere in the middle.   ☐

4. Using a wire stripper, identify the size of the wire you are going to repair.   ☐

5. Wire gauge size. _____

6. Using the wire stripper, remove about 1/2″ of insulation from each end of the wire.   ☐

7. Cut a section of heat shrink and insert it on the wire.   ☐

    Caution: Keep heat shrink away from the area you are going to solder! If the heat shrink gets warm, it will shrink in the wrong spot.

8. Connect the two sections of wire together. Use a splice clip or use the twist method. Describe the method you used to connect the two wires.

    _____

9. Heat the soldering gun up. When the tip starts to smoke, clean the tip with a small screwdriver.   ☐

10. Describe the type of solder you are using. _____

11. The soldering gun is heated up and ready to use. Use the tip of the soldering gun to touch the bare wires and melt the solder onto the copper wires.   ☐

    NOTE: You don't need a lot of melted solder to make a good connection.

12. Let the solder cool down, check the strength by pulling on both ends. It should not pull apart or be loose.   ☐

    DO NOT COVER SOLDER JOINT UNTIL INSPECTED BY INSTRUCTOR!!

Solder joint #1	Solder joint #2

13. Slide the heat shrink over the repaired section and use a heat gun the shrink and seal the connection.   ☐

14. Using a DMM check the resistance of the repaired wire. Record your reading.

    _____

15. Perform the same repair as before, but this time use a butt connector. See your instructor for a connector.   ☐

Butt splice #1	Butt splice #2

Repair five different terminals on five different connectors.

1. Obtain a wire harness with a connector that is in a GM vehicle. Have your instructor assign a connector.   ☐

2. Have your instructor choose a wire that is inside a harness with a connector.   ☐

3. Inspect the wire harness for frayed wires, cracks, damage from fluids and any faulty insulation. Describe what you found. _____

Task Completed

4. Determine the type of terminal locking mechanism for the connector or terminal. ☐
Select the correct terminal tool.

5. Carefully remove the terminal from the connector and inspect it. Check for signs of
damage or burning from excessive heat due to current flow or loose connection on the
wire. Record your findings. _____

6. Locate the proper terminal for replacement. Only after you have found the correct ☐
terminal, cut the terminal from the wire as close as possible where the terminal and
wire connect.

7. Determine the wire gauge and, using the appropriate wire stripping tool,
remove approximately ½″ of insulation from the end of the wire.
Record wire gauge size. _____

8. Using the proper crimping tool install the new terminal onto the wire. If the terminal ☐
has a weather insulator install a new insulator.

9. With the terminal installed on the wire show your instructor **before** reinstalling into ☐
the connector.

Terminal #1	Terminal #2	Terminal #3	Terminal #4	Terminal #5
_____	_____	_____	_____	_____

10. Reinstall the terminal into the connector and check to make sure the connector in ☐
installed securely.

Connector #1	Connector #2	Connector #3	Connector #4	Connector #5
_____	_____	_____	_____	_____

Procedure

Task Completed

Using bulk wire from your instructor, make the following jumper wires.

SEE DEMO JUMPER WIRES FOR EXAMPLES!

1. 12″ jumper wire with alligator clips on both ends. ☐

2. 3′ jumper wire with alligator clips on both ends. ☐

3. 6″ jumper wire with male spade terminals on both ends. ☐

4. 6″ jumper wire with female terminals on both ends. ☐

5. 12″ jumper wire with one male terminal and one female terminal. ☐

6. 24″ jumper wire with a fuse and male terminals on both ends. ☐

7. Make two more jumper wires that you want. ☐

NOTE: TEST ALL FINISHED WIRES WITH A DMM FOR RESISTANCE! ☐

8. Why is it important to use heat shrink on soldered connections?

    _____

9. What is a better connection: butt terminal or soldering? Explain.

    _____

    _____

Instructor's Response

_____

_____

_____

CHAPTER 7

ENGINE INSPECTION AND PREVENTIVE MAINTENANCE

Upon completion and review of this chapter, you should understand and be able to:

- Perform an oil and filter change.
- Reset the engine oil life indicator or reminder.
- Inspect under hood fluids.
- Change air and fuel filters.
- Inspect and service the cooling system.
- Inspect and replace a serpentine drive belt.
- Inspect the timing chain or belt.
- Remove and replace the timing chain or belt.
- Inspect and service exhaust systems.

- Inspect and service emission control devices.
- Use a scan tool to check for diagnostic trouble codes (DTC).
- Perform engine vacuum and pressure tests.
- Interpret engine vacuum and pressure tests.
- Perform an express service.
- Recognize special procedures for servicing hybrid electric vehicles.
- Remove an engine cover and reinstall it with a new gasket or chemical sealant.

Terms To Know

Backlash	Filter gasket	Muffler
Cooling fan	Filter wrench	Petcock
Cylinder leakage test	Front pipe	Serpentine belt
Diagnostic trouble code (DTC)	Heat-resistant gloves	Slime
Diesel exhaust fluid (DEF)	Malfunction indicator lamp (MIL)	Spout
Diesel particulate filter	Micro-vee	Tailpipe
Diverter valve	Monitor	Top off

INTRODUCTION

The technician's first jobs will probably be lubrication service and tires. While performing a routine oil change, the technician should observe the general condition of the vehicle and the engine. The customer may not realize that there is a potential problem. Driving the vehicle to the lift gives the technician a chance to get a feel for the brakes, steering, and engine operating condition. When a defect is suspected, notify the customer or the service writer. Reassembling automotive components requires the installation of different types of gaskets, seals, and chemical sealants. This chapter will give you a short explanation of these tasks and some items that need to be checked as the repairs or servicing are completed.

⚙️ **SAFETY TIP** The services performed in this chapter may require the technician to work on or with hot components or liquids. **Heat-resistant gloves** that cover the hands and forearms are available at local parts stores. Remember, heat absorbed by the outside of a material can burn the skin if touched. If the glove material is wet, it will have no resistance and the heat will travel straight through to the skin. Even when using protective clothing, the technician must be careful. Eye protection should always be worn. Always use the locks on a lift or use jack stands when working under the vehicle.

Most entry-level technicians with sufficient supervision can perform the services covered in this chapter and the remainder of this book. Keep in mind that shop managers or service writers will not assign more complicated jobs to a new technician until they feel comfortable with the technician's skill and knowledge.

CHANGING THE ENGINE OIL AND FILTER AND CHECKING FLUID LEVELS

Almost every technician starts his or her first job as a lube technician and will continue to perform this basic service throughout his or her career. If any task is considered to be a basic and very necessary vehicle service, it is the scheduled engine oil change.

The brand and type of oil are strictly a customer's choice. However, the technician may advise the customer on the type of engine oil recommended by the vehicle manufacturer.

Before starting the service, get the oil and filter. Usually the type of oil is determined by the engine specifications, but the customer may have a preference for a brand or type (**Figure 7-1**). Check the repair order for customer preference or any additional tasks the customer may have requested. The customer may have heard a noise or other symptom that she or he would like to have checked.

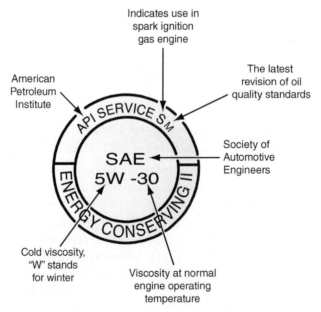

Figure 7-1 Definition of the ratings shown on an oil container.

Figure 7-2 The oil filter can be easily removed.

Move the vehicle to the bay. As the vehicle is operated, note the performance of the engine, brakes, and other systems. This short drive may point to something that requires a closer inspection. Raise the hood and make a quick inspection for the oil filter. Every engine has the oil filter in a different location, so look carefully (**Figure 7-2**). Lift the vehicle to a comfortable working height. Refer to **Photo Sequence 13** for the typical procedure for changing oil and filter.

⚠️ WARNING **Do not remove the wrong drain plug. A few vehicles have a drain plug for the automatic transaxle. Accidentally draining the wrong fluid will cost the shop some money, but if the mistake is not caught before releasing the vehicle to the customer, damage to the engine and transaxle could occur.**

Classroom Manual
pages 211–214

 Special Tools
Lift
Drain pan
Filter wrenches

PHOTO SEQUENCE 13
Typical Procedure for Changing the Oil and Oil Filter

P13-1 Open the hood and pull the dipstick out slightly. Remove the oil cap.

P13-2 Raise the vehicle, locate the oil drain plug, and position the oil drain pan so that it will contain the oil. NOTE: The oil will come out with light pressure, and it will cause a stream to go to the side. Adjust for this by moving the oil drain pan to the side.

P13-3 Remove the oil drain plug, using a wrench. NOTE: Always use a properly sized wrench. You will find that over time the oil drain plug head will get rounded by the use of incorrect sized tools. It will need replacement if it is difficult to remove.

PHOTO SEQUENCE 13 (CONTINUED)

P13-4 Watch the oil coming out of the pan. The content, color, and clarity will give you clues to how often it is maintained and if there are any problems.

P13-5 Adjust the oil drain pan location as needed when the oil flow becomes less.

P13-6 After the oil is done draining, reinstall the oil drain plug with a torque wrench. Use the service manual to find the specification.

P13-7 Remove the oil filter. You may need to use an oil filter wrench if it is tight. There are several types to use. Oil filters are located at different places on different vehicles.

P13-8 Make sure that the oil filter gasket is removed with the old filter. Double-check the engine side to make sure it is not there. CAUTION: If you do not remove the old oil filter gasket from the engine, when you install the new one, it will have two gaskets. This is called "double gasketing" an engine. This can cause oil to leak from the space in between the two gaskets at a very high rate and pressure.

P13-9 Lubricate the new oil filter gasket with new oil before installing it. Tighten the oil filter one-half turn past hand tight. Do not use any tools to tighten it.

P13-10 Some vehicles have a cartridge style oil filter that requires a socket to remove.

P13-11 Similarly, lubricate the new oil filter O-ring if it is a cartridge style. Use a torque wrench to tighten the housing cover of a cartridge filter (if equipped).

P13-12 With the drain plug and filter tight, add the correct amount and type of oil to the engine.

P13-13 Check the level on the dipstick before starting the engine. If needed, add oil.

P13-14 Start the engine. Immediately watch the oil pressure light or gauge. The engine should have oil pressure within 5 seconds.

P13-15 If it does not have pressure, then turn the engine off and inspect. With the engine still running, walk around and look under the engine for any leaks.

P13-16 Turn the engine off, wait 30 seconds, and recheck the oil level. It should have gone down just a little bit, because oil is filling into the new filter. Adjust the oil level as needed.

In many cases, the vehicle is hot and the customer is waiting, so speed is essential. Do not, however, forget basic safety procedures. Use heat-resistant gloves and eye protection. With the vehicle raised, move the drain container under the oil drain plug (**Figure 7-3**). Use the correct wrench to loosen the plug. A typical drain plug is not made of the strongest metal and can be rounded with the wrong wrench. A boxed-in wrench is suggested.

While the oil is draining, make a quick inspection of the undercarriage of the vehicle. Look for any leaks and bent or damaged components. On some vehicles, it is possible to see part of the brake disc and pads. Inspect the tire treads for wear. The tread should show even wear across the tire.

Check the drain plug for a gasket. Many plugs have a brass, paper, or plastic washer to help seal the drain hole (**Figure 7-4**). It is usually best to replace the gasket if one is used. Once the oil has completed draining, install and torque the drain plug. Do not overtighten the plug. The drain hole goes through thin metal and the threads can be stripped out.

> ⚙ **INTERNET TIP** For proper disposal and control procedures for repair shop waste, information is available at the Web site for the Coordinating Committee for Automotive Repair. The Web site is http://www.greenlink.com.

Locate the filter and determine the best type of **filter wrench** to be used. There are several types of wrenches available and more than one may work (**Figure 7-5**). The location of the filter and the components around it will determine which wrench will be the easiest to use. Position the drain pan directly under the filter and loosen it. At this point, the technician may have to work around a very hot exhaust and other hot

⚠ **Caution**

Use hand and forearm protection when changing oil on a hot engine. Burns could result from the oil, filter, and engine components.

⚠ **Caution**

Always capture the used oil from the engine. Clean any spills immediately. Failure to do so may result in an injury to the technician or charges by the EPA for illegal waste disposal.

Filter wrenches are specifically designed to remove oil filters.

Figure 7-3 An oil drain container.

Figure 7-4 The drain plug seal may be made of brass, copper, or neoprene.

The **filter gasket** is usually an O-ring. A cartridge-style oil filter requires a new O-ring as well (**Figure 7-6**). There is a specific amount of torque from the container to the filter housing.

components. Many filters are placed so the draining oil will flow down over engine and vehicle components. Be alert with the drain pan and watch for splashing oil. Work carefully and safely.

While the filter is draining, double-check that the old oil **filter gasket** came off with the old oil filter. Double-check the engine side to make sure that it was removed. Leaving it on will be catastrophic! If the old gasket is left on, you will end up with two gaskets (because the new oil filter has one). This is known as double gasketing the filter. It will cause a major leak and oil will spill everywhere when the engine is started, if you are lucky. In some cases, the double gasket will not leak until after the vehicle has left the shop, causing a major oil leak and possibly losing oil pressure and causing major engine damage.

Figure 7-5 Examples of oil filter wrenches.

Figure 7-6 You will encounter different style oil filters. This is an example of the two most common types.

Next open the box for the new filter and a container of oil. Open the new oil container and use a finger to smear a light coat of oil on the filter's gasket (**Figure 7-7**). This will help remove it during the next service.

Wipe the engine oil filter mounting surface with a clean rag and install the new filter. Hand tighten the filter to the specifications shown on the filter. In almost all cases, the filter is hand tightened about one-half of a turn after the seal contacts the mount. Do not use a wrench to tighten the filter or it will be very difficult to remove during the next service.

With the plug and filter in place, check the area for tools and remove the drain pan from under the vehicle. Lower the vehicle to the floor.

There are different methods used to install oil into an engine. Some shops have overhead drops that are plumbed to large oil containers in a storeroom (**Figure 7-8**). Follow the equipment instructions and shop policy to select the amount and type of oil. Many shops use quart containers, and the technician must pour each quart individually into a filler hole located in a valve cover.

Figure 7-7 This gasket goes around the outer edge of the filter, and it should be lightly lubed before the filter is installed.

The offset position of the container's **spout** allows air to enter the container. It is suggested that the container not be squeezed to speed the process. This will block the airflow.

⚠ Caution

Do not open a hot coolant system. Serious burn injuries could result.

Use a clean funnel to prevent oil from dripping on the engine. In the few situations when a funnel is not available, oil can be poured directly from the container. Most quart containers are shaped to reduce the amount of spillage. Notice where the pouring **spout** is attached to the container body (**Figure 7-9**). It is usually to one side at the top. Hold the

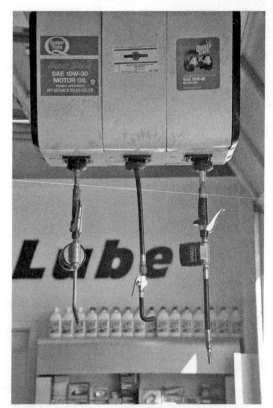

Figure 7-8 Fleet repair and specialty shops have bulk containers and the oil or grease is pumped to each bay as needed.

container so the pouring spout is the highest point, placing the oil below the spout. The oil can be poured slowly and air can enter the container as the oil exits (**Figure 7-10**). Turning the container so the spout is down will cause a burping of the oil as air is drawn into the emptying container. Install the amount of oil specified and then replace the filler cap.

Many modern vehicles have a computer program that makes the driver aware of the need to change oil (**Figure 7-11**). If this is the case, you must reset the oil life indicator or reminder. Some vehicles require a special tool, connected like a scan tool. But most can be reset using the driver's controls on the instrument cluster or the information center. Even if the vehicle has a maintenance reminder, many shops still install the windshield sticker (**Figure 7-12**).

Figure 7-9 The spout of a quart container of engine oil is offset to facilitate pouring.

Figure 7-10 The offset spout should be held as shown when adding oil to the engine.

Figure 7-11 Many newer vehicles have a computer program that makes the driver aware of the need to change the oil.

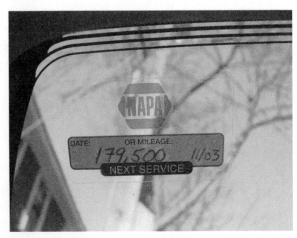

Figure 7-12 Always put an oil sticker on the windshield or in the driver's door jam to remind the customer when to have the oil changed again.

Conduct a check of the fluid levels available under the hood. The most common are the brake master cylinder, windshield washer, coolant, and transaxle or transmission (**Figure 7-13**). The transaxle and transmission must be checked with the engine running, but the fluid can be checked for burning or age. This fluid should be reddish. If it is burned or dark in color, a service recommendation may be made to the customer.

The coolant may be hot, so do not open the radiator. Check the coolant level in the overflow reservoir. It should be between the hot (high) and cold (low) marks (**Figure 7-14**).

Figure 7-13 In most shops, an oil change includes a check of the fluid level at each location shown here.

Figure 7-14 The coolant should be visible through the reservoir and between the hot and cold mark on a warm engine.

Removing a cap or observing the level through the translucent reservoirs provides a means to check the master cylinder and windshield washer fluids. A low brake fluid level may be correct based on the amount of brake wear. As the brakes wear, the fluid level will drop. The technician or service writer may consult with the customer if the brake fluid is low.

Some shops will **top off** some fluids without charge to the customer. Generally, antifreeze and brake fluids are charged. With all the fluids checked, start the engine and observe the oil warning light. It should go out within a few seconds after the engine is running. If it does not, shut the engine down immediately and correct the problem. When the light is out, shut off the engine and recheck the engine oil level. Some of the oil is used to fill the filter and the level may drop slightly. The level, with the filter full, should be at the full mark or slightly below the mark. Top off the oil as needed. Check under the vehicle for leaks. When the oil level is correct and no leaks are apparent, clean the vehicle of any oil drips. Complete the repair order and return the vehicle to the customer. Properly dispose of waste oil and the old filter.

Top off is a term that means fluid is added to a system to bring it to the correct level.

Classroom Manual
pages 226–227

CHANGING THE AIR AND FUEL FILTERS

Air Filter

The air filter is usually checked during an oil change service. However, it should be checked any time the customer complains of poor engine operation. The filter will be located in a housing on top of the engine or in an area near a fender behind a headlight. The housing cover may be bolted down or retained with spring clips (**Figure 7-15**).

Select the correct replacement filter and open the housing. There may be electrical or vacuum connections that must be disconnected before the cover can be removed. Be sure that neither is damaged during removal. Remove the filter and inspect the housing for dirt, leaves, and oil. Dirt and leaves on the inlet side of the housing is normal. Clean the housing. If oil is present, this may indicate a problem with the PCV system. Further inspection of the PCV system should be accomplished.

Figure 7-15 Snap clamps typically hold the air filter housing cover in place. Some use threaded fasteners for the same purpose.

Special Tools

Line wrenches

Quick-disconnect
  tools

Catch pan for fuel

Hose-clamping tools

⚠ **Caution**

Do not allow any hot
components or
flame-producing
devices near the vehi-
cle when the fuel sys-
tem is being serviced.
Dangerous fires could
result.

⚙ **SERVICE TIP** Many of the vacuum hoses are made from hard plastic and break easily. Use both hands, one to pull the hose and one to hold the component, to pull the hose straight from the component. Do not bend or flex the hose. A small, pocket-size, flat-tipped screwdriver may be used to unlock electrical connections.

In some vehicles, a separate air filter for the PCV is located in the engine air filter housing. This filter should also be replaced at this time. Install the new engine air filter and cover. If bolts are used, do not overtighten them. On the round units mounted directly to the top of the engine, the tightening of the bolts can be critical. In many cases, the cover on this type of housing applies force to the air filter. Overtightening can depress the filter and restrict airflow to the engine.

Vehicles that have a PCV valve should have it inspected every time the air filter is inspected. Remove the PCV valve (**Figure 7-16**) and shake it. If it rattles, it is still good and not stuck. If no noise is heard, replace it.

In other vehicles, no PCV valve is used. These vehicles use an orifice tube that goes from the valve cover to the intake manifold. You must inspect this hose by squeezing it to see if it is softened from being oil soaked. Often these hoses will collapse from oil saturation over time. Removing this hose is important to inspect the inside of it for damage or blockage.

Fuel Filters

Changing the fuel filter presents some safety and mechanical problems. Opening the line to remove the filter will allow some fuel to escape. A suitable catch container should be used, and the waste fuel must be stored in an appropriate container. The oil drain pan should *not* be used. Consult with a supervisor for instructions. The vehicle should be

Figure 7-16 Some PCV systems have an additional air filter within the engine's air filter housing. Change it at the same time the engine filter is changed.

allowed to sit until the exhaust system is cool. It is preferable that the entire vehicle has cooled down.

The technical end involves relieving the fuel pressure on electronic fuel injection (EFI) systems and using tools to disconnect and close the lines. Consult the service manual to determine how the fuel pressure is relieved. It usually involves disabling the fuel pump and starting the engine. When the engine stalls, the fuel pressure is removed but there will still be some fuel in the lines.

Some fuel filters are connected to the fuel lines with compression fittings requiring line wrenches for loosening (**Figure 7-17**). Others use quick-release fittings that can be unlocked with special tools or a small, flat-tipped screwdriver (**Figure 7-18**). Almost all filters have a small mounting bracket that is removed or loosened with a socket or screwdriver.

Before starting work, select the correct filter and check it for the bracket and replacement quick-release locks if used (**Figure 7-19**). Some types of quick-release locks are damaged during removal. Most EFI systems have fuel filters mounted on their frame rails and use the quick-release connections used for this discussion. Other types follow the same procedures except for filter location and type of connection.

Figure 7-17 Twist the fuel lines, if possible, at the quick-disconnect. This will loosen the connection for easier release.

Figure 7-18 There are many types of quick-disconnect tools available.

Figure 7-19 Most fuel filters for EFI are mounted to the frame near the fuel tank. Some are mounted in the engine compartment.

Follow the instructions to relieve the fuel pressure and locate the fuel filter. Do not enable the fuel pump until the filter is replaced. If the filter is under the vehicle, raise the vehicle to a good working height. Locate the filter and determine the tools needed. If the filter is connected to flexible hoses, some may be clamped shut with special tools. Check the service manual.

Clamp flexible hoses if possible and position the pan securely as two hands will be required to remove the filter (**Figure 7-20**). Remove or loosen the bracket. Loosen the fittings by prying the locks open with the tool. Twist the filter once or twice to loosen it on the lines. Fuel may be allowed to drain at this point. Allow as much as possible to drain. Grasp a fuel line to hold it in place as the filter is pulled from the line. The locks will have to be held open while separating the filter and hose. Allow the fuel to drain before separating the filter from the other line. There will still be fuel in the filter, so handle accordingly.

Ensure that the filter is mounted in the correct direction. The filter should be marked "inlet" and "outlet" or may only have one end labeled (**Figure 7-21**). The label may be just an arrow showing flow. The filter is installed with the outlet end toward the engine. Install one end of the filter into the fuel line and be sure the lock slips in place. The short pipe at each end of the filter has a flange or collar. The lock must lock between the collar and the filter body. The connection should be checked by attempting to pull the filter from the line. The filter may move a short distance. Repeat the procedures with the other end. Install and tighten the bracket. Remove any hose clamps installed.

Figure 7-20 This type of quick-disconnect does not require a tool. It can be done by hand.

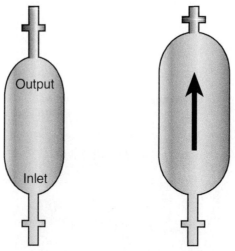

Figure 7-21 The fuel filter may be labeled in several ways. Two examples are shown.

Lower the vehicle just enough to gain access to the door. Do not start the engine at this time. Enable the fuel pump by turning the ignition key from OFF to RUN two or three times. This will pressurize the system. Check the fuel filter connections for leaks. If dry, start the engine and allow it to run long enough to settle down and operate smoothly. Shut off the engine and check the filter for leaks. This procedure is easier if a second person is available to check for leaks or operate the running engine while the filter area is checked. With the service complete, check for tools and lower the vehicle.

Drain all the fuel from the old filter. Properly dispose of the filter and waste fuel. Complete the repair order and return the vehicle to the customer.

The newest light vehicle EFI system is a *return less fuel arrangement.* Instead of a pressure regulator situated near the injectors, pressure is controlled by the speed and time of the fuel pump. This is done by varying the voltage supply to the pump. For instance, during idling conditions, the pump's supply voltage is reduced, thereby lowering the fuel pressure and volume. As the engine load increases, more voltage is supplied, and fuel flow is increased to meet demand. Systems of this type usually do not have external fuel filters that need periodic replacement. The filter is part of the pump and sending unit assembly inside the tank. Most newer Chrysler vehicles have this type of EFT.

COOLING SYSTEM INSPECTION AND SERVICE INSPECTIONS

Inspection of the cooling system consists of checking for external coolant leaks, hoses, belts if applicable, and the visual condition of the coolant. The coolant is usually a 50/50 mixture of water and antifreeze, which can be green, orange, or red in color. Check the inside of the cap for rust, corrosion, contamination, or slime. If any of these three conditions is present, the cooling should be flushed, and new coolant installed.

⚡ WARNING **Engine manufacturers specify exactly which antifreeze to use in their engines. Many antifreezes are chemically designed for a specific brand or model of vehicle engine. Even many of the so-called one-size-fits-all antifreezes will not work well in many modern engines. They prevent freezing; however, they can corrode the coolant system or cause excessive electrolysis to build. Use the antifreeze recommended by the vehicle or engine manufacturer.**

External cooling leaks are easy to spot and can be easily identified by the color of the antifreeze. An experienced technician should locate and repair internal cooling leaks. Check under the vehicle and inside the engine compartment for coolant. Do not forget to check the heater hoses and heater. A leaking heater will deposit coolant in the passenger compartment on the floorboard. The owner will usually complain of a coolant smell.

Hoses and Coolant Changes

⚡ WARNING **Some new engines require special procedures for refilling the coolant systems. Follow the manufacturer's instructions. Damage to an engine may occur if the system has air or is not completely full.**

Most radiator hoses are easily replaced. The coolant must be drained, captured, and recycled or disposed of properly. Before starting the replacement process, read the service manual to determine the correct method for draining the system, refilling the system, and choosing the amount and type of antifreeze required (**Figure 7-22**). Many engines, including diesels, require special antifreeze.

Antifreeze is normally slick and oily. Metal dissolving in the coolant, however, causes it to feel like slime. You can also tell a lot about coolant by its color. Green is the color of traditional coolant. This type of coolant will last about 3 years. Extended life coolants are very common on modern vehicles. They will last anywhere from 3 to 6 years. Extended life coolants are available in many colors: red, pink, orange, yellow, and so on. But they are not green.

⚡ Caution

Collect and dispose of the waste coolant according to shop policy. Wrongful disposal of waste coolant could damage the environment, injure or kill wild or domestic animals, and bring legal charges filed by the EPA against the shop owner.

Figure 7-22 This underhood label indicates that the vehicle requires an extended-life coolant. Always use the proper type of coolant.

Special equipment is available to make draining and refilling a coolant system easy or easier than the following procedures. One very simple system uses shop air pressure to create a vacuum within the coolant system and withdraw the coolant. The same system is used to force coolant back into the system. Other than the radiator cap, the system is not opened. This speeds up the work, eliminates replacing clamps, and sometimes time-consuming gaskets.

Drain the coolant into a suitable container using the **petcock** located on the lower side of the radiator (**Figure 7-23**). Do not loosen the cap at this time. Some vehicles have petcocks on the engine block. If this is to be a complete coolant change, those petcocks will have to be opened to drain all the coolant. When the amount of draining coolant slows, remove the radiator cap. Initially leaving the cap on allows the coolant recovery tank to drain through the radiator.

Release the hose clamp using pliers, wrenches, or a screwdriver. The hose will probably be stuck to the mounts and will have to be twisted and pulled off. A possible solution is to split the hose where it fits over the mounts and slide a flat-tipped screwdriver between the hose and mount (**Figure 7-24**). Use a small piece of sandpaper or a scraper to remove any corrosion from the mounts. Slip a new clamp over the new hose and slide the hose onto the mount. Place the clamp about halfway between the end of the mount (feel it through the hose) and the end of the hose. Tighten the clamp until bits of the hose extend into the clamp's screw slots. Repeat the procedures for the other end and any other hoses. The same basic procedures also can be used to replace the heater hoses. Do not use

A **petcock** is a drain plug for the coolant system.

Figure 7-23 Most light vehicles do not have a coolant drain plug on the engine block, but many heavy trucks and equipment do have this plug.

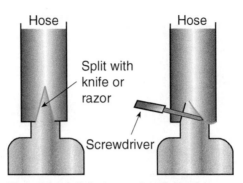

Figure 7-24 Split the hose and pry it open for removal. Do not do this if the hose is to be reused.

excessive force to remove a hose from the heater. Heater hose connections are thin and can be damaged while trying to remove the hoses. Use a knife or another blade to slice the hose as previously mentioned.

Close the petcocks and pour in the correct amount of antifreeze. Completely fill the cooling system with water. Follow the manufacturer's instructions to bleed air from the system. Some vehicles have a small bleeder screw at the upper radiator hose/engine connection for this purpose. Others require increasing the engine speed to 1,200–1,500 rpm, adding additional coolant, and capping the radiator before reducing the engine speed to idle. The heater temperature control should be placed in the HOT position during refilling. This will flush air from the heater and completely fill the system with coolant.

Once the system is full, allow the engine to idle while checking for leaks. If the repair is satisfactory, complete the repair order, dispose of the waste, clean the area, and return the vehicle to the customer.

SERVICE TIP The condition of the antifreeze can be checked using a digital voltmeter. The red voltmeter lead is inserted into the coolant, and the black lead is attached to a good engine ground. If the voltmeter reads 0.2 volt or less, the antifreeze is good. A reading between 0.2 and 0.6 volt indicates the antifreeze is deteriorating and is borderline. A voltage value above 0.7 volt indicates the antifreeze must be changed. High voltage values can also indicate a blown cylinder head gasket. The voltage reading is the result of the chemical reaction between the engine block materials and the coolant. As the additives in the antifreeze deplete, the antifreeze becomes a weak acid. The result is the cooling system becoming a weak battery.

Coolant Inspection

The antifreeze content in the coolant may be tested with a coolant hydrometer, refractometer (**Figure 7-25**), or coolant test strips. The vehicle manufacturer's specified antifreeze content must be maintained in the cooling system because the antifreeze contains a rust inhibitor to protect the cooling system. Always check the vehicle or service manual for the type of coolant to add to the system. Inspect the coolant for rust, scale, corrosion, and other contaminants such as engine oil or automatic transmission fluid. Aside from a visual check of coolant condition, a paper test strip can be used to test the coolant pH level. This will give you a visual indication of whether the coolant has become too acidic and requires flushing. If the coolant is contaminated, the cooling system must be drained and flushed (**Figure 7-26**). If oil is visible floating on top of the coolant,

Figure 7-25 This refractometer can be used to determine the antifreeze protection level of any type of coolant.

Figure 7-26 This coolant reservoir is covered in sludge; the system should be flushed and refilled with clean coolant.

the automatic transmission cooler may be leaking. This condition also contaminates the transmission fluid with coolant. If the vehicle has an external engine oil cooler, it may also be a source of oil contamination in the coolant. These coolers may be pressure tested to determine if they are leaking. Testing or repairing radiators and internal or external oil coolers is usually done by a radiator specialty shop. Inspect the air passages through the radiator core and external engine oil cooler for contamination from debris or bugs. These may be removed from the heater core with an air gun and shop air, or water pressure from a water hose.

Cooling System Component Testing

Testing the Thermostat. If a customer brings in a vehicle with an overheating problem, it is possible the thermostat is not opening. In addition, an engine that fails to reach normal operating temperature can be caused by a faulty thermostat. A thermostat that is not opening will cause the coolant temperature gauge to rise steadily into the red zone after several minutes of operation. Sometimes the thermostat will stick closed intermittently. The customer may report that the coolant gauge spiked into the red and then came back down to a normal reading one or more times. The thermostat is inexpensive; problems caused by engine overheating are not. It is wise to replace the thermostat if you suspect that it is sticking. Thermostats have a high failure rate and are often replaced simply based on the symptoms described or as part of routine maintenance.

To attempt to verify a sticking thermostat, start the engine while it is relatively cool. Hold the upper radiator hose toward the radiator as the engine warms up. Periodically check the coolant temperature gauge to make sure the engine does not overheat. If the thermostat is working properly, the upper radiator hose should get quite hot and feel pressurized as the temperature gauge rises into the normal zone. If the gauge starts to read hot and you have not felt a distinct temperature increase in the hose temperature, the thermostat is sticking closed. Replace a faulty thermostat. You can also test for a stuck-open thermostat using a similar method. Start the engine after it has cooled. Hold the upper radiator hose toward the radiator end. Let the vehicle run for a few minutes. If the hose starts to get warm within the first few minutes of operation or before the engine reaches a normal temperature, it is sticking open (**Figure 7-27**).

Sometimes you may have to remove the thermostat to try to verify that it is faulty and causing the system overheating. Check the rating of the thermostat, and confirm it is the proper one for the engine application. Also confirm it was installed in the right direction. It is possible to test thermostat operation by submerging it in a container of water and heating the water while observing the thermostat. Use a thermometer so the temperature when the thermostat opens can be determined. At the rated temperature of the thermostat, it should begin to open.

Cooling Fan Inspection. The fan can be driven by a drive belt off of the crankshaft, or it can be operated electrically (**Figure 7-28**). Regardless of the design used, inspect the fan blades for stress cracks. The fan blades are balanced to prevent damage to the water pump bearings and seals. If any of the blades are damaged, replace the fan.

> **Cooling fans** force airflow draw the radiator to help in the transfer of heat from the coolant to the air.

⚡ **WARNING An electric fan may start at any time, even after the engine has stopped. Either let the engine cool down or disconnect the fan electrically before touching the fan.**

Upper radiator   Thermostat
hose   housing

Figure 7-27 The thermostat sits on the top front of this engine.

Figure 7-28 (A) Belt driven fan. (B) Electric fan.

Electric fans are also inspected for damage and looseness. If the fan fails to turn on at the proper temperature, the problem could be the temperature sensor, the fan motor, the fan control relay, the circuit wires, or the powertrain control module (PCM). To isolate the cause of the malfunction, attempt to operate the fan by bypassing its control. On many modern vehicles, this can be done by using a scan tool to activate the fan. If the fan operates, the problem is probably in the coolant sensor circuit. Individual electrical component testing will be covered in later chapters of the book.

To direct airflow more efficiently, many manufacturers use a shroud (**Figure 7-29**). Proper location of the fan within the shroud is also required for proper operation. Generally, the fan should be at least 50 percent inside the shroud. If the fan is outside the shroud, the engine may experience overheating due to hot under hood air being drawn by the fan instead of the cooler outside air. If the shroud is broken, it should be repaired or replaced. Do not drive a vehicle without the shroud installed.

Figure 7-29 The fan shroud is used to control airflow around the engine.

INSPECTING AND REPLACING THE SERPENTINE DRIVE BELT AND PULLEY SYSTEM

Modern vehicles use a **micro-vee** belt called a **serpentine belt**. A micro-vee belt is like a vee belt used on older cars, but the vee grooves are much smaller and there are multiple grooves. A serpentine belt gets its name from the configuration it is in. Micro-vee belts are not always found in serpentine configuration.

Belts will stretch over time and start to crack with age. A stretched-out belt does not have the tension of a new belt, and it will cause a squeaking noise because it is not tight. It is the slipping on the pulleys that causes the squeaking noise. Some micro-vee belts are installed so that they are a stretch fit. These belts contain a chemical that shrinks the belt after installation. No tensioner or tightening is required. These belts wear out too, and you must cut them off to remove them.

Replacing a serpentine belt can be both complicated and simple. The method of removing the belt tension or even locating the tensioner is usually the most difficult. Before beginning the repair, locate a drawing of the belt's routing (**Figure 7-30**). This is sometimes located on an under hood label but can usually be found in the owner's manual or the service manual. If a drawing is not available, it may be advisable to draw a diagram of the old belt's routing. It would appear to be a simple operation, but a single-belt system has many bends as the belt goes around each pulley. Without a drawing, errors in routing the new belt may cost the technician valuable time. **Photo Sequence 14** highlights the typical sequence of tasks needed to remove and replace a serpentine drive belt.

> **SERVICE TIP** Before attempting to move the tensioner, note how the belt is threaded over it. If the belt contacts the top of the tensioner pulley, the tensioner will probably have to be swung down and away from the belt. Just the opposite is true if the belt touches the pulley on the lower side.

Use a tool to force the tensioner back from the belt. A great deal of force is usually required. With the tension released, remove the belt from the pulley that is the easiest to access. It is also easiest to choose to remove the belt from a pulley without ribs on it. It

A **micro-vee** belt is a belt composed of multiple small vee grooves in it.

The **serpentine belt** is so named because of its winding route among the various pulleys at the front of the engine.

Classroom Manual pages 313–314

⚠ **Caution**

Ensure that the tensioner and tool are securely mated to each other before attempting to relieve tension from the belt. The tensioner is under a great deal of force. A tool that is not placed properly could slip and cause injury or damage.

⚠ **Caution**

Do not stand in the rotating plane of a belt. It can come off and cause injury.

Locate the tensioner and determine the type of tool or wrench needed to remove tension from the belt. In some cases, a box-end wrench or a 3/8" ratchet is all that is needed. A special tool is available for this task and will work in most cases (**Figure 7-31**).

Figure 7-30 Note that on this engine the alternator pulley would be the best place to make the initial and final step in replacing this belt.

PHOTO SEQUENCE 14
Removing and Replacing a Serpentine Belt

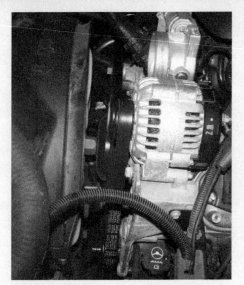

P14-1 This photo shows the limited amount of workspace available for replacing a typical serpentine belt.

P14-2 The tensioner tool is used to move the automatic tensioner to put slack into the belt. The belt is removed from the easiest pulley to access, usually the alternator.

P14-3 (Engine is out-of-vehicle for clarity.) The belt is removed and bent backwards. This will reveal any cracks or wear and tear on the belt.

P14-4 (Engine is out-of-vehicle for clarity.) The belt is installed starting with the crankshaft pulley and around, under, or over each of the other pulleys except for the last one (same as the first from which removed).

P14-5 With the tensioner moved to provide sufficient slack in the belt, the belt can be looped over the last pulley. Before starting the engine, recheck the belt for proper placement around each pulley.

Figure 7-31 Pictured are a few of the different belt tensioner tools available on the market.

will slide off instead of being lifted off. The alternator is usually the best option. This pulley is small and is usually best placed for removing and installing the belt. Once the belt is loose, slowly allow the tensioner to move back into place. Sometimes the tool will hang on an engine component and need to be worked loose. Do not allow the tool to hold the tensioner as the belt is removed. It could come loose and cause an injury.

Remove the belt from the rest of the pulleys and from the vehicle. When the belt is removed you should inspect the tensioner and pulleys. Spin the pulleys by hand and listen for a grinding or squeaking bearing. Check each pulley for looseness and bearing movement. If the tensioner is a spring-loaded tensioner (**Figure 7-32**), use the removal tool to move it back and forth to check for movement and stickiness. Do not get in the habit of overlooking these items as they frequently require replacement.

With the belt removed, inspect the belt for cracks. The rule of thumb for replacement is three cracks in 3 inches. However, many newer belts are made of EPDM, a chemical used in the belt's construction. This type of belt will stretch and require replacement, but it will not crack. You must use a special tool (**Figure 7-33**) to check the depth of the belt.

Install the new belt starting at the crankshaft pulley. Maintain hand tension on the belt as it is routed and installed over all but the last pulley. The last pulley should be the easiest to access. Secure the belt while the tool is replaced on the tensioner.

Figure 7-32 Typical automatic belt tensioner.

Figure 7-33 Special belt depth tool is used to check a belt's condition.

This step is easier and safer to do with two people, but one person can do it. Move the tensioner until it is at the furthest end of travel. Pull the belt over the last pulley. Before allowing the tensioner to move, take a quick check to ensure that the belt is properly positioned on each pulley. Release the tensioner and recheck the belt. If the belt is not positioned properly on any pulley, remove the tension and place the belt properly. Don't start the engine until you are sure that the belt is on properly. Even if the belt is slightly off, it will not correct itself; instead it will get cut on the edge of the pulleys.

Remove the tensioner tool. Have a second technician crank the engine once or twice while observing the belt from the side. If the belt routing is correct, start the engine and observe the belt operation. If the repair is correct, remove and store the tools, complete the repair order, and return the vehicle to the customer.

INSPECTING THE TIMING BELT OR CHAIN

Symptoms of a Worn Timing Mechanism

We mentioned that a worn timing chain may slap against the cover or against itself on a quick deceleration and make a rattling noise. Use a stethoscope in the area of the chain to confirm your preliminary diagnosis, and then proceed with the checks described in the next section to confirm the extent of the damage (**Figure 7-34**).

A customer may also report that the engine seems to have poor acceleration from starts but runs well, perhaps even better, at higher rpm. When there is slack in the chain, the camshaft timing is behind the crank. This is called retarded valve timing. This improves high-end performance while sacrificing low-end responsiveness.

Figure 7-34 Listen for a slapping or rattling noise on deceleration.

SERVICE TIP A noisy timing chain slapping against the timing cover should be replaced right away to avoid failure and the need for more costly repairs.

Timing gears may clatter on acceleration and deceleration. The engine does not have to be under a load; just snap the throttle open and closed while listening under the cover for gear clatter. The customer is unlikely to notice the reduced low-end performance before the gears break from too much **backlash**.

Customers with worn or cracked timing belts will notice nothing until their engines stop running or begin running very poorly. What you and they should be paying close attention to is the mileage on the engine and the recommended belt service interval. Skipping a recommended replacement is at the very least an invitation to meet a tow truck driver.

Backlash is the small amount of clearance between the gear teeth that allows room for expansion as the gears heat up.

Symptoms of a Jumped or Broken Timing Mechanism

When a timing chain, belt, or gear breaks and no longer drives the camshaft(s), you can often tell by the sound of how the engine turns over. It cranks over very quickly, and no sounds even resembling firing occur (**Figure 7-35**). The engine has no compression. Hopefully, it has been towed to your shop; that's the only way it'll run again.

When a timing chain, belt, or gear skips one or more teeth, or jumps time, it sets the valve timing off significantly. Some engines will run when the timing is off one or even two teeth but not well. Usually you can hear the problem as the engine cranks over. It turns over unevenly; the engine speeds up and then slows down as you crank. Idle quality, acceleration, and emissions will be affected. If the timing is off significantly, it will pop out of the intake or exhaust, or backfire under acceleration. Often the engine will not even start.

Timing Belt Inspection

To inspect a timing belt, you really need to pull the top timing cover off of the engine. This is usually a plastic cover with an upper section and lower section. For inspection purposes, it only will be necessary to remove the upper cover (**Figure 7-36**). You will be looking for cracks like those in serpentine belts. Rotate the engine around to be able to view the whole belt. Look on the underside of the belt at the cogs or teeth that hold the belt in the sprocket. Any broken or missing teeth dictate immediate replacement. **Figure 7-37** shows some typical timing belt failures. If the belt shows signs of oil or coolant contamination,

Figure 7-35 This snapped timing belt left the customer stranded on the highway.

Clip

Figure 7-36 Remove the upper timing cover to get a look at the belt. This cover is easily removed with accessible clips.

Figure 7-37 Any fault with the belt warrants replacement.

the belt should be replaced, and the leak must also be fixed as part of the repair. Be sure to diagnose the cause of the leak and include that procedure and part(s) in your estimate to the customer. As we've discussed, there is no reason to try to get a few more months out of a damaged belt; it could result in a very expensive repair.

CUSTOMER CARE Many maintenance manuals provided with new vehicles have a long list of components and systems that are supposed to be inspected at regular intervals. It may be, for example, that at every 60,000 miles when the spark plugs are replaced, the timing belt is also supposed to be inspected. A careless technician may ignore this and other portions of the 60,000-mile service and just replace the new components. A true professional will perform all the tasks recommended by the manufacturer; they are listed for a reason. By performing full service, you will be providing your customers with the excellent quality repair work they are paying for.

Tensioner Inspection

The pulley-type tensioner used with belts is also commonly replaced when the service interval is long, every 90,000 miles, for example. If the timing belt is being replaced for the second time and the tensioner was not replaced the first time, you should definitely recommend replacement (**Figure 7-38**). An engine that requires belt replacement every 60,000 miles and is being serviced at 120,000 miles deserves a new tensioner to prevent premature failure. To inspect the tensioner for faults, check for scoring on the

Belt tensioner pulley

Figure 7-38 Replace the belt tensioner pulley if it has excessive mileage or if you can feel any roughness or wobbling in the bearing.

pulley surface. Scratches, sharp edges, or grooves can damage a new belt rapidly. Also check the pulley's bearing. Rotate it by hand and feel for any roughness. If you can hear it scratching or feel coarseness, replace it. Also check the pulley for sideplay; grasp it on its ends and check for looseness. If it wobbles at all, replace it.

A timing belt automatic tensioner is also typically replaced during belt replacement. If there are any signs of oil leakage, replacement is mandatory. If the belt replacement interval is relatively frequent, it may be possible to use the automatic tensioner through the life of two timing belts. The best advice is to follow the recommendation of the manufacturer and carry out any inspections or measurements indicated.

Timing Mechanism Replacement

The actual repair procedures for service of timing chains, belts, and gears vary immensely. Some overhead cam (OHC) timing chains can be replaced in a few hours with the engine in place; others require the engine to be removed from the vehicle. It is common to have to remove the cylinder head or the timing cover. Timing belts can generally be replaced with the engine in the vehicle. They can be simple and straightforward or time-consuming and a little tricky. A timing belt on a transversely mounted V6 engine in a minivan, for example, may leave you little room to access the front end of the engine. Replacing the timing mechanism on cam-in-the-block engines using either a chain or a gear set is usually uncomplicated. In every case, it is critical that you set all the shafts into perfect time with the crankshaft. As we've discussed, setting the valve timing off by one tooth will impact engine drivability and emissions. Similarly, an engine with balance shafts not timed properly will vibrate noticeably. Often it is too easy to miss the timing of a shaft by one tooth if you aren't paying close attention and double-checking your work. This is important and time-consuming service work; you want to do it only once! Many technicians will replace the camshaft and crankshaft seals while they have access to them during the job. If the water pump is driven by the timing belt, you will usually replace that as well. Always check idle quality and engine performance after a replacement job to be sure the engine is operating as designed. This text cannot offer a replacement for the manufacturer's service procedures. Instead it offers a couple of specific examples of the procedures for replacing different timing mechanisms. You'll still need manufacturers' specific information to achieve proper timing mark alignment and to perform the job in the most efficient manner. You may also need special tools to hold the camshaft sprockets in position while replacing the belt. **Photo Sequence 15** shows how a timing belt is replaced.

PHOTO SEQUENCE 15
Typical Procedure for Replacing a Timing Belt on an OHC Engine

P15-1 Disconnect the negative cable from the battery prior to removing and replacing the timing belt.

P15-2 Carefully remove the timing cover. Be careful not to distort or damage it while pulling it up. With the cover removed, check the immediate area around the belt for wires and other obstacles. If some are found, move them out of the way.

P15-3 Align the timing marks on the camshaft's sprocket with the mark on the cylinder head. If the marks are not obvious, use a paint stick or chalk to mark them clearly.

P15-4 Carefully remove the crankshaft timing sensor and probe holder.

P15-5 Loosen the adjustment bolt on the belt tensioner pulley. It is normally not necessary to remove the tensioner assembly.

P15-6 Slide the belt off the camshaft sprocket. Do not allow the camshaft pulley to rotate while doing this.

P15-7 To remove the belt from the engine, the crankshaft pulley must be removed. Then the belt can be slipped off the crankshaft sprocket.

P15-8 After the belt has been removed, inspect it for cracks and other damage. Cracks will become more obvious if the belt is twisted slightly. Any defects in the belt indicate it must be replaced. Never twist the belt more than 90 degrees.

P15-9 To begin reassembly, place the belt around the crankshaft sprocket. Then reinstall the crankshaft pulley.

PHOTO SEQUENCE 15 (CONTINUED)

P15-10 Make sure the timing marks on the crankshaft pulley are lined up with the marks on the engine block. If they are not, carefully rock the crankshaft until the marks are lined up.

P15-11 With the timing belt fitted onto the crankshaft sprocket and the crankshaft pulley tightened in place, the crankshaft timing sensor and probe can be reinstalled.

P15-12 Align the camshaft sprocket with the timing marks on the cylinder head. Then wrap the timing belt around the camshaft sprocket, and allow the belt tensioner to put a slight amount of pressure on the belt.

P15-13 Adjust the tension as described in the service manual. Then rotate the engine through two complete turns. Recheck the tension.

P15-14 Rotate the engine through two complete turns again; then check the alignment marks on the camshaft and the crankshaft. Any deviation needs to be corrected before the timing cover is reinstalled.

CUSTOMER CARE When repairing an engine with a broken timing mechanism, it is important to know whether the engine is an interference engine or not. If it is, advise the customer of the potentially very high added cost of removing the cylinder head and replacing valves. Also make sure the customer understands that even more serious damage, such as a valve stuck into a piston, could have occurred. Even on freewheeling engines, it's wise to inform the customer that valve damage is possible. Discuss these issues with the customer or service advisor before offering an estimate for the work.

USING A CHEMICAL SEALANT AS A GASKET

This section is intended as a practice exercise. If cost is a factor, a tube of cheap toothpaste can be used instead of Room Temperature Vulcanizing (RTV). Find a scrap engine, transmission, differential cover, or a plain piece of flat metal. If a flat piece of metal is used, a marker can be used to track and simulate bolt holes. Any of the items can be cleaned after the practice is completed. The technician will have to use RTV in many applications.

> **⚙ SERVICE TIP** Cutting the tube at an angle allows only a small portion of the tube to rest on the cover. This helps prevent spreading the RTV.

Remove the cover from the RTV tube and cut it at an angle until its opening is about 1/8 inch in diameter (**Figure 7-39**). Once the bead is started, do not break it. If it is necessary to move a hand, keep the tube end in place and begin to squeeze the tube again from that point. Also, when going by a bolt hole, lay the bead to the inside (lubricant side) of the hole (**Figure 7-40**). Do not run the bead over the hole.

Place the end of the tube at a convenient starting point. The tube should be angled at 45 degrees to the cover (**Figure 7-41**). Squeeze from the bottom and continue squeezing as the tube is moved. The tube should be moved with an even spread. The RTV should come out in a smooth, even flow without humps or dips in the bead. Continue until the

Figure 7-39 Cut the RTV tube at an angle.

Figure 7-40 Lay the RTV bead in the manner shown. Keep the bead to the lubricant side of the bolt holes.

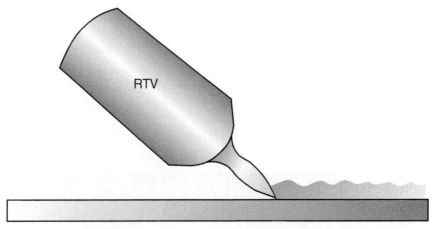

Figure 7-41 Keep the point of the tube to the metal at about a 45-degree angle. The angled cut will allow for a smooth bead.

bead is laid completely around the cover. Finish the bead by rolling the tube tip into the starting point of the bead. This will mesh the two ends together and provide a leak-resistant seal.

> ⚙ SERVICE TIP Do not place RTV on a gasket of any material unless directed to do so by the gasket installation instructions. The RTV forms a rubber-like bead and will not stick to cork and most other types of gaskets, thus forming a leak between the RTV and gasket. Typically, the only time RTV and gaskets are used together is when RTV is used to seal the pieces of a multipiece gasket.

Before reinstalling the cover or pan or applying sealant to a cover/pan, it should be checked for straightness. It is not unusual for a thin stamped piece of metal like the typical cover or pan to become deformed around the clamping flange. This may be caused by overtightening on the last installation or the stresses of operation. Either way, a warped cover or pan will not properly seal. Any time a pan or cover is removed, it should be checked for straightness along its flange. For the purpose of this discussion, the term *pan* will be used because it is the most accepted. An oil pan is used for this discussion.

There are several means for straightening the clamping flange of a pan, but all require a solid, flat surface such as a heavy steel work table. The one discussed here is probably the most common technique. Place the oil pan right side up on the table and use a straight-edge to locate high and low spots along the flange. Emphasis should be directed along the line between the bolt holes. Use some type of marker to mark the start and end of each low or high spot. Compare these spots to determine if the flange is warped downward here or if the spot is caused by an area that has been warped upward at another spot. Any high-warped areas can be flattened more easily than trying to raise warped low areas.

Place the oil pan upside down on the table with the worst side most easily accessible. Now the low areas are the high areas. Locate or make a 3- to 4-inch strip of metal or hard wood (such as ash or oak) block with a width equal to that between the oil pan body and the outer edge of the flange and several inches tall. It should be of strong enough metal or wood to transfer hammer blows or clamping force to the flange without deforming. It depends on the thickness of the flange and its metal composition.

Secure the pan or have an assistant hold it in place. Place the block onto the first area to be straightened. The block should be laid lengthwise along the flange (**Figure 7-42**). Use a ball-peen hammer to strike the block at different places along its length. Do not use excessive force with the hammer. Repetitive midforce blows will do a better job than hard blows. A couple of hard blows may cause even more warping. Continue to work the area until it is flat for the length of the working block. Check it with the straightedge before continuing. With one area straightened, move the block along the flange, taking care to have the block overlap the previously straightened area. Continue in this fashion completely around the pan or as necessary. The corners will provide some difficultly. However, the

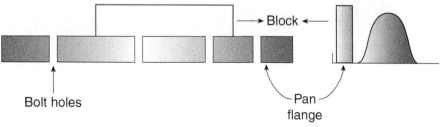

Figure 7-42 Lay the block so the hammering force is transmitted over a larger area.

use of a shorter piece of metal or a flat nose punch with a diameter approximately the width of the block can be used at those points. Skill in this type of job is gained with experience.

> ⚙️ **SERVICE TIP** Check with a machine shop. Sometimes they will have a metal jib that can do the straightening job much easier—or at the least they could machine a block for use in this type of repair. Most vehicle pans have about the same width of flange.

When removing an engine cover, you may need to use RTV. Its use is becoming very popular. When installing a new gasket or sealant, make sure to follow the manufacturer's service directions. **Photo Sequence 16** outlines the typical procedures for replacing a valve cover gasket.

PHOTO SEQUENCE 16
Removing and Installing an Engine Valve Cover with a New Gasket

P16-1 Consult the service information for instructions on how to replace the valve cover gasket properly.

P16-2 Remove the air distribution ductwork to gain access to the valve cover.

P16-3 Remove and label connectors and vacuum lines as needed.

P16-4 With the intake ducts out of the way, remove the spark plug wire loom and wires that will obstruct removal.

P16-5 Loosen the valve cover bolts using the proper sequence if one is specified.

P16-6 Maneuver the valve cover off the cylinder head.

P16-7 Remove the old gasket material, and thoroughly clean and dry the valve cover.

P16-8 Install the new valve cover gasket securely. Use sealant only if it is specified.

P16-9 Always properly torque the valve cover to the specification to ensure a quality repair.

P16-10 Reinstall the ductwork securely.

P16-11 Reconnect all wiring and vacuum lines. Wipe down the valve cover and surrounding components. Finally, start the vehicle and check to be sure it runs properly and has no leaks.

CASE STUDY

A technician performed a routine job of replacing the differential fluid. She sealed the cover with RTV as recommended by the service manual. The customer returned some weeks later with a small but unsightly leak around the cover. The technician removed, cleaned, and installed the cover. RTV was used. Within a few days, the customer returned with the same problem except the leak was now worse. A careful inspection of the cover and housing revealed no visible problem. As a matter of curiosity, the technician screwed in some of the cover bolts. They bottomed out with about half of the bolt threads exposed. Digging into the holes with a pick revealed that every hole was filled about halfway full with RTV. The technician had laid the bead straight over the hole on the cover, and the installed bolt pushed the RTV into the hole. The bolts were tightened with a quarter-inch-drive impact wrench, so the technician did not have a feel for the bolt as it bottomed on the RTV. Cleaning the holes completely, installing RTV properly, and torqueing the fasteners properly corrected the problem. Rushing the "simple" tasks on a routine job can result in a loss of profits and customers.

EXHAUST SYSTEM INSPECTION

The **front pipe**, **muffler**, and **tailpipe** are the three main components used to channel exhaust gases from the engine to the rear bumper and reduce exhaust noises.

A **diesel particulate filter** is used to capture soot and incinerate it during regeneration cycles.

The engine computer uses a strategy to spray a fine mist of **diesel exhaust fluid (DEF)** in the exhaust during certain periods to help clean up the exhaust emissions and burn the soot in the filter.

Most parts of the exhaust system, particularly the **front pipe**, **muffler**, and **tailpipe**, are subject to rust, corrosion, and cracking. Broken or loose clamps and hangers can allow parts to separate or hit the road as the car moves.

Some newer diesel engines use a different type of exhaust system that is essential to its emission system. This is known as an after-treatment system. This type of exhaust uses a **diesel particulate filter** that captures soot to incinerate it during regeneration cycles. The engine computer uses a strategy to spray a fine mist of **diesel exhaust fluid (DEF)** in the exhaust during certain periods. This helps clean up the exhaust emissions and burn the soot in the filter. This reduces the black smoke typically found in the exhaust of a diesel engine.

⚠ **WARNING** If the engine has been running, exhaust system components may be extremely hot. Wear protective gloves when working on these components and keep flammable objects away from hot areas.

⚠ **WARNING** Exhaust gas contains poisonous carbon monoxide (CO) gas. This gas can cause illness and death by asphyxiation. Exhaust system leaks are dangerous for customers and technicians.

Any exhaust system inspection should include listening for hissing or rumbling that would result from a leak in the system (**Figure 7-43**). An on-lift inspection should pinpoint any of the following types of damage:

1. Holes, road damage, separated connections, and bulging muffler seams.
2. Kinks and dents.
3. Discoloration, rust, soft corroded metal, and so forth.
4. Torn, broken, or missing hangers and clamps.
5. Loose tailpipes or other components.
6. Bluish or brownish catalytic converter shell, which indicates overheating.

Figure 7-43 Exhaust system for an OBD II vehicle.

INSPECTING THE PRESENCE AND CONDITION OF EMISSION CONTROL DEVICES

Each of the emission systems outlined in the Classroom Manual needs periodic servicing and replacement. In state and local areas requiring emissions testing, a vehicle will not pass inspection if emissions components are missing or damaged, even if the exhaust emissions are within standards. In this section, basic inspection procedures on some emission systems are discussed. Systems not listed usually require an experienced technician and electronic testers. All systems will require an experienced technician for detailed testing.

Secondary Air Injection

The drive belt for the secondary air injection is inspected as other belts. As the engine is operated, there should be no exhaust noises in or near the secondary air **diverter valve**. If there is exhaust noise, check the flexible hose running from the valve to the metal line going to the exhaust manifold or catalytic converter (**Figure 7-44**). There is a check valve inside the metal line that prevents exhaust from entering the flexible hose and valve. It may become stuck in the open position over a period of time and gases can enter the hose. If it allows exhaust to back up, pulsing or heat may be felt in the flexible hose. The check valve requires replacing if this condition is noted. Accelerate the engine several times and quickly release the throttle. If there is a backfire through the exhaust, the diverter valve may not be shutting off the airflow during deceleration. Vacuum hoses leading to the diverter valve should be flexible, crack-free, and respond to squeezing much like any other rubber-type hoses. Replace any that show signs of wear or dry rotting.

Evaporative Systems

Removing the fuel tank filler cap is the first check of the evaporative (EVAP) system. If the sound of moving air occurs as it is loosened, the cap is probably working. The other end of the system is located in the engine compartment and has vacuum hoses that need to be checked (**Figure 7-45**). The remainder of the system can only be inspected for leaks. Further tests to diagnose the EVAP system require electronic test equipment. Always replace the fuel filler cap with one of identical design and operation.

Classroom Manual
pages 72–73

 Caution

Do not attempt to touch or repair hot engine components. Serious burn injury could result. It is best to allow the vehicle to cool before beginning the tasks in this section.

Diverter valve A valve designed to supply air to the exhaust system to help control emissions.

 Caution

Use care when working around an operating converter. The catalytic converter may reach temperatures of 1,400°F. Second- and third-degree burns on hands and arms could result.

Figure 7-44 On the secondary air injection system, if the check valve fails it often leads to a complete failure of the system.

Figure 7-45 An older but typical EVAP system. The biggest difference between the old system and the newer ones is that the various valves are now computer controlled.

Catalytic Converters

The catalytic converter (**Figure 7-46**) cannot be checked completely, but some simple checks can be made. Start the engine and allow it to warm. While the engine is operating, check the end of the tailpipe. If there is soot in the pipe or the exhaust gas is black, an overly rich air/fuel mixture is indicated. Too much fuel in the exhaust will clog a converter and restrict the exhaust flow. After the engine has warmed to operating temperature, attempt to observe the exhaust pipe between the converter and exhaust manifold. A pipe that is beginning to turn or has turned red also indicates a clogged exhaust system. Do not touch the pipe or any other component of the exhaust. A noise from near or inside the converter may indicate a loose heat shield or the material in the converter may be coming apart or loose.

The remainder of the emission systems are checked at this point for their presence. Some of the components may not be visible and may require specialized testing equipment and training. As you grow into an experienced technician and student, you will learn how to diagnose these components.

Figure 7-46 Monolithic-type catalytic converter.

SCANNING FOR DIAGNOSTIC TROUBLE CODES (DTC) AND MONITORS

The use of a scan tool like the ones discussed in Chapter 3 can aid the technician during the diagnostics process. Since the late 1980s, vehicle computers have been designed to provide technicians with some insight into the electronic world on the vehicle. The first system was known as On-Board Diagnosis I (OBD I), and it grew into the present system known as OBD II. OBD II first arrived on model year 1996 vehicles. Not all vehicles were OBD II-compliant in 1996, most notably Honda. Certain characteristics of an OBD II vehicle are the under hood certification label, the location and shape of the 16-pin Diagnostic Link Connector (DLC) (**Figures 7-47** and **7-48**), having a post-catalytic converter oxygen sensor, having heated oxygen sensors, monitoring systems, saving freeze frame data, and having one of a few chosen computer languages and programs to convey the information to a scan tool. Another common characteristic with OBD II vehicles is the MIL. This stands for **malfunction indicator lamp**, also known as the "check engine" or "service engine soon" light on the dash (**Figure 7-49**).

Classroom Manual
pages 72-73

MIL (malfunction indicator lamp) is a dash-mounted warning light commonly referred to as the check engine light.

Figure 7-47 The DLC on an OBD II vehicle is located under the driver's side dash.

Pin 1
Pin 16
Figure 7-48 An OBD II DLC.

MIL light

Figure 7-49 The malfunction indicator light (MIL) will come on when the PCM has detected a problem that could allow emissions to increase; this certainly includes misfire.

Monitor The vehicles computer observes engine data and adjusts the controls to keep emissions as low as possible.

OBD II provides more accurate data and has the ability to **monitor** certain electronic components and systems for wear or damage. To retrieve this information, a technician must use a scan tool.

> ⚙️ **SERVICE TIP** Some vehicles, such as those manufactured by Ford, will go through an OBD self-test. This self-test will turn on and actuate certain components under the hood. Don't be alarmed when you hear this happening, and always watch the scan tool screen for what is happening.

Diagnostic trouble codes (DTCs) are alphanumeric fault codes that represent a circuit failure in a monitored system.

A scan tool is capable of retrieving **diagnostic trouble codes (DTCs)**, parameter identification data (PID), freeze frame data, and inspecting the monitor status. Scan tools come in two different forms. A generic scan tool will work on all OBD II vehicles, but it will only scan the generic side of OBD II. This will get you the minimum amount of data needed to diagnose and repair the vehicle (**Figure 7-50**). A manufacturer scan tool (**Figure 7-51**) can only connect and work with the manufacturer's vehicles that it was designed for. However, a manufacturer scan tool can get you more data than a generic scan tool (also called Global OBD II scan tool) and can perform some tasks that no other scan tool can.

With the scan tool connected to the DLC, power it on. Some scan tools use the vehicle's battery, and some use their own battery. If you will be using the scan tool for a while, you may want to connect a battery charger to the vehicle or disable the daytime running lights. The two computers will communicate, and any stored and pending DTCs can be displayed on the scan tool's display screen. You can now record the codes shown on the screen. OBD II will have five digits. There will be a letter followed by four numbers. All OBD II codes are the same for all manufacturers. To decipher the codes, you will want to consult the service manual.

> ⚙️ **SERVICE TIP** If a NO COMMUNICATION message is displayed, check the vehicle setup and entry on the scan tool and tool/vehicle connection and the vehicle's battery voltage.

Figure 7-50 Scan tool communicates with the PCM and displays DTCs and engine data.

Figure 7-51 Laptop computers are now often used as a scan tool.

Here is an example of deciphering a code: If the code is P0304, the "P" stands for powertrain (meaning engine and transmission). The "0" stands for a generic OBD II code, meaning a generic OBD II scan tool can see this code, and it is the reason the MIL is illuminated. The "3" is the subsystem. In this case, the "3" means it is an ignition or misfire code. The "04" is the variable code. In this case, the "04" means cylinder #4. So this code is telling the technician that there is a misfire on cylinder #4. **Photo Sequence 17** shows the typical procedure for retrieving a DTC using a generic (Global) OBD II scan tool.

PHOTO SEQUENCE 17
Typical Procedure for Using a Scan Tool to Retrieve DTCs

P17-1 Roll the driver's side window down, turn the ignition to the off position, and block the wheels (or set the parking brake).

P17-2 Select the proper scan tool data cable for the vehicle you are working on.

P17-3 Connect the scan tool data cable to the DLC on the vehicle.

P17-4 Turn the ignition key to the KOEO (key on, engine off) position. Some scan tools will automatically turn on at this point; others may need to be turned on.

P17-5 Enter the model year, make, and selected VIN components into the scan tool.

P17-6 Maneuver through the scan tool's menu, and select "read/retrieve DTC."

P17-7 Record the DTCs on the repair order.

ENGINE PRESSURE TESTING

Cranking Compression Test

A cranking compression test is the most commonly used type of engine pressure testing. It will test the pressure output of each individual cylinder. It can clearly determine if the valves, rings, and head gasket(s) are sealing the combustion chamber properly (**Figure 7-52**). When any of those components are faulty, the engine may not be able to develop adequate compression. The manufacturer will provide a specification for the appropriate cranking compression pressure. Low compression will cause less productive combustion. You and the customer will notice that the engine lacks power or misfires on a cylinder(s). An OBD II code that indicates a misfire may lead to this test series.

To perform a cranking compression test (**Photo Sequence 18**):

1. Let the engine warm up to normal operating temperature. (This ensures that the pistons have expanded to properly fit and seal the cylinder.)
2. Let the engine cool for 10 minutes so you don't strip any threads; then remove the spark plugs. Remember to keep them in order so you can read them.
3. Block the throttle at least part way open so the engine can breathe.

Figure 7-52 A cranking compression test will detect leaks at the valves, rings, pistons, and head gasket.

4. Disable the fuel system by removing the fuel pump fuse or relay to prevent contamination of the engine oil.
5. Disable the ignition system so you do not damage the coils. You can either remove the ignition fuse or remove the low-voltage connector to the coil(s). Do not simply remove the plug or coil wires and let them dangle. The coil can be damaged as it puts out maximum voltage trying to find a ground path for the wires.
6. Install a remote starter or have an assistant help you.
7. Carefully thread the compression tester hose into the spark plug hole and attach the pressure gauge to the adaptor hose (**Figure 7-53**).
8. Crank the engine over through five full compression cycles. Each cycle will produce a puffing sound as engine compression escapes through the spark plug ports.
9. Record the final pressure.
10. Repeat the test on each of the other cylinders.

PHOTO SEQUENCE 18
Typical Procedure for Performing a Cranking Compression Test

P18-1 Tools required to perform this task: compression tester and adapters, spark plug socket, ratchet, extensions, universal joint, remote starter button, battery charger, service manual, and squirt can of oil.

P18-2 Follow the service manual procedures for disabling the ignition and fuel systems.

P18-3 Clean around the spark plugs with a low-pressure air gun, and remove all of the spark plugs from the engine.

P18-4 Install the remote starter button.

P18-5 Carefully install the compression tester adapter into the first cylinder's spark plug hole.

P18-6 Connect the battery charger across the battery, and adjust to maintain 13 to 14.5 volts. This is needed to allow the engine to crank at constant speeds throughout the test.

P18-7 With the throttle plate held or locked into the wide-open throttle position, crank the engine while observing the compression reading.

P18-8 Continue to rotate the engine through four compression strokes as indicated by the jumps of the needle.

P18-9 Record the reading after the fourth compression stroke. Continue until all cylinders have been tested and compare the test results. Compare test results with the manufacturer's specifications.

Figure 7-53 This style compression tester has adapters that thread into the spark plug hole.

⚙ **SERVICE TIP** A "blown" head gasket that is allowing coolant into the combustion chamber may be doing an adequate job of sealing compression. If an engine passes a compression test, it does not prove that the head gasket is sealing properly.

Analyzing the Results

With all the pressure readings in front of you, compare them to the manufacturer's specifications. Typical compression pressures on a gas engine range between 125 and 200 psi. The readings should be close to the specification and within 15 to 20 percent of each other. Some manufacturers specify only that the compression be over 100 psi and that each reading is within 20 percent of the others. To calculate 20 percent easily, take the normal reading (say, 160 psi) and drop the last digit (16 psi). Multiply by 2 (32 psi), and that is 20 percent. If the readings were 160, 150, 100, and 155, the third cylinder is more than 32 psi different than the other three. This indicates a significant problem with cylinder number three that must be identified and corrected.

If one or more cylinders are below specification, it is likely caused by worn rings, burned valves, or valves sticking open from carbon on the seats, a faulty head gasket, or a worn or broken piston. Perform a wet compression test to help find the probable cause. When all the cylinders are slightly low, it usually indicates a high mileage engine with worn rings. Two adjacent cylinders with low compression readings may indicate a blown head gasket leaking compression between two cylinders. If all the cylinders are near zero or very low, suspect improper valve timing. A timing belt, chain, or gears that have "jumped" time by slipping a tooth will allow the valves to be open at the wrong time, causing low compression readings.

Wet Compression Test

To help determine the cause of a weak cylinder, perform a wet compression test (**Figure 7-54**). Use an oil can and put two squirts of oil into the weak cylinder. Reinstall the compression gauge, crank the engine over five times, and record the reading. If the low reading cylinder increases to almost normal compression pressure, it is likely that worn rings are causing the cylinder leakage. The extra oil in the cylinder helps the rings to seal better during the test. If a valve is burned or being held open by carbon or if there is a hole in the piston, a film of oil will not dramatically improve the compression pressure. Note that a wet compression test may not be as effective on horizontally opposed engines, because the oil will seal only the lower half of the cylinder wall.

Look at the examples that follow:

Compression test results					
Cyl. #1	Cyl. #2	Cyl. #3	Cyl. #4	Cyl. #5	Cyl. #6
175 psi	165 psi	170 psi	80 psi	170 psi	170 psi
Wet compression test					
Cyl. #4					
160 psi					

This is a clear indication that the rings are severely worn on cylinder number four. If the wet compression test had shown a rise to only 105 psi, you would suspect a burned valve or leaking valve. To definitively determine the cause of low cranking compression, perform a cylinder leakage test (**Figure 7-55**).

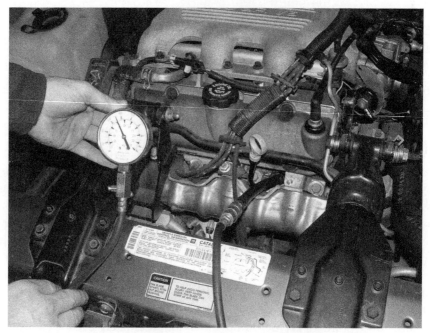

Figure 7-54 This cylinder showed low compression of only 100 psi while cranking dry and wet. A leakdown test will verify a leaking valve.

Figure 7-55 Listen for air in the coolant reservoir for indications of a leaking head gasket.

Running Compression Test

It is possible that the cranking compression can show good results but the engine still is not mechanically sound. A cranking compression test does not do a good job of testing the valvetrain action. There is so much time for the cylinder to fill with air when the engine is spinning only at cranking speeds of about 150 to 250 rpm. A more accurate way to test the loaded operation of the valvetrain is to perform a running compression test. Running compression readings will be quite low compared to the cranking compression pressures because there is much less time to fill the cylinders with air.

If you have identified a weak cylinder through power balance testing, check that cylinder and a few others to gain comparative information. In most cases, it is worth the time to check all the cylinders. To perform the running compression test:

1. Remove the spark plug only from the test cylinder.
2. Disable the spark to that cylinder by grounding the plug wire or removing the primary wiring connector to the individual coil.
3. Disable the fuel to the test cylinder by disconnecting the fuel injector connector.

4. Install the compression gauge into the spark plug port.
5. Start the engine and let it idle.
6. Release the pressure from the gauge whose reading reflects cranking compression. Record the new pressure that develops at idle.
7. Rev the engine to 2,500 rpm. Release the idle pressure, and record the new pressure that develops at 2,500 rpm.
8. Remove the gauge and reinstall the spark plug. Remember to reattach the plug wire or coil and the fuel injector connector.
9. Repeat for several or all cylinders.

Analyzing the Results

Generally, the running compression will be between 60 and 90 psi at idle and between 30 and 60 psi at 2,500 rpm. The most important indication of problems, however, is how the readings compare to each other.

Look at the following readings:

	Cyl. #1	Cyl. #2	Cyl. #3	Cyl. #4
Idle	70 psi	75 psi	45 psi	70 psi
2,500 rpm	40 psi	40 psi	20 psi	45 psi

Cylinder number three is barely below the general specifications, but it is clearly weaker than the other cylinders. These results warrant investigation of the problem in cylinder number three. The most likely causes of low running compression readings are:

- Worn camshaft lobes
- Faulty valve lifters
- Excessive carbon buildup on the back of intake valves (restricting airflow) (**Figure 7-56**)
- Broken valve springs
- Worn valve guides
- Bent pushrods

Use a stethoscope, vacuum testing, and visual inspection under the valve covers to help locate the reason for the low reading.

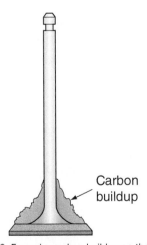

Carbon buildup

Figure 7-56 Excessive carbon buildup on the back of intake valves caused by a bad valve seal or guide leakage prevents the engine from pulling in enough air to make adequate power. The added weight can also cause the valve to float at higher rpm.

CYLINDER LEAKAGE TESTING

A **cylinder leakage test** places air pressure into a combustion chamber so you can listen for leaks from possible faulty components.

Once you have identified a cylinder(s) with low cranking compression, you can use a **cylinder leakage test** to pinpoint the cause of the problem. To perform a cylinder leakage test, you put pressurized air in the weak cylinder through the spark plug hole when the valves are closed. You can then listen for leakage of air from the combustion chamber at various engine ports to determine the source of the leak. This is an excellent way to gather more information before disassembling the engine or when deciding whether a replacement engine should be installed. A cylinder leakage test is also a very thorough way of evaluating an engine when someone is considering purchasing a vehicle. You can even perform a cylinder leakage test on a junkyard engine to assess its value before purchasing it.

> **⚙ SERVICE TIP** If the piston and connecting rod are not exactly at TDC, the engine may rotate as you apply air pressure. Inline engines need to be almost exactly at TDC, whereas V-type engines can be a little off. If the engine starts to turn while you apply air pressure, try to move the cylinder closer to TDC. If you still cannot get the cylinder to stop moving, then try other things like holding the ratchet (watch your hand), using a piston position locater, or checking the timing marks on the crankshaft pulley. Each vehicle is very different, but you may be able to use a bar in a location where it stops a pulley from moving.

> **CUSTOMER CARE** A customer was anxious to buy a good-looking, performance-modified Honda. He was finally convinced to allow us to perform a cylinder leakage test after we noticed a bit of blue smoke coming from the tailpipe. A leakdown test showed very serious ring wear on two cylinders. Although he still bought the car, he paid $1,500 less for it than he had planned.

Cylinder Leakage Test

The cylinder leakage test can help you evaluate the severity of an engine's mechanical problem and pinpoint the cause. You can test one low cylinder to find a problem, or you can test each cylinder to gauge the condition of the engine. Like a compression test, the cylinder leakage test will detect leaks in the combustion chamber from the valves, rings, or pistons and the head gasket. From this test, however, you can clearly identify which is the cause of the leak.

To perform a cylinder leakage test (**Photo Sequence 19**):

1. Let the engine warm up to normal operating temperature.
2. Let the engine cool for 10 minutes, and then remove the spark plug from the cylinder to be tested.
3. Remove the radiator cap and the PCV valve. This prevents damage to the radiator or engine seals if excessive pressure leaks into the radiator or the engine's crankcase.

PHOTO SEQUENCE 19
Typical Procedure for Performing a Cylinder Leakage Test

P19-1 Loosen the ignition cassette to access the spark plugs.

P19-2 Disconnect power to the ignition cassette to avoid damaging the coils.

P19-3 Remove the ignition cassette.

P19-4 Remove the spark plugs.

Warm for piston to rise to TDC

Rotate engine over slowly with crank pulley

P19-5 Rotate the engine to TDC compression for the test cylinder so that the piston is at TDC compression.

P19-6 Hook up shop pressure to the leakage tester and set to 0 percent leakage.

P19-7 Connect adapter to test cylinder and read leakage. This cylinder is leaking slightly over 30 percent.

P19-8 Listen for air at the intake manifold for leakage past an intake valve.

P19-9 Listen for air at the oil fill to detect leakage past the rings. In this case you could feel the air blowing out the fill pipe.

P19-10 Listen for air in the coolant reservoir for indications of a leaking head gasket.

P19-11 Listen for air at the tailpipe to find leakage past an exhaust valve.

P19-12 The moderate blue smoke from the tailpipe substantiates the diagnosis of worn rings on this high-mileage vehicle.

4. Turn the engine over by hand until the cylinder is at TDC on the compression stroke. This ensures that the valves will be closed, and the rings will be at their highest point of travel. The cylinders wear the most at the top because of the heat and pressure of combustion, so if the rings are leaking, they will show the greatest evidence at the top. There are a few different methods of getting a cylinder to TDC compression, depending on the engine.

   a. Have an assistant turn the engine over slowly while watching the piston come up to TDC. You should be able to feel and hear blowing out of the spark plug hole if it is coming up on the compression stroke. If you begin the leakage test and show 100 percent leakage, you are on TDC exhaust. Rotate the engine 360 degrees and retest.

   b. Remove the valve cover and watch the rockers or camshaft as you turn the engine over. On the intake stroke, the intake valve will open and then close. Then both the valves will remain closed while the piston comes up to TDC. If the engine has rocker arms, move them up and down to be sure they both have lash (clearance) between the arm and the pushrod. If the engine has an overhead camshaft(s), the followers will be on the base circle, not the lobes of the camshaft.

   c. If the vehicle has a distributor, remove the cap and rotate the engine until the rotor points at the test cylinder's firing point.

5. With the cylinder at TDC compression, thread the cylinder leakage tester adaptor hose into the spark plug hole.

6. Apply air pressure to the leakage tester and calibrate it according to the equipment instructions. On the tester shown, you turn the adjusting knob until the gauge reads 0 percent leakage without it attached to the cylinder.

7. Connect the leakage tester to the adaptor hose in the cylinder while watching the crankshaft pulley. If at all the engine rotates at all, reset it to TDC. When the engine is right at TDC, it will not rotate with air pressure applied.

8. Read the percentage of leakage on the tester and gauge as follows:
   - Up to 10 percent leakage is excellent; the engine is in fine condition.
   - Up to 20 percent leakage is acceptable; the engine is showing some wear but should still provide reliable service.
   - Up to 30 percent leakage is borderline; the engine has distinct wear but may perform reasonably well.
   - More than 30 percent leakage shows a significant concern that warrants repair.

9. With the air pressure still applied, use a stethoscope and listen at the following ports:
   - At the tailpipe: Leakage here indicates a burned or leaking exhaust valve.
   - At the throttle or a vacuum port on the intake manifold: Leakage here indicates a burned or leaking intake valve.
   - At the radiator or reservoir cap: Leakage here proves that the head gasket is leaking.
   - At the oil fill cap or dipstick tube: Leakage here indicates worn rings.
   - If the leakage at the oil fill is near 100 percent, suspect damage to a piston (**Figure 7-57**).

10. Release the pressure, remove the adaptor, and rotate the engine over to the next test cylinder. Repeat as indicated. Refer to **Photo Sequence 19** for detailed pictures and directions on performing a cylinder leakage test.

Figure 7-57 This destroyed piston showed 100 percent leakage.

SERVICE TIP There will always be some leakage past the rings; a light noise is to be expected. Excessive ring wear will cause distinct blowing out of the cap or dipstick hole. You will be able to hear the noise and feel the pressure.

CUSTOMER CARE Although a cylinder leakage test can identify the primary cause of low compression readings, remember to keep your mind and eyes open to other problems when repairing the engine. Do not automatically inform a customer that an exhaust valve is leaking and a valve job will cure the problem. Consider the engine mileage, oil condition, pressure and usage, and degree of leakage past the rings. If the engine has high miles, it is very likely that the rings are significantly worn. By replacing the faulty valve(s) and restoring the others to like-new condition, you will increase the compression and combustion pressures. The change may be significant enough that the old rings will not seal as well as they had been before the valve job. In this case, the customer would be back complaining about blue smoke from oil consumption or about the engine still lacking adequate power. This may very well be an example of when a replacement engine would be the most cost-effective repair for the customer. Evaluate the whole situation, and make your recommendation to the customer. If you offer the customer repair options, be very clear about explaining the possible consequences.

ASE-STYLE REVIEW QUESTIONS

1. *Technician A* says a plugged PCV may be indicated by oil in the air filter housing.
   Technician B says that some vehicles do not have a PCV valve.
   Who is correct?
   A. A only
   B. B only
   C. Both A and B
   D. Neither A nor B

2. OBD II scan tools are being discussed.
   Technician A says a generic (Global) OBD II scan tool will connect to all OBD II vehicles and be able to read the stored DTCs.
   Technician B says that a manufacturer scan tool can perform more tasks than a generic (Global) OBD II scan tool can.
   Who is correct?
   A. A only
   B. B only
   C. Both A and B
   D. Neither A nor B

3. Belts are being discussed.
   Technician A says that a serpentine belt is really just a micro-vee belt that is in a certain configuration.
   Technician B says that some belts do not crack as they age and wear out.
   Who is correct?
   A. A only
   B. B only
   C. Both A and B
   D. Neither A nor B

4. The fuel system is being discussed.
   Technician A says the pressure must be released on all vehicles before removing the fuel filter.
   Technician B says poor engine operation may result if the inlet end of the fuel filter is toward the fuel tank (filter is installed backwards).
   Who is correct?
   A. A only
   B. B only
   C. Both A and B
   D. Neither A nor B

5. Coolant that is green in color is:
   A. Considered a traditional coolant.
   B. Commonly found on new cars.
   C. A long-life coolant.
   D. None of the above.

6. The lubrication system is being discussed.
   Technician A says an oil filter gasket should have a light coat of RTV applied.
   Technician B says the old oil filter gasket may stick to the filter mounting on the engine.
   Who is correct?
   A. A only
   B. B only
   C. Both A and B
   D. Neither A nor B

7. The cooling system is being discussed.
   Technician A says coolant must be mixed with antifreeze and water.
   Technician B says a slimy feel to the coolant indicates the coolant should be changed and the system flushed.
   Who is correct?
   A. A only
   B. B only
   C. Both A and B
   D. Neither A nor B

8. *Technician A* says the oil drain plug must be torqued to specifications.
   Technician B says the oil pan plug may require a new gasket.
   Who is correct?
   A. A only
   B. B only
   C. Both A and B
   D. Neither A nor B

9. An OBD II vehicle is being discussed.
   Technician A says all OBD II vehicles have a 10-pin DLC.
   Technician B says the OBD II DLC is located under the dash.
   Who is correct?
   A. A only
   B. B only
   C. Both A and B
   D. Neither A nor B

10. An engine test that can tell you the pressure of each individual cylinder is called a:
    A. compression test.
    B. vacuum test.
    C. power balance test.
    D. cylinder leakage test.

ASE CHALLENGE QUESTIONS

1. Two technicians are discussing an oil change.

   Technician A says the only time you should use an oil filter wrench is to remove an old oil filter.

   Technician B says you should use an oil filter wrench to remove and install an oil filter.

   Who is correct?

   A. A only   C. Both A and B

   B. B only   D. Neither A nor B

2. A vehicle is being serviced for an oil change.

   Technician A says all the fluids should be checked.

   Technician B says air filter should be replaced with every oil change.

   Who is correct?

   A. A only   C. Both A and B

   B. B only   D. Neither A nor B

3. Replacing a fuel filter is being discussed.

   Technician A says before removing the filter you must relieve the fuel pressure by cracking the line open.

   Technician B says that if the vehicle is not running there is no fuel pressure in the line.

   Who is correct?

   A. A only   C. Both A and B

   B. B only   D. Neither A nor B

4. Timing belt inspection is being discussed.

   Technician A says using a stethoscope can determine if further diagnosis is needed.

   Technician B says you must remove the timing cover to perform a belt inspection.

   Who is correct?

   A. A only   C. Both A and B

   B. B only   D. Neither A nor B

5. A cooling system is being discussed.

   Technician A says that a sticking thermostat can cause the engine to overheat.

   Technician B says an intermittent stuck thermostat can cause a coolant gauge to spike.

   Who is correct?

   A. A only   C. Both A and B

   B. B only   D. Neither A nor B

Name _____   Date _____

PERFORMING AN ENGINE OIL AND FILTER CHANGE

Upon completion and review of this job sheet, you should be able to perform a routine engine oil and filter change service.

ASE Education Foundation Correlation

This job sheet addresses the following **MLR** tasks:

1.A.3   Inspect engine assembly for fuel, oil, coolant, and other leaks; determine necessary action. **(P-1)**

1.C.5   Perform engine oil and filter change; use proper fluid type per manufacturer specification; reset maintenance reminder as required. **(P-1)**

This job sheet addresses the following **AST/MAST** tasks:

1.A.4   Inspect engine assembly for fuel, oil, coolant, and other leaks; determine needed action. **(P-1)**

1.D.10   Perform engine oil and filter change; use proper fluid type per manufacturer specification. **(P-1)**

We Support
ASE | Education Foundation

Tools and Materials

Drain pan

Filter wrenches

Box-end wrenches

Service manual

Protective Clothing

Goggles or safety glasses with side shields

Describe the vehicle being worked on:

Year _____ Make _____ Model _____

VIN# _____ Engine type and size _____

Procedure

Task Completed

1. Move the vehicle to the lift. Record any noise or obvious defects found during the movement. _____

   _____

   _____

   _____

2. Engine Identification:

   Displacement _____ Block _____ Valvetrain _____

3. Determine the type and amount of oil and the filter:

   Oil _____   Amount_____

   Filter/customer preferred brand

4. Raise the vehicle and place the drain pan under the oil pan plug.   ☐

5. Remove the plug and drain the oil. Note the condition of the oil.

   _____

6. Check the plug gasket and replace if necessary.   ☐

7. Install the plug after the oil has drained. Torque to specifications.   ☐

8. Move the drain pan if necessary. Remove and drain the oil filter. Did the gasket come off with the old filter?

_____

9. Clean the oil filter mount, lube the new gasket, and install the new filter.   ☐

10. Lower the vehicle.   ☐

11. Install the proper amount of oil into the engine.   ☐

12. Start the engine. Watch the light or gauge. If no oil pressure is indicated within a few seconds, shut down the engine and locate the problem.   ☐

13. If the oil pressure is correct, shut down the engine.   ☐

14. Check for leaks at the oil filter and oil pan plug.   ☐

15. Check the oil level and top off as needed.

    Was oil required? _____ If yes, how much? _____

Instructor's Response

_____

_____

_____

NAME _____   DATE _____

COOLING SYSTEM FLUSH

Upon completion of this activity, you will be able to understand how to properly determine if the cooling system needs to be flushed or drain and filled. You will also perform a power coolant flush using a coolant exchanger.

ASE Education Foundation Correlation

This job sheet addresses the following **MLR** tasks:

1.C.1   Perform cooling system pressure and dye tests to identify leaks; check coolant condition and level; inspect and test radiator, pressure cap, coolant recovery tank, and heater core and galley plugs; determine necessary action. **(P-1)**

1.C.4   Inspect and test coolant; drain and recover coolant; flush and refill cooling system; use proper fluid type per manufacturer specification; bleed air as required. **(P-1)**

This job sheet addresses the following **AST/MAST** tasks:

1.D.1   Perform cooling system pressure and dye tests to identify leaks; check coolant condition and level; inspect and test radiator, pressure cap, coolant recovery tank, and heater core and galley plugs; determine needed action. **(P-1)**

1.D.4   Inspect and test coolant; drain and recover coolant; flush and refill cooling system; use proper fluid type per manufacturer specification; bleed air as required. **(P-1)**

We Support
ASE | Education Foundation

Tools and Materials

Safety glasses

Recommended coolant

Coolant hydrometer/refractometer

Coolant pressure tester

DMM

Coolant power flush machine

Describe the vehicle being worked on:

Year _____ Make _____ Model _____

VIN# _____ Engine type and size _____

Procedure	Task Completed

1. Prepare the vehicle on the hoist. Set the hoist arms so the vehicle can be lifted if necessary.   ☐

2. Check for any related technical service bulletins for cooling system on this vehicle.   ☐

3. List all bulletins found.

_____

_____

Let the engine cool down if needed. Then carefully remove the radiator cap or recovery cap.   ☐

⚡ **WARNING** **Serious injury can occur if the cap is removed while the vehicle is still hot!!**

4. Using service manuals, determine the type of coolant required by the manufacturer.   ☐

5. What is the proper coolant mixture for this vehicle? (coolant-to-water mix)
   _____

6. Test the condition of the coolant with a hydrometer or a refractometer. Record coolant strength. _____

7. Perform a visual inspection of the coolant. Does the coolant look dirty or cloudy? Record your findings. _____

8. Perform an electrolysis test using a DMM. Record your voltage. _____

9. Is the voltage reading you recorded in step 6 normal? Explain.

   _____

10. Based on your coolant inspection, what type of coolant service do you recommend?

   _____

11. Perform a cooling system pressure test. Inspect for cooling system leaks. Record what pressure you tested the system. _____

12. Did you find any leaks? _____

Perform a Power Coolant Flush

13. Determine where the thermostat is located on this vehicle. Use service manuals to locate thermostat position. Describe location.

   _____

14. You will need to lower the coolant level to install the power flush adaptors.   ☐

15. Open the petcock on the radiator or use the flush machine to lower the coolant level. Use a clean drain pan to save the coolant.   ☐

16. With the coolant level lowered, install the coolant flush machine adaptors onto the vehicle's radiator hose.   ☐

17. With the flush machine and adaptors installed, reinstall the coolant you drained in step 13. Coolant level should be full.   ☐

18. Install the radiator cap.   ☐

19. Prepare the flush machine. Fill the flush machine with the proper mix of coolant.   ☐

20. Connect the coolant flush machine to the vehicle's battery terminals.   ☐

21. Place an exhaust hose on the vehicle and start the engine. Allow the engine to warm up to operating temperature.   ☐

   The flush process will be a lot faster if you let the thermostat open before starting the flush machine.

22. With the engine at operating temperature, turn on the flush machine and flush the old coolant out of the system. Watch as the new coolant container empties and the old coolant container fills.   ☐

23. When the new coolant container is empty, turn off the flush machine and turn off the   ☐
engine. Disconnect the coolant flush machine hoses.

⚠ Caution

Use caution because the hoses will be very hot! Use gloves or shop rags to remove the quick-connects.

24. Allow the engine to cool down. Then carefully remove the radiator cap.   ☐

**⚠ WARNING Serious injury can occur if the cap is removed while the vehicle is
still hot!!**

25. Remove the flush machine adaptors from the vehicle. You will need to use a hose   ☐
pinch off tool so you don't lose coolant from the radiator hose. Work quickly as
possible to reinstall the radiator hose to the engine.

26. Secure the hose clamps. Top off the radiator with coolant mixture. Rinse off the engine   ☐
compartment with water.

27. Start the engine with the radiator cap off and make sure that the cooling system has no   ☐
air trapped in the system. Adjust the coolant level as needed. You may need to idle the
engine until the thermostat opens.

28. Install the radiator cap.   ☐

29. With the engine running and the cooling system sealed, perform a cooling system   ☐
inspection to ensure there are no coolant leaks. Raise the vehicle on the hoist!

This is a good time to clean the floor with soap and water.

30. If the cooling is operating properly, return the vehicle to the parking lot.   ☐

Instructor's Response

_____

_____

_____

NAME _____   DATE _____

INSPECT, TEST, AND REPLACE A THERMOSTAT

Upon completion of this activity, you will be able to understand how to inspect, test, and replace a thermostat.

ASE Education Foundation Correlation

This job sheet addresses the following **MLR** tasks:

1.A.3   Inspect engine assembly for fuel, oil, coolant, and other leaks; determine necessary action. **(P-1)**

1.C.3   Remove, inspect, and replace thermostat and gasket/seal. **(P-1)**

8.A.6   Verify engine operating temperature. **(P-1)**

This job sheet addresses the following **AST/MAST** tasks:

1.A.4   Inspect engine assembly for fuel, oil, coolant, and other leaks; determine needed action. **(P-1)**

1.D.7   Remove, inspect, and replace thermostat and gasket/seal. **(P-1)**

8.A.10   Verify engine operating temperature; determine needed action. **(P-1)**

We Support
ASE | **Education Foundation**

Tools and Materials

Safety glasses
Hand tools
New thermostat and gasket
Recommended coolant
Scan tool
Inferred gun

Describe the vehicle being worked on:

Year _____ Make _____ Model _____

VIN# _____ Engine type and size _____

Procedure   **Task Completed**

1. Using service manuals, search for technical service bulletins related to cooling system issues. Record any bulletins.

   _____

   _____

2. Using service manuals, find the correct engine operating temperature. Record specification.

   _____

3. Start the engine and let it warm up. Check the temperature of the engine using a scan tool. Record the engine coolant temperature (ECT). _____

4. Use an IR gun or dash gauge (if equipped) and compare to scan tool reading. Record temperature. _____

Task Completed

5. Check the operation of the thermostat by watching the scan tool data or by using the IR gun on the upper radiator hose. Is the thermostat opening at the correct temperature? Record thermostat opening temperature. _____

6. Turn off the engine. Let the engine cool down. Then carefully remove the radiator cap or coolant recovery reservoir cap. While you're waiting for the engine to cool, print procedures and answer question at the bottom of work sheet. ☐

⚡ **WARNING Serious injury can occur if the cap is removed while the vehicle is still hot!!**

7. Using service manuals, **print out and attach** the service procedure for thermostat removal. ☐

8. With the radiator cap off, install a coolant drain pan under the radiator, carefully open the petcock, and drain the coolant from the radiator until the level is below the thermostat. ☐

   Note: Save as much coolant as possible. You will need to reuse the coolant.

9. Remove the radiator hose from the thermostat housing. ☐

10. Carefully remove the thermostat housing from the engine. ☐

11. Remove the thermostat from the engine and inspect the thermostat. Do you see anything wrong with the thermostat visually? _____

12. Clean the thermostat housing gasket surface and its mating surface on the engine. ☐

13. What is the torque specification for the thermostat housing bolts? _____

14. Reinstall the same thermostat using a new gasket. See instructor for gasket. ☐

15. Install thermostat housing and gasket and torque bolts to specifications. ☐

⚡ **WARNING Overtightening the housing bolts will crack the housing. Use extreme caution!!!**

16. Reinstall the upper radiator hose and secure the clamp. ☐

17. Make sure that the petcock is closed and not leaking coolant. ☐

18. Top off the cooling system with the coolant you saved in step 8. ☐

19. Start the engine with the radiator cap off and make sure that any air trapped in the radiator is out before putting the radiator cap back on. ☐

20. Bring the engine temperature back up to manufacturer's specifications. Is the thermostat opening properly? Use scan tool to verify. Record temperature. _____

21. Inspect for any coolant leaks. Did you find any leaks? _____

22. Clean the engine off with water, rinse the floor and squeegee! ☐

23. Explain the function of the cooling system.

_____

_____

_____

24. The thermostat uses a _____ pellet to move a piston.

25. Typically, thermostats are fully open by what temperature? _____

26. Explain the function of the thermostat. (Why do we need one?)

_____

_____

_____

27. Explain the operation of the thermostat. (How does it work?)

_____

_____

_____

28. What would the customer's complaint be if the thermostat is stuck open?

_____

29. How do you open a two position radiator cap?

_____

_____

30. What is a surge tank? Explain.

_____

_____

Instructor's Response

_____

_____

_____

CHAPTER 8

TRANSMISSION AND DRIVETRAIN SYSTEMS INSPECTION, MAINTENANCE, AND SERVICE

Upon completion and review of this chapter, you should understand and be able to:

- Diagnose driveline noise.
- Adjust a clutch linkage.
- Inspect and service a manual transmission or transaxle.
- Inspect and service an automatic transmission or transaxle.
- Inspect and service a drive shaft or axle.
- Inspect and service a differential.

Terms To Know

Axle stub	Front-wheel drive (FWD)	Special tools
Center bearing	Grease fitting	Suction pump
Dipstick	Limited slip	Yoke
Free travel	Sealing tool	

INTRODUCTION

Like the engine, most of the powertrain components need routine servicing for providing the best performance. Many times, a driver will routinely change the engine oil and filter but fails to do the same for the transmission or transaxle and differential. Noise is sometimes the first indication that the driveline is failing. In some cases, the customer will not bring the vehicle in for transmission or driveline service until it is too late. At this point, major repairs or replacement are sometimes the only option. It is your job as a technician to help prevent this by inspecting the transmission and driveline properly and helping the customer (or service advisor) make informed decisions about the maintenance.

DRIVELINE DIAGNOSIS BY NOISE

Most driveline noises are easy to separate from the noise of an engine failure. However, the noise can echo or vibrate through the vehicle, leading the technician to look in the wrong place. If the noise can be associated with a type of component, it can help eliminate some of the possible sources. Some of the driving conditions needed during a test drive can be produced on a lift in the shop. Only an experienced technician operating within the shop's policy should perform or supervise a test drive on the road or on the lift.

Classroom Manual
pages 243–244

⚡ **WARNING** Care must be taken when operating the vehicle in an attempt to isolate a driveline noise. The vehicle must be operated to load, unload, and stress the driveline. It is best to use an empty parking lot or a low-traffic road. Failure to pay attention to other motorists or traffic conditions can result in serious consequences.

⚠ **WARNING** Do not conduct a test drive without permission of the owner and managers, and only in accordance with shop policy. An experienced technician should perform or supervise any test drive. Failure to follow proper procedures could result in damage to the vehicle, injury, and legal actions.

⚠ **WARNING** Use extra care in placing lift pads and lifting and operating a vehicle while raised on the lift. Do not raise the vehicle any more than necessary to clear the wheels off the floor. It is best to keep the nondriven wheels chocked on the floor. Clear the areas of equipment and people in front of and behind the vehicle. Damage to the vehicle or injury could occur if the operation is not done in a safe and cautious manner.

Manual Transmissions or Transaxles

Noises from the transmission and transaxle usually include growling from the gears or whining from the bearings. Gear noise will change or disappear when another gear is selected. Sometimes a damaged gear, synchronizer, or maladjusted clutch will result in grinding or grating noises as the damaged teeth engage with the teeth on the mating gear (**Figure 8-1**). Bearing whine will change in pitch as the vehicle speed changes.

Noises can be further diagnosed by operating the vehicle under different conditions. Usually, transmission noise will not change because of road conditions or the turning of the vehicle. It will appear as the load is applied and released. If the clutch is disengaged as the vehicle is driven and the noise does not change, it is probably caused by driveline components behind the transmission.

Disengaging the clutch unloads the driveline and the noise will change in pitch, volume, or both. Placing the transmission in neutral as the vehicle is moving will cause most transmission noise to disappear or at least change drastically. A vibration may be felt in the shift lever as a certain gear is engaged.

⚙ **SERVICE TIP** All the internal components of an automatic transmission or transaxle are heavy, continuously lubed, and tend not to fail unless abused.

Automatic Transmissions or Transaxles

Normally, automatic transmissions or transaxles do not produce major noises. The most common noise is a whine from the hydraulic pump or possibly the torque converter. If there is a clunking or other noise coming from the transmission area, the flexplate (also known as a flywheel) may be cracked (**Figure 8-2**). A cracked flexplate will also result in noise during

Figure 8-1 Gears can make noise if they are chipped or missing parts. Proper lubrication and maintenance are crucial.

Figure 8-2 Inspect the flexplate (or flywheel) for cracks around the mounting bolt holes.

cranking as the starter tries to turn the flexplate and torque converter. Any other noise from the automatic transmission or transaxle usually means it must be replaced or rebuilt.

Drive Shafts

A bent or unbalanced drive shaft will cause a continuous vibration, which gets worse as the shaft's speed is increased. The vibration may lessen but will not go away completely as the load changes. This vibration should not be confused with bad or unbalanced tires. Tires will usually produce a vibration on an axle. Front tires will cause a vibration or jerking of the steering, whereas rear tire vibration is felt in the seat. Drive shaft vibration may resound throughout the vehicle.

U-joints can create a clunking noise as the drive shaft is loaded and unloaded (**Figure 8-3**). This very closely resembles noise from worn differential gears. Closer inspection is required. A squeak at speeds of 3 to 6 mph usually means the U-joints are dry. If a **grease fitting** is

Most original equipment U-joints do not have **grease fittings**. Replacement U-joints will include a fitting with the joint.

Figure 8-3 A common customer complaint is a squeak at low speeds. This noise indicates a dry (not lubricated) U-joint(s).

Move back and
forth to check
backlash

Dial indicator

Figure 8-4 A roaring or whining sound traced to the differential could indicate that the gear backlash is incorrect.

present on the joint, lubricating it may help. It is best to replace a squeaking U-joint as soon as possible. If the joint breaks apart during vehicle operation, the drive shaft assembly can come through the floorboard, causing injury to the passenger. A drive shaft that drops to the road and digs into the road surface can flip a car over.

Differentials

As stated earlier, clunking noise from under the car may be caused by worn U-joints or differential gears. Other differential noises can be just as difficult to isolate. A high-pitched whine from the differential can mean the pinion gear and ring gear are set too tightly or one of the bearings may be going bad (**Figure 8-4**). Operating the vehicle in tight, slow turns can isolate axle bearing noises. As the vehicle leans, its weight loads and unloads the bearings sited near the drive wheels. The bad bearing will make more noise when it is on the outside of the turn and loaded. Rear-wheel-drive (RWD) axles usually do not make any noise until they break. Then there is no drive or noise.

Front-Wheel-Drive (FWD) Axles

Most **front-wheel-drive (FWD)** axle bearings are one-piece, double ball bearings that are permanently lubed and sealed.

The noise from **front-wheel-drive (FWD)** axles will originate in the constant velocity (CV) joints. They are fairly easy to diagnose and isolate. From a dead stop, accelerate hard straight ahead. A fast, clicking noise is usually related to a bad inner CV joint. Operating the vehicle in a tight, slow circle will produce a slow-to-fast clicking noise from bad outer CV joints. The noise will be worse when the bad joint is loaded or on the outside of the turn. Bad outside CV joints may also be felt in the steering wheel as a turn is made.

A bearing mounted in the wheel hub supports the outer end of the drive axle (**Figure 8-5**). A common noise from a damaged bearing is a loud crushing or cracking sound. This may also produce a noticeable feel in the steering wheel that increases when the bearing is loaded.

Clutches

Clutch problems can be easy to diagnose because the system is so simple. Clutch noises must be separated from transmission noises.

Figure 8-5 To isolate axle bearing noise (roaring), test drive and change vehicle direction. The noise will change as the bearing on the outside of the turn is loaded and unloaded.

Diagnosis

A rattling or whining sound with the clutch disengaged can be traced to the release bearing. If the noise is audible at all times but changes in pitch or volume as the clutch is operated with the transmission in neutral, the noise points to a continuously running release bearing that is dry of lubrication. A transmission input shaft's bearing noise will be most prominent with the clutch engaged and the transmission in neutral (**Figure 8-6**). It will disappear within a short time when the clutch is disengaged and held that way for a few

A clutch friction disc can get overheated and have hot spots like those found in brake friction material. It usually results in chattering or grabbing during clutch engagement.

Figure 8-6 Noise from the input shaft bearing is most noticeable when the transmission is under load. It will cease when the clutch is disengaged (pedal depressed) and a few seconds allowed until the shaft stops spinning.

seconds. A chattering noise and jerking as the clutch is engaged usually indicates a bad clutch disc, pressure plate (clutch cover), or flywheel. A leak in the engine's rear main or transmission input seal will also cause chattering and jerking during clutch operation. The clutch assembly must be removed for an exact diagnosis.

> ✿ **SERVICE TIP** Problems engaging first or reverse gears may also be caused by low fluid level in the transmission. Low fluid may cause the gears and synchronizers to bind. They can completely block the first or reverse shift, particularly after the vehicle has sat idle for a time. Check the fluid before beginning removal of the transmission and clutch. A new customer may be found when informed that the suspected $300.00 clutch replacement will only cost $40.00 or less.

A worn clutch disc will slip during load conditions, most notably in the higher gears. A racing engine as the engine load goes up in high gear is the first sign of a worn clutch disc. Mechanical clutch linkage that has not been properly adjusted will not allow the pressure plate to clamp correctly and will cause the same problem. This type of linkage requires periodic adjustment as the clutch disc wears (**Figure 8-7**). An early sign of pressure plate wear or maladjusted linkage is a difficulty in shifting into first or reverse gears while the vehicle is halted. The pressure plate in this instance is not completely releasing the clutch disc. There may not be a slipping condition, but other gears may also be hard to select when the vehicle is under load.

Figure 8-7 A typical adjusting point for mechanical clutch controls.

Because most hydraulic clutch linkages are self-adjusting, slipping normally indicates bad components in the clutch assembly. If the hydraulic clutch system should leak, the pressure plate will not release the clutch disc. Topping off and bleeding the system may temporarily correct the problem, but the only proper repair is to replace the leaking component.

Classroom Manual
pages 248–253

INSPECTING AND ADJUSTING THE CLUTCH LINKAGE AND FLUID

Mechanical clutch linkages must be adjusted periodically due to normal clutch lining wear. Before discussing the mechanical-type adjustment, we will discuss inspecting and changing the fluid in a hydraulic clutch. The fluid can be replaced easily by attaching a rubber hose to the slave cylinder bleeder screw and draining the old fluid into a container, preferably a clear one (**Figure 8-8**). Add fluid to the master cylinder and operate the pedal until clean fluid comes out of the slave. Keep the end of the hose below the level of fluid in the container and watch for air bubbles as the system is bled. Bleed the air out of the system by closing the bleeder screw and pumping the pedal several times. Hold the pedal to the floor and slightly open the bleeder. Continue this process until there are no air bubbles in the fluid. This process should be done at about the same interval as that for the coolant and brake fluid, about every 2 years. Consult the service or operator manual for the maintenance schedule. This process is very similar to brake bleeding. When completed, make sure to check the level of fluid in the clutch (**Figure 8-9**). A low level may indicate a leak. Clutch fluid is like brake fluid, so it must be inspected in a similar manner.

> ⚙ **SERVICE TIP** Bleeding a hydraulic clutch can be aggravating on some vehicles. If the system cannot be bled using the manual method, use a pressure bleeder.

Special Tools

Lift *or* floor jack and jack stands

Many newer models of cars and light trucks have hydraulically controlled clutches, but there are many older vehicles still on the road. A heavy-truck or equipment fleet technician will perform repairs and adjustments as a routine maintenance service on heavy vehicles (on- and off-road).

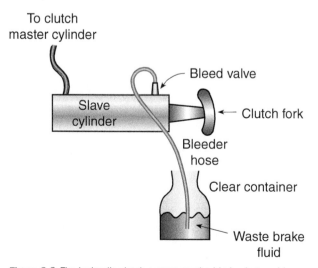

Figure 8-8 The hydraulic clutch system can be bled using a rubber hose and a catch container.

Figure 8-9 Fluid for a hydraulic clutch is stored in a small reservoir that may be attached to the clutch master cylinder or remotely mounted on the firewall.

Very few hydraulic control clutch systems have pedal **free travel**. Almost all manual clutch linkage systems have a specific free travel.

Adjusting the pedal's **free travel** sets the mechanical linkage engagement point for the clutch. Free travel is the distance the pedal moves before the release bearing contacts the release levers of the pressure plate (**Figure 8-10**). The vehicle will probably have to be lifted some amount to gain access to the adjustment. If the vehicle is lifted to a normal (stand up) working height, a second person will have to ride the vehicle up the lift to operate the clutch and measure the pedal free travel. If only one technician is available, it is probably quickest and easiest to jack and block the front of the vehicle. This will allow the technician to use a creeper to make the adjustment, slide out, measure the free travel, and if necessary, slide back under the vehicle to make additional adjustment.

Clutch pedal height

Clutch pedal free play

Distance between the clutch pedal and the cowl when the clutch is disengaged

Figure 8-10 The adjustment shown is for adjusting pushrod length on a hydraulic-controlled clutch. Remember to check the brake switch operation whenever the clutch linkage, mechanical or hydraulic, is adjusted.

> ⚙️ **SERVICE TIP** Always make sure to check the manufacturer's Technical
> Service Bulletins for issues related to the transmission servicing, maintenance, and
> inspection. Some manufacturers use a dye in their automatic transmission fluids,
> and that dye will change color as it ages. The color is misleading and can cause an
> improper maintenance suggestion for service.

The adjustment device will be located on the clutch linkage at the left side of the bell
housing near the release bearing's fork. It usually consists of one or two nuts screwed on
a threaded rod. Adjust the linkage using the adjusting nuts until the pedal free travel is the
correct distance. Most clutches have a free travel of 3/4 to 1 inch. Consult the service
manual for the correct specifications.

⚠️ **WARNING** **Ensure that the parking brake is set and the areas to the front and
rear of the vehicle are cleared before running the engine to check clutch operation.
Damage to the vehicle, shop equipment, or injury to persons may result if the
clutch fully engages unintentionally.**

With the adjustments completed, check the clutch and transmission operation on the
lift. All gears should be checked at a complete stop with the engine running. First and
reverse should be selected easily at a complete stop. All other gears should be smooth with
no grinding or hanging. The vehicle will have to be tested on the road to completely check
the adjustment. Consult the instructor or supervisor.

Classroom Manual
pages 272–276

TRANSMISSION, TRANSAXLE, AND DIFFERENTIAL SERVICE

Both manual and automatic transmissions and transaxles need routine servicing and
maintenance. Most of the time, it is only a fluid and filter change. The service manual
specifies the interval, type of fluid, and capacity.

Manual Transmissions or Transaxles

This type of transmission does not need service very often, other than a fluid change.
A few manual systems use a filter. The fluid can be drained through a drain plug located
at the bottom of the transmission housing. Manual transmission fluid comes in different
blends. Older vehicles use a 85W-90 thick oil with few additives, but newer systems
require lighter, thinner fluid. Some smaller cars use an automatic transmission fluid. The
current classification is listed as GL-4 or GL-5 (**Figure 8-11**). GL fluids may have additives
that lubricate, clean better, and are designed for specific temperature ranges. Many manufacturers also used special lubricants for their manual transmissions. Most newer light
trucks require a special fluid for their manual transmissions. Other manufacturers use
automatic transmission fluid or a lightweight engine oil for their manual transmission or
transaxle systems. Most manual transmissions require 3 to 5 quarts of fluid. Always check
service manuals for proper fluid types and capacities.

To change the fluid in a manual transmission, lift the vehicle to gain access to the drain
plug (**Figure 8-12**). The transmission should be hot enough for the thick fluid to flow
freely. Cold transmission fluid is very slow to flow and will not remove all the debris that
may be inside the bottom of the transmission. The transmission should be at the same

Figure 8-11 Manual transmission and differential fluids range from a thick 85W-90 to thin automatic transmission fluid or a special blend fluid.

Drain plug

Figure 8-12 A typical drain plug for a manual transmission or transaxle.

Special Tools

Drain pan

Air- or hand-operated pump *or* suction gun

angle as it is when the vehicle is setting on its tires. Remove the drain plug, capture the old fluid, and dispose of it according to the shop's policies. While the fluid is draining, check for leaks at the transmission's extension housing seal where the drive shaft mounts or the drive axle seals on a transaxle. Repair as necessary.

SERVICE TIP Attempting to service a transmission that is not at the same angle as an operating vehicle may not allow all the fluid to drain.

The **suction pump** is used to draw fluid from a container and force the fluid into the transmission or transaxle by hand.

Reinstall and torque the drain plug once all the fluid has drained. Fluid can be installed with an air-operated pump or using a hand **suction pump**. The suction pump is slow and can be messy.

SERVICE TIP Always refer to the service manual when checking the fluid level on a manual transmission, transfer case, or differential. Some manufacturers call for a special tool to check the fluid level, and the level is not supposed to be all the way up to the fill plug.

WARNING Overfilling a transmission could cause high pressure within the housing and rupture a seal or gasket. Damage could result from an overfilled transmission.

Locate the fill plug. It is normally about halfway up on the side of the transmission. Remove the plug and insert the pump's hose tip into the hole. If an air- or hand-operated pump is used, the amount being installed is measured with a dial on the pump (**Figure 8-13**). Suction pumps will hold about a pint of fluid and will have to be refilled until the transmission is full (**Figure 8-14**). Most manual transmissions do not have a dipstick to check fluid

Figure 8-13 This type of lubrication gun is attached to a large bulk container of lubricant.

Figure 8-14 A suction gun is a slower method of transferring fluids, but it still works.

level. The most common method of checking the fluid involves the use of the technician's finger (**Figure 8-15**). Insert the little finger into the fill hole to the first knuckle. Bend the fingertip downward inside the transmission. If the fluid touches the finger, the level is correct. **Special tools** can be purchased or made to check the fluid level. A quick tool can be made from a welding rod. Bend the rod 90 degrees, about the length of the last little finger joint from one end (**Figure 8-16**). About an inch from this bend, curve the rod so the bent end can be fitted into the fill hole. Place the bent end into the fill hole and the rod against the bottom of the hole (**Figure 8-17**). If the fluid level is correct, there will be fluid on the end of the rod inserted into the fluid. A tool is not an absolute necessity unless the technician performs this type of service several times a day or week.

During filling, the fluid is added until it barely begins to run from the fill hole. If it is overfull, drain and capture the excess. Reinstall and torque the fill plug.

Automatic Transmissions or Transaxles

Automatic transmission or transaxle service includes a fluid and filter change. The fluid removed will not be all of the fluid in the system. A typical fluid change will only remove the fluid in the transmission pan. The torque converter and pump will retain some fluid. Before starting the service, check the manual for the correct amount of fluid used for a filter change service. Ensure the amount used is not for a full flush. It is best if the transmission is warm. Transmission and transaxle services are performed using similar procedures.

> **SERVICE TIP** Check the used transmission fluid using feel, sight, and odor. If the fluid is burnt or very dirty, changing the filter and 3 quarts of fluid will not help the transmission. Many times, a full-system flush will help restore most of the transmission's performance, but a flush may not help and may even decrease performance more if there is a lot of mileage on the fluid. **Photo Sequence 20** highlights the procedures for changing the filter and gasket.

Special tools: Tools that are needed to perform a special operation the complete a repair of task.

Special Tools

Large drain pan
Funnel

Fill the transaxle to the bottom of the filler plug hole

Figure 8-15 Typically the transmission or transaxle/differential check plug is on the side of the housing. The fluid level can be checked with a finger.

Approx.
½ to ¾
inch

Metal
rod

Figure 8-16 A straight piece of small bar stock can be made to check the fluid level in a manual transmission or differential.

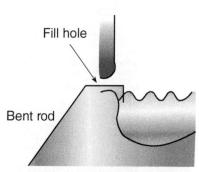

Fill hole

Bent rod

Figure 8-17 Place the rod on the bottom of the fill hole and allow the short portion to extend down into the fluid.

P20-1 To begin the procedure for changing automatic transmission fluid and filter, raise the car on a lift. Before working under the car, make sure it is safely positioned on the lift.

P20-2 Before continuing, compare the new pan gasket with the shape of the oil pan.

P20-3 Locate the transmission pan. Remove any part that may interfere with the removal of the pan.

P20-4 Position the oil drain under the transmission pan. A large-diameter drain pan helps prevent spills.

P20-5 Loosen all of the pan bolts and remove all but three at one end. This will cause fluid to flow out around the pan into the drain pan.

P20-6 Supporting the pan with one hand, remove the remaining bolts and pour the fluid from the transmission pan into the drain pan. Be careful, there may still be up to a quart of fluid left in the pan.

P20-7 Carefully inspect the transmission pan and the residue in it. The condition of the fluid is often an indication of the condition of the transmission and serves as a valuable diagnostic tool.

P20-8 Remove the old pan gasket and wipe the transmission pan clean with a lint-free rag.

P20-9 Unbolt the fluid filter from the transmission's valve body. Keep the drain pan under the transmission while doing this. Additional fluid may leak out of the filter or the valve body.

P20-10 Install the new filter onto the valve body and tighten the attaching bolts to proper specifications. Then lay the new pan gasket over the sealing surface of the pan and move the pan into place.

P20-11 Install the attaching bolts and tighten them to the recommended specifications. Carefully read the specifications for installing the filter and pan. Some transmissions require torque specifications of inch-pounds rather than the typical foot-pounds.

P20-12 With the pan tightened, lower the car. Pour automatic transmission fluid (ATF) into the transmission. Start the engine. With the parking brake applied and a foot on the brake pedal, move the shift lever through the gears. This allows the fluid to circulate throughout the entire transmission. After the engine reaches normal operating temperature and with transmission in park, check the transmission fluid level and correct it if necessary.

Before proceeding, compare the pan gasket in the filter kit with the pan. There are many different pan shapes, none of which are particular to a certain brand or model. If the pan is removed and the gasket is incorrect, the lift and bay will be tied up until another gasket is available.

Lift the vehicle to gain access to the transmission's oil pan. Most automatics do not have drain plugs. The pan must be removed (**Figure 8-18**). Care must be taken to prevent spills.

Drain plug (optional)

Figure 8-18 A drain plug for an automatic transmission or transaxle is uncommon but very handy if present.

Remove the pan bolts from the lowest side of the pan. Place a wide drain pan under the transmission and remove bolts along each side, starting at the point where the first bolts were removed. At some point, the pan will tilt, and fluid will begin to drain. Loosen the other pan bolts, but do not remove all of them until most of the fluid is removed. Keep the last two or three bolts loose until the pan can be supported by hand. Hold the pan against the transmission and remove the last several bolts. Control the oil pan and pour the remaining fluid into the drain. There will be fluid dripping from the filter and transmission. Keep the drain pan in place until the oil pan is reinstalled. Note how the filter is installed. Automatic transmission filters come in many different shapes and their installation location may not be clear once the old one has been removed. Remove the filter fasteners and filter.

Usually, there is a small magnet at the bottom of the oil pan (**Figure 8-19**) with a few metal shavings on the magnet. This is normal. Excessive shavings indicate gear or bearing damage. Look for black friction material at the bottom of the pan as well. There will probably be some, but like the metal shavings, this is normal. Pieces or chucks of friction material indicate that the hydraulic clutches are bad. The lower side of the filter needs to be checked for the same items. Excessive shavings or friction material normally requires a transmission overhaul or rebuild. Consult the instructor, supervisor, or service manual for advice as needed.

Clean the oil pan of all materials, including any gasket material. The oil pan mounting area on the transmission also must be cleaned of all gasket materials. Be sure to check the bolt holes in the pan and transmission for damage or gasket material. Some manufacturers use room temperature vulcanizing (RTV) instead of a gasket to seal the pan, but RTV can enter and block the hole. Like the manual system, take the time now to check the seals at the drive shaft or drive axles.

Install the filter and torque the filter bolts (**Figure 8-20**). If RTV was the original sealing material but an aftermarket gasket is available, use the gasket. Place the gasket on the oil pan. Do not use RTV on the gasket unless the gasket manufacturer recommends it. If RTV must be used in place of a gasket, it should be laid in a 1/8-inch bead and applied evenly around the pan. Remember to lay the bead on the fluid side of the bolt hole and **do not** allow RTV to cover the hole. Ensure that the RTV meets the vehicle manufacturer's specifications and is allowed to cure for 10 to 15 minutes before mounting the pan to the transmission. Before installing the oil pan, wipe the fluid from the mounting area on the transmission. Place the pan in place and start the bolts. Overtightening the bolts can strip the hole threads very easily. It is important to use a good torque wrench and tighten the

Figure 8-19 The magnet located in the transmission oil pan will attract metal shavings. Check it to help determine the mechanical condition of internal components.

Figure 8-20 Always torque the filter fasteners. The filter can allow air to enter the hydraulic system if not properly torqued.

Figure 8-21 Always torque the oil pan fasteners. A too-tight condition will cause just as much leakage as a too-loose condition.

Figure 8-22 Dipsticks usually have a "hot" and a "cold" mark. Note that the amount of fluid needed to move the level from the add mark to full is usually just 1 pint.

In automotive applications, a **dipstick** is a thin flexible rod used to check fluid levels in certain components.

Special Tools

Drain pan
Air- or hand-operated
  pump **or**
Suction gun

Classroom Manual
page 273

A **limited slip** differential is designed to reduce wheel spin using some type of clutch mechanism. Using incorrect differential fluid will damage the clutches.

Classroom Manual
pages 274–276

bolts, in sequence, to torque specification (**Figure 8-21**). With the pan installed, lower the vehicle and add the proper amount of fluid. The fluid should register at the cold level mark or slightly below the full mark on the **dipstick**.

With the fluid installed, start the engine and select each gear in turn to fill the pump and clutches. The fluid in most automatic transmissions or transaxles is checked with the engine running and the gear selector in park or neutral and hot. The first check after a service will be with cold fluid, but the level should be nearly correct but not be overfull (**Figure 8-22**). With the proper level for cold fluid, the vehicle may be driven to heat the fluid, so it can be checked at the hot fluid level. Next, check the oil pan for leaks. Remove any fluid over the full mark and repair any leaks if necessary.

Differentials

The differential is serviced by changing the oil. Some differentials have a drain plug, whereas others have a rear cover that must be removed. The fill plug is usually about halfway up the side, like the manual transmission. The fluid is drained, refilled, and checked the same way as a manual transmission and uses the same type of fluid on older differentials (**Figure 8-23**). Lockup or **limited slip** differentials use special additives for the clutches. Using regular fluid can seriously damage a lockup differential. Newer differentials use special blended fluids. Final drives on transaxles are serviced as part of the transaxle.

DRIVE SHAFT AND DRIVE AXLE REPAIR

Drive shaft repair usually consists of replacing a U-joint or replacing the entire drive shaft assembly. On the FWD axle, CV joints or boots can be replaced, but the most common task is to replace the entire axle. CV joint boots can be replaced, but because of the CV joint's highly machined and matching mechanism, damage to the joint can occur very quickly when a boot is torn, and the grease is flung outward. Axle replacement is usually the cheapest and quickest method for the customer and shop.

Replacing just a boot or one joint may save some time or the cost of the part, but the labor will offset most of that savings. In addition, the shop will have to warrant the repair component and the labor. Replacing the entire drive axle allows the shop to give the

Less than
5 mm (0.20 in.)

Figure 8-23 The differential fluid level is checked the same way as a manual transmission.

customer a new axle or one that was rebuilt in a specialty driveline shop at roughly the same total cost and a better warranty.

Drive Shafts

We will cover a basic U-joint replacement. Some vehicles may have a **center bearing** on the drive shaft and more than two U-joints. The procedures will be the same except for removing the center bearing. Before removing the drive shaft, mark the position of the shaft to its companion flange on the differential (**Figure 8-24**). The companion flange is the part where the drive shaft's rear U-joint mounts to the differential. There may be another companion flange at the center bearing or the transmission depending on vehicle design. Use a **sealing tool** to slide over the transmission's output shaft once the drive shaft is removed. This will stop any fluid from leaking.

⚡ WARNING **Do not overtighten the vise on the drive shaft and do not place the vise over the balancing weight. It may bend or dent the tube, knock a weight off, and unbalance the assembly.**

⚡ WARNING **Do not use tools for purposes for which they were not designed. Damage to components, tools, or injury to a person may occur.**

🔧 **Special Tools**
Vise
Marker
U-joint bearing press
Grease gun
Drip pan

The **center bearing** is used on two-piece drive shafts.

A **sealing tool** is a pliable cover shaped like the slip joint. The plug slides over the output shaft to prevent oil from leaking out during servicing.

Propeller
shaft

Shaft flange

Differential flange

Reference
marks

Figure 8-24 Failure to mark the drive shaft to the companion flange could result in an out-of-phase drive shaft with vibration problems.

Figure 8-25 Special tools are available for pressing the U-joints in and out of the drive shaft yoke.

The yoke of a light drive shaft (small car) can be easily distorted. The correct use of the proper tool will help prevent this distortion.

Place the drive shaft in a vise near the U-joint being replaced. Remove the rear U-joint by first removing the retaining clips and pressing the bearing cups out. The clips may be internal or external. Some original U-joints do not use clips but have plastic retainers that will be sheared when the cups are pressed out. The replacement U-joints for this type will have internal clips. When pressing the cups, use an adapter to support the drive shaft's yoke. This prevents the **yoke** from spreading. A common practice is to use two sockets, one slightly larger than the cup hole and the other slightly smaller than the cup itself. The smaller socket is used to push the cups away from the yoke while the larger socket acts as a standoff between the yoke and the vise. However, in Chapter 3, "Tools and Equipment Usage," there is no mention of using sockets as pressing or standoff tools. Another common practice is to use a hammer and punch. Although the hammer and punch can be used as drivers in some instances, this is not one of them. There are tools designed for removing and installing U-joints (**Figure 8-25**). See **Photo Sequence 21** for U-joint replacement with the correct tools.

PHOTO SEQUENCE 21
Replacing a Single Universal Joint

P21-1 Ensure the vise is not placed over a drive shaft weight and is not overtightened.

P21-2 Remove the bearing cup retainers, if equipped. The clips may be internal or external on the yoke.

P21-3 Select the proper adapters and position the removal clamp over the adapters.

P21-4 Use a box-end wrench to turn the screw that applies force to the adapters. The adapters will have to be switched from one side to the other to press the second cup out.

P21-5 When installing the cups, first position one cup hand tight into the cavity.

P21-6 Install one end of the cross into the partially installed cup, then maneuver the cross so the other end fits into the opposite cavity. The hole for the grease fittings should face toward the drive shaft.

P21-7 Hold the cross in place to retain the needle bearings in place of the first cup as the second cup is fitted over the cross end and hand-pressed into the cavity.

P21-8 Position the clamp and adapters over the cups, ensuring the cross does not slide out of one of the cups. This could cause a bearing to drop inside the cup.

P21-9 Use the box-end wrench to press the cups in place. At about three-fourths of the way in, check to make sure the cross is centered and free in the bearing cups. If not, remove the cups and cross, inspect the bearings, and reinstall.

P21-10 Once the cups are completely seated, install the cup retainers. Sometimes the aftermarket retainers may be installed differently from the original parts.

P21-11 Install the grease fitting. Do not lube until all four cups are fitted and locked in place.

Pressing on one cup will force the opposite cup out of the yoke. With the first cup out, try to remove the U-joint cross by moving it into the empty bore and twisting it to the side. If the space is not sufficient, the yoke will have to be rotated and force applied to the end of the cross. This will move the second cup outward enough to remove the cross. The second cup can be knocked out with a rod installed through the empty bore once the cross is removed.

During the installation of the U-joints, the bearing cups must be kept squarely in the bore as force is applied. The needle bearings can be displaced if the cup is dropped or not aligned correctly on the cross (**Figure 8-26**). The cross has a **grease fitting** hole that must be placed correctly as well. The hole is positioned to allow the insertion of the grease fitting and allow the connection of a grease gun with the drive shaft on the vehicle.

Place the cross in a bore and align a cup on the cross to keep the bearings in place as the cup is pressed. Press the first cup just enough to position the other end of the cross to align the second cup. With both cups being guided by the cross, apply force to each side so the cups are pressed into the yoke. One cup will probably be in position before the other. Install the retaining clip on the cup and then press and clip the second cup into position. If the second cup does not go all the way in, one or more of the needle bearings may have dropped to the bottom of one of the cups. The joint will have to be removed and checked. Install the grease fitting.

Some rear U-joints have the second set of cups in a flange, which bolts to the companion flange (**Figure 8-27**). Replace the cups on this type of arrangement at this time. Many systems have outer cups that connect to the companion flange by U-bolts. If this is the case, move the two new cups out of the way. They will not be needed until the drive shaft is reinstalled. Rotate the U-joint to ensure that it is free. It should be stiff but can be moved by hand.

Mark the position of the slip joint to the drive shaft. This U-joint is removed and installed the same as the rear one; however, it may require several attempts to clamp the shaft in the proper position in the vise. Ensure that the slip joint is positioned correctly as the U-joint is installed. Install the grease fitting and grease the joint. If all four cups are installed on the rear joint, it can also be greased. If only two cups are installed, the joint cannot be greased correctly until the drive shaft is installed into the vehicle.

The **grease fitting** on some aftermarket U-joints are placed in the bearing cup instead of the joint cross.

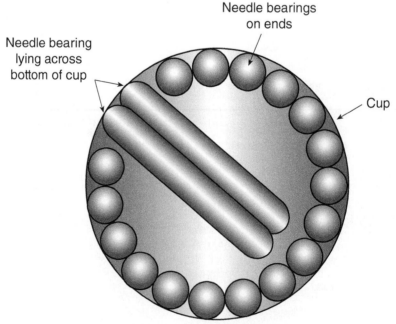

Figure 8-26 Needle bearings stand on end and can be easily knocked over during installation.

Figure 8-27 Flanges like the two shown must be bolted together.

Remove the sealing tool from the transmission and slide the slip joint into position. Place the two rear bearing cups on the cross and in position on the companion flange. Ensure that the marks are aligned. Install and torque the bolts. If necessary, grease the rear U-joint. The vehicle can now be lowered, the area cleaned, the repair order completed, and the vehicle returned to the customer.

Drive Axles

Replacing a FWD axle is not normally assigned to an entry-level technician, but he or she may be asked to assist in the replacement. The replacement process is simple and can be learned quickly.

FWD axles consist of two CV joints and a shaft. If the shaft is bent, the entire axle is always replaced. If a CV joint is damaged or the boot is torn, either can be replaced, but as mentioned earlier, it is usually cheaper in the long run for the customer to replace the axle as an assembly. The replacement of a typical drive axle will be discussed here. The following is typical of manual or automatic transaxles and should be supervised by the instructor or an experienced technician until proficiency is achieved by the new technician.

⚡ **WARNING Use extreme care when working on vehicles with antilock brake systems (ABSs). Damage to the system or components may occur if improper tools or procedures are used.**

Lift the vehicle to gain access to the inner end of the shaft. A drain pan should be available to catch any fluid that may drain from the transaxle's final drive housing. Remove the wheel assembly. Locate, remove, and hang the brake caliper and rotor out of the way. Before removing any other components, inspect the area around the steering knuckle. Closely inspect for antilock system components. If a sensor is present, consult the service manual for procedures to protect it during axle replacement. Also check the outer CV joint for a tone ring (**Figure 8-28**).

There are several methods used to attach the hub to the suspension. If a wheel alignment is performed at the strut, the vehicle will have to be checked for alignment before returning it to the customer. **Photo Sequence 22** shows the typical procedures.

🔧 **Special Tools**
Torque wrench, 3/8 and 1/2 inch
Small pry bar

Figure 8-28 This tone ring on the CV joint can be damaged if care is not taken as the axle is removed and installed.

PHOTO SEQUENCE 22
Typical Procedure for Removing and Installing Drive Axles

P22-1 With the tire on the ground, remove the axle hub nut and loosen the wheel nuts.

P22-2 Raise the vehicle and remove the tire and wheel assembly.

P22-3 Remove the brake caliper and rotor. Be sure to support the caliper so that it does not hang by its brake hose.

P22-4 Remove the bolts that attach the lower ball joint to the steering knuckle. Then pull the steering-knuckle assembly from the ball joint.

P22-5 Remove the drive axle from the trans-axle by using the special tool.

P22-6 Remove the drive axle from the hub and bearing assembly using a spindle remover tool. Most times, the drive axle can be pushed out by hand.

P22-7 Remove the drive axle from the car.

P22-8 Install joint in a soft-jawed vise to make necessary repairs to the shaft and joints and prepare to reinstall the shaft.

P22-9 Loosely install the axle shaft into the transaxle.

PHOTO SEQUENCE 22 (CONTINUED)

P22-10 Tap the splined end into the transaxle or install the flange bolts and tighten them to specifications.

P22-11 Pull the steering knuckle back and slide the splined outer joint into the knuckle.

P22-12 Fit the ball joint to the steering-knuckle assembly. Install the pinch bolt and tighten it.

P22-13 Install the axle nut and tighten it by hand. Do not reuse the old axle nut.

P22-14 Install the rotor and brake caliper.

P22-15 Torque the caliper bolts to specifications.

P22-16 Install the wheel and tire assembly.

P22-17 Lower the car.

P22-18 Torque the axle nut to the specified amount.

Figure 8-29 With either the top or bottom steering knuckle's fasteners removed, the steering knuckle can be twisted to one side, so the axle stub will slide out. The bolts removed here will require the wheel alignment to be checked after axle installation.

The first step after inspection is to remove the axle nut. The cotter key and locking plate, or tab, are removed followed by the nut. There are two ways to set up the knuckle for removal of the axle. The first is to remove the steering linkage from the knuckle and then remove the strut or knuckle mounting bolts. The freed knuckle, swinging on the lower ball joint, can be swung out of the way as the **axle stub** is pushed from the hub (**Figure 8-29**). The wheel alignment will probably have to be checked in this instance.

A second method is to separate the steering linkage from the knuckle and unbolt the lower control arm. When this is done, the knuckle will hang from the strut (**Figure 8-30**). This is the best method because it will not change the wheel alignment.

> **Axle stubs** are the outer ends of the axle.

⚡ WARNING **Do not drive directly on the end of the axle. Damage to the axle threads will occur. Use the axle nut or a piece of wood between the driver and the axle.**

Figure 8-30 Separating the steering knuckle from the lower control arm prevents misalignment of the steering following a drive axle or CV boot installation.

Separate the axle from the hub by pulling out the steering knuckle to the side. In most cases, the axle will slide from the hub. If the axle will not slide out, press the axle stub from the hub using a special puller tool. Once loosened, the axle should slide out. Allow the hub and knuckle to hang to one side as the axle is separated from the transaxle. Do not just pull on the axle shaft. This will cause the inner CV joint to separate.

⚠ WARNING **Always use a drive axle that has the correct locking device on the transaxle end of the axle. An undersized clip may allow the axle to come out of the transaxle during operation. An oversized clip may damage transaxle gears during axle installation and hinder removal if it does lock in place.**

⚠ WARNING **Do not pry or hit the CV boot. Damage to the boot may occur and allow the CV grease to be flung out during operation.**

A small pry bar is fitted between the inner end of the inner CV joint and the transaxle housing to unlock the inner end of the axle (**Figure 8-31**). A special tool is also available to unlock the axle. Slide the axle out. Check the end of the axle's inner stub for a retaining clip or lock. Ensure that the replacement axle has the same type and size of lock or clip.

Place the new axle into the transaxle until it locks or is fully positioned against the transaxle housing. The axle must not slide out when using only hand force. A brass hammer can be used to tap the axle in place. If it does not lock in, inspect and correct the problem. Line the outer axle stub with the hub and slide the hub onto the axle. Install the washer and loosely tighten the axle nut. Install the lower ball joint and the steering linkage. As the ball joint and steering linkage nuts are tightened to torque, line up the nuts to the cotter key's hole, if used. If the nut passes the hole, do not loosen the nut. Tighten it until the hole aligns again. If the nut must be loosened, it must be removed completely and the two components broken apart before reinstalling.

Install the brake components and wheel assembly. Lower the vehicle until the tires are resting on the floor. Torque the axle nut and install the locking collar and cotter key. If locking tabs are used, tap or crimp the tabs in place. Operate the brakes to reseat the pads. Finish the repair by completing the repair order, cleaning the area, storing the tools, and returning the vehicle to the owner.

Figure 8-31 A flat or curved pry bar can be used to "pop" or unlock the inner axle stud from the transaxle's side gear. Do not pull on the axle; the CV joint could be pulled apart.

CASE STUDY

An automatic transaxle was rebuilt and installed in a vehicle. A rebuilt pump and torque converter were also installed. During the test drive, a serious fluid leak was detected at the bottom of the bell housing. Knowing that the only possible transaxle components that could leak in this area were the pump gasket, pump seal, or torque converter, the technician removed the transaxle. Inspection of the seal and torque converter showed no damage to either part. Fluid was present around the inside of the bell housing and over the torque converter. The technician removed the pump and found the gasket to be dry. A closer inspection of the suspected components was conducted. When the pump seal was removed, it was found that the pump bushing behind the seal cavity was missing. This allowed the converter pump lug to move within the pump and break contact with the seal. A replacement pump seal and bushing were replaced, and the transaxle reinstalled. A test drive revealed no leaks and the vehicle was returned to the customer.

ASE-STYLE REVIEW QUESTIONS

1. Drive axles are being discussed.

   Technician A says some drive axles have center bearings.

   Technician B says the CV joint or boot may be replaced individually.

   Who is correct?

   A. A only
   B. B only
   C. Both A and B
   D. Neither A nor B

2. Drive shafts are being discussed.

   Technician A says the drive shaft must be marked before removal.

   Technician B says the U-joint does not need to be marked before removal.

   Who is correct?

   A. A only
   B. B only
   C. Both A and B
   D. Neither A nor B

3. Transmission fluid is being discussed.

   Technician A says most manual transmissions use automatic transmission fluid.

   Technician B says the level of fluid in an automatic transaxle should be checked while hot for an accurate measurement.

   Who is correct?

   A. A only
   B. B only
   C. Both A and B
   D. Neither A nor B

4. Transmission servicing is being discussed.

   Technician A says the oil pan bolts must be torqued.

   Technician B says the fill plug is on the side of a manual transmission.

   Who is correct?

   A. A only
   B. B only
   C. Both A and B
   D. Neither A nor B

5. A vehicle has a rattling noise from the area of the clutch. The noise goes away when the clutch is disengaged. This noise could be caused by any of the following EXCEPT:

   A. Dry release bearing.
   B. Input shaft bearing.
   C. Worn friction (clutch) plate.
   D. Worn input gear.

6. A whining noise is heard as the engine is running and the transmission is in neutral.

   Technician A says this may be caused by a torque converter.

   Technician B says the output shaft bearing may be the cause.

   Who is correct?

   A. A only
   B. B only
   C. Both A and B
   D. Neither A nor B

7. During servicing of an automatic transmission, metal shavings are found on the magnet.

   Technician A says a small amount of shavings is normal.

   Technician B says excessive shavings may indicate a problem with the overrunning clutch.

   Who is correct?

   A. A only
   B. B only
   C. Both A and B
   D. Neither A nor B

8. A whining sound is present at all times when the engine is operating. The vehicle is equipped with an automatic transmission.

   Technician A says this could be caused by low transmission fluid.

   Technician B says the transmission pump could be worn.

   Who is correct?

   A. A only
   B. B only
   C. Both A and B
   D. Neither A nor B

9. U-joint replacement is being discussed.

   Technician A says the bearing cups are pressed out.

   Technician B says a cup that fails to seat completely may be caused by a misplaced needle bearing.

   Who is correct?

   A. A only
   B. B only
   C. Both A and B
   D. Neither A nor B

10. A clunking noise is heard during gear changing.

    Technician A says the noise is probably in the differential or U-joints on an automatic-transmission-equipped vehicle.

    Technician B says the noise may be caused by a bad inner CV joint.

    Who is correct?

    A. A only
    B. B only
    C. Both A and B
    D. Neither A nor B

ASE CHALLENGE QUESTIONS

1. Driveline diagnosis is being discussed.

   Technician A says road testing should be performed in a empty parking lot with a experienced technician.

   Technician B says some driveline diagnosis can be performed on a vehicle lift.

   Who is correct?

   A. A only
   B. B only
   C. Both A and B
   D. Neither A nor B

2. A clutch pedal adjustment is being discussed.

   Technician A says the only time an adjustment is needed is after replacing a clutch disc and pressure plate.

   Technician B says the clutch linkage must be adjusted periodically due to normal wear.

   Who is correct?

   A. A only
   B. B only
   C. Both A and B
   D. Neither A nor B

3. A automatic transmission is being serviced.

   Technician A says you should use RTV with the new transmission pan gasket.

   Technician B says you should not use RTV with a new transmission pan gasket.

   Who is correct?

   A. A only
   B. B only
   C. Both A and B
   D. Neither A nor B

4. A customer is complaining about a clicking noise when accelerating from a stop sign.

   Technician A says a bad CV joint could be the cause.

   Technician B says a bad CV joint may be felt in the steering wheel.

   Who is correct?

   A. A only
   B. B only
   C. Both A and B
   D. Neither A nor B

A rattling or whining noise is being discussed.
Technician A says if the noise is only when the clutch pedal is engaged it could be a possible release bearing.

Technician B says if the noise is heard at all times but changes volume as the clutch is operated the release bearing could be dry or no lucubration.

Who is correct?

A. A only

B. B only

C. Both A and B

D. Neither A nor B

Name _____   Date _____

CHANGING FLUID IN AN AUTOMATIC TRANSMISSION

Upon completion and review of this job sheet, you should be able to change the fluid in an automatic transmission or transaxle.

ASE Education Foundation Correlation

This job sheet addresses the following **MLR** tasks:

2.A.2   Check fluid level in a transmission or a transaxle equipped with a dipstick. (**P-1**)

2.A.3   Check fluid level in a transmission or a transaxle not equipped with a dipstick. (**P-1**)

2.A.4   Check transmission fluid condition; check for leaks. (**P-2**)

This job sheet addresses the following **MLR/AST/MAST** task:

2.B.4   Drain and replace fluid and filter(s); use proper fluid type per manufacturer specification. (**P-1**)

This job sheet addresses the following **AST/MAST** tasks:

2.A.3   Diagnose fluid loss and condition concerns; determine needed action. (**P-1**)

2.A.4   Check fluid level and condition in a transmission or a transaxle equipped with a dipstick. (**P-1**)

2.A.5   Check fluid level and condition in a transmission or a transaxle not equipped with a dipstick. (**P-1**)

We Support

ASE | Education Foundation

Tools and Materials

Goggles or safety glasses with side shields

Drain pan

Hand tools

Service manual

Torque wrench 1/4-inch drive, inch-pounds

Describe the vehicle being worked on:

Year _____ Make _____ Model _____

VIN _____ Engine type and size _____

Procedure

Task Completed

1. Identify the transmission and specifications

    Transmission _____

    Type of fluid _____

    Amount of fluid _____ L/Qt

    Fluid to be checked _____ hot/cold _____ park/neutral

    Filter bolt torque _____ inch/foot-pounds

    Pan bolt torque _____ inch/foot-pounds

2. Lift the vehicle to gain access to the oil or fluid pan.   ☐

Task Completed

3. Position the drain pan and remove the lowest pan bolts. ☐

4. Remove other bolts slowly until the fluid begins to drain. ☐

5. Support the pan and remove all bolts. Drain any remaining fluid. ☐

6. Inspect the magnet. Describe any materials that adhered to the magnet.

   _____

   _____

7. Remove the filter. Remove any filter gasket or O-ring if present. ☐

8. Remove all gaskets or RTVs from the pan and pan-mounting area on the transmission. ☐

9. Install the filter and gaskets or O-ring. Torque the filter bolts. ☐

10. Place the gasket or RTV on the pan. Wipe the pan mounting area on the transmission clean. ☐

11. Position the pan and gasket and install the bolts. ☐

12. Torque the pan bolts in a sequence. ☐

13. Lower the vehicle. Keep the wheels off the floor. ☐

14. Add the specified type and amount of fluid. Conduct an initial fluid level check. Top off as needed. ☐

15. Set the parking brake if it has not already been set. Start the engine and engage each gear for 5 to 10 seconds. Ensure that the transmission engages reverse and drive. ☐

16. Select the park (neutral) gear. ☐

17. Check the transmission fluid for cold level. Top off as necessary. ☐

18. Test drive the vehicle until the engine is at operating temperature. ☐

19. Check the fluid level. Is the fluid hot? _____ Cold? _____

20. Top off as necessary. ☐

Instructor's Response

_____

_____

_____

Name _____   Date _____

SERVICING A DIFFERENTIAL

Upon completion and review of this job sheet, you should be able to service a differential on an RWD vehicle.

ASE Education Foundation Correlation

This job sheet addresses the following **MLR** tasks:

3.A.1   Research applicable vehicle and service information, fluid type, vehicle service history, service precautions, and technical service bulletins. (**P-1**)

3.A.2   Drain and refill manual transmission/transaxle and final drive unit; use proper fluid type per manufacturer specification. (**P-1**)

3.A.3   Check fluid condition; check for leaks. (**P-2**)

3.E.1   Clean and inspect differential housing; check for leaks; inspect housing vent. (**P-1**)

3.E.2   Check and adjust differential case fluid level; use proper fluid type per manufacturer specification. (**P-1**)

3.E.3   Drain and refill differential housing. (**P-1**)

This job sheet addresses the following **AST/MAST** tasks:

3.A.2   Research applicable vehicle and service information, fluid type, vehicle service history, service precautions, and technical service bulletins. (**P-1**)

3.A.3   Check fluid condition; check for leaks; determine needed action. (**P-1**)

3.A.4   Drain and refill manual transmission/transaxle and final drive unit; use proper fluid type per manufacturer specification. (**P-1**)

3.E.1.1   Clean and inspect differential case; check for leaks; inspect housing vent. (**P-1**)

3.E.1.2   Check and adjust differential case fluid level; use proper fluid type per manufacturer specification. (**P-1**)

3.E.1.3   Drain and refill differential case; using proper fluid type per manufacturer specification. (**P-1**)

We Support
ASE | **Education Foundation**

Tools and Materials

Goggles or safety glasses with side shields
Hand tools
Service manual
Torque wrench, 3/8- or 1/4-inch drive
Drain pan with screen or magnet or wide flat pan
Air- or hand-operated pump **or** suction gun

Describe the vehicle being worked on:

Year _____ Make _____ Model _____

VIN _____ Engine type and size _____

Procedure

1. Identify specifications:

   Transmission _____ manual/automatic

   Type/amount of fluid _____

   Cover bolt torque _____

2. What is your reason for servicing this differential? (diagnosis, symptoms)

_____

3. Lift the vehicle to a comfortable working height. ☐

4. Position the drain pan under the differential. If possible, use a pan that will capture and hold metal shavings with a screen or magnet or use a wide, flat pan in which fluid can be examined after draining. ☐

5. If the differential has a drain plug, remove it and drain the fluid. ☐

6. Once the fluid is completely drained, install and torque the plug to specifications. Go to step 21. ☐

7. Remove the lower bolts on the cover. Loosen the side bolts starting at the bottom on each side. Do not remove bolts. Fluid may start to drain as the bolts are loosened. ☐

8. If necessary, use a screwdriver to pry the bottom of the cover from the housing. Use caution because fluid will begin to flow as soon as the seal is broken. ☐

9. Allow the fluid to drain. ☐

10. With most of the fluid drained, remove the remainder of the bolts and the cover. ☐

11. Examine the waste fluid for any metal shavings. There should be none. Inspection results:

_____

_____

12. If shavings are found, further diagnosis of the gears and bearings is required. ☐

13. Clean the cover and housing of all sealing material. Clean the inside of the housing as much as possible. ☐

14. If a gasket is available, position it on the cover and insert a bolt through the cover. ☐

15. Place the gasket and cover on the housing. Start two bolts to hold them in place. Start the remaining bolts and hand tighten all of them. ☐

16. Torque the cover bolts to specifications. Go to step 21. ☐

17. If a gasket is not used, lay a bead of RTV on the cover. Ensure that the bead is about 1/8-inch thick, evenly applied around the cover edges, and attached to the fluid side of the bolt holes. Let it stand for 10 to 15 minutes. ☐

18. Align the cover's bolt holes with the holes in the housing. ☐

19. Align the cover with the housing. Ensure that the cover does not slide as the bolts are started. ☐

20. Start and hand tighten all bolts. Torque all bolts to specifications. ☐

21. Remove the fill plug. It is normally about halfway up on the side of the housing just behind the drive shaft connection. ☐

22. Secure the correct fluid and install it through the fill plug. Use the available fluid pump to install the fluid. ☐

23. As the fluid nears specified capacity, reduce the fluid flow. ☐

24. Cease filling just as soon as the fluid begins to flow from the fill hole. ☐

25. Check the fluid level using the first joint of the little finger or a measuring tool. Add or drain fluid as necessary to the correct level. ☐

26. Check the cover for leaks. ☐

27. If the repair is complete, clean the area of tools and waste fluid and lower the vehicle. ☐

28. Complete the repair order and return the vehicle to the owner. ☐

Instructor's Response

_____

_____

_____

Name _____   Date _____

REPLACING A U-JOINT

Upon completion and review of this job sheet, you should be able to replace a U-joint in a drive shaft assembly.

ASE Education Foundation Correlation ——————————————————

This job sheet addresses the following **MLR** tasks:

3.D.1   Inspect, remove, and/or replace bearings, hubs, and seals. (**P-2**)

3.D.2   Inspect, service, and replace shafts, yokes, boots, and universal/CV joints. (**P-2**)

This job sheet addresses the following **AST/MAST** tasks:

3.D.3   Inspect, remove, and/or replace bearings, hubs, and seals. (**P-1**)

3.D.4   Inspect, service, and replace shafts, yokes, boots, and
               universal/CV joints. (**P-1**)

We Support

ASE | Education Foundation

Tools and Materials

Goggles or safety glasses with side shields
Lift **or** floor jack and stands
U-joint bearing press
Grease gun
Marker
Transmission sealing plug
Drain pan
Vise

Describe the vehicle being worked on:

Year _____ Make _____ Model _____

VIN _____ Engine type and size _____

Procedure

Task Completed

1. Identify the vehicle:

   Transmission type _____ U-joint bolt torque _____

   Are all the U-joints being replaced? _____

   Is there a center bearing on this shaft? _____

   Is it being replaced? _____

   If yes, what is the bearing bolt torque? _____

2. Lift the vehicle to a good working height.   ☐

3. Compare the replacement U-joint with the installed joint.

   Are they the same in size of bearing cup? _____

   Do they have the same locking devices? _____

4. Mark the rear end of the drive shaft to the differential companion flange.   ☐

5. Position a drip pan under the rear of the transmission.   ☐

6. If present, remove the center bearing fasteners. Support the shaft.   ☐

Task Completed

7. Remove U-joint or flange fasteners at the differential. Support the shaft. ☐

8. Slide the slip joint from the transmission. Set the shaft aside. ☐

9. Install the sealing plug. ☐

10. Position the shaft in a vise with the vise jaws near the U-joint being removed. ☐

11. Remove all locking clips, if present, from the joint. ☐

12. Position the U-joint bearing press. ☐

13. Tighten the press until one bearing cup is forced out. ☐

14. Remove the press, twist the joint cross sideways, and remove the cross. If it will not clear, position the press in an opposite direction to the first setup. Force the second bearing cup out. Remove the cross. ☐

15. Rotate the shaft in the vise to more fully expose the second set of bearing cups. ☐

16. Repeat steps 12 through 14. ☐

17. Remove the bearing cups from the replacement joint and place them aside. ☐

NOTE: Read steps 18 through 25 before starting the installation.

18. Start one cup into one hole of the yoke. ☐

19. Insert the cross into the yoke by placing one end of the cross into the seated cup. ☐

20. Twist the cross until the opposite end is fitted into the second and opposite hole of the yoke. Hold in place. ☐

21. Start a second bearing cup over the exposed cross end and into the yoke. ☐

22. Position the cross so the two ends are started slightly into their respective cups and the grease fitting is facing in the correct direction. Hold in place. ☐

⚡ **WARNING: Do not apply excessive force while pressing the cups. They will require some force, but if it becomes difficult, the bearing cups will have to be removed to check for a dropped needle bearing. If a bearing is dropped, excessive force could damage it, the cup, and the yoke.**

23. Position the bearing press and slowly press the cups into the yoke. Rotate the cross as the cups are pressed. ☐

24. At about the halfway point, stop pressing the cups. Check the position of the cross within the cups. If the cross is positioned properly, continue pressing the cups. ☐

25. When the cups are in position, move the cross as far as it will go in each direction. Install the locking clips if the cross can be moved by hand. If it will not move, the cups and cross will have to be removed to check for dropped bearings. ☐

26. Install the grease fitting. ☐

27. Repeat steps 18 through 25 to remove and install the other cups on this joint. ☐

28. Repeat steps 8 through 25 to remove and install the other U-joints. ☐

29. Grease all U-joints that have all bearing cups installed and locked. The other U-joints will be greased after the drive shaft has been installed on the vehicle. ☐

Task Completed

30. Move the shaft to the vehicle. ☐

31. If required, replace the center bearing onto the shaft. ☐

32. Remove the sealing plug and slide the slip joint into the transmission. ☐

33. If this is a two-piece drive shaft (with a center bearing), support the shaft as the rear bolts are being installed. ☐

34. Install the bearing cups, if necessary, onto the cross. ☐

35. Position the rear U-joint into the companion flange and start the bolts. Align the marks by turning one rear wheel to rotate the flange. ☐

36. Torque the fasteners. ☐

37. Install the center bearing, if present, and torque the fasteners. ☐

38. If necessary, grease the U-joints that have not been lubricated. ☐

39. If necessary, top off the transmission fluid at this time (manual) or after the vehicle has been lowered (automatic). ☐

40. Secure all tools and clean the area of any oil spots. ☐

41. Lower the vehicle, complete the repair order, and return the vehicle to the customer. ☐

Instructor's Response

_____

_____

_____

Name _____   Date _____

REPLACE A WHEEL STUD

Upon completion of this activity, you will be able to properly remove and replace a wheel stud.

ASE Education Foundation Correlation

This job sheet addresses the following **MLR** task:

5.F.6   Inspect and replace wheel studs. **(P-1)**

This job sheet addresses the following **AST** task:

5.F.7   Inspect and replace wheel studs. **(P-1)**

This job sheet addresses the following **MAST** task:

We Support

5.F.8   Inspect and replace wheel studs. **(P-1)**

ASE | Education Foundation

Tools and Materials

Goggles or safety glasses with side shields

Basic hand tools

Flat washers

Describe the vehicle being worked on:

Year _____ Make _____ Model _____

VIN _____ Engine type and size _____

Procedure

	Task Completed
1. Place a vehicle on the hoist. Remove one of the rear wheels.	☐
2. Remove the rear drum.	☐

3. Describe the condition of the wheel stud you are about to replace. Are the threads damaged? Is it cracked, rusted, or broken?

_____

4. Remove any additional components that may be in the way. This may include brake components.   ☐

5. Inspect the drive axle flange, is it damaged, bent, or cracked?

Record findings. _____

** INSTRUCTOR CHECK POINT ** _____

6. Use a hammer to drive out the old stud. **Caution: Do not hit the stud next to the one you are removing because you will damage the threads!**   ☐

7. Select 3 or 4 correctly sized washers/spacers to go over the stud. **** Note**** The washers should be thick enough so that the locking ridges of the stud are covered.   ☐

8. Insert the new stud from the rear of the flange.   ☐

9. Screw in a standard lug nut of the same diameter and tread pitch with the cone (taper) side out.   ☐

Wheel lug nut

Wheel hub

Washers

Task Completed

10. Ensure that the stud is straight and tighten the nut down with a ratchet and socket or combination wrench. ****WARNING** DO NOT USE AN AIR-POWERED TOOL!!**   ☐

11. Check the back side of the stud to ensure it was pulled through all the way to the flange.   ☐

   INSTRUCTOR CHECK POINT _____

12. Remove the lug nut and washers. Inspect the new stud.   ☐

13. Did the new stud go in correctly? Was it damaged? Is it straight?

   _____

   If NO, remove stud and replace with a new one.

14. Replace the brake drum, any other components you had to remove, and the wheel.   ☐

15. What is the lug nut torque spec?

   _____

Instructor's Response

_____

_____

_____

_____

CHAPTER 9
SUSPENSION AND STEERING SERVICE

Upon completion and review of this chapter, you should understand and be able to:

- Inspect and rotate tires.
- Inspect and replace a tire pressure monitoring system (TPMS) sensor.
- Perform a TPMS relearn procedure.
- Repair a tire.
- Replace a damaged wheel stud.
- Perform a prealignment check and set up the alignment lift and machine.
- Diagnose and inspect suspension and steering components.
- Inspect and replace a shock absorber.
- Replace inner tie-rod on a rack-and-pinion system.
- Diagnose tire wear.
- Remove and install wheel assembly.
- Remove, balance, and mount tires on wheels.

Terms To Know

bounce	Gas-charged shocks	Steering knuckle
Camber wear	Isolators	Steering wheel free play
Castellated nut	Joint stud	Toe wear
Coupler	Negative camber	TPMS relearn procedure
Cupping	Positive camber	tread depth gauge
Drifting	Pulling	vehicle ride height
Dry-park check	Seepage	Wave

INTRODUCTION

The driver may not realize that a steering or suspension problem exists until a part fails. The technician should make a thorough inspection of any vehicle being serviced. The inspection may only be a quick ride into the shop or a check of the tread on the tires. Spotting a potential hazard may bring business to the shop but more important is the recognition and correction of a fault that could prevent accidents and injury.

CHECKING FRAME AND SUSPENSION MOUNTING POINTS

As stated in the Classroom Manual, the condition of the frame affects the suspension and steering operations. A technician performing suspension and steering work should inspect the frame. The frame should not show any unusual bends or cracks. Mounting fasteners that show any sign of movement indicate that a portion of the frame has shifted.

Classroom Manual
page 295

Attention should also be directed to the point where brackets are welded to the frame. Any break in a weld bead could indicate a twisted or bent frame. Abuse of the vehicle through overloading or driving can result in distorted frame members, particularly in unibody vehicles. Check the door, hood, and trunk alignment and hinge fittings. A misaligned door, hood, or trunk does not necessarily mean the frame is bent, but a bent or twisted frame could deform the body.

Classroom Manual
pages 309–316

Caution

When placing the lift pads on a jack or lift under the vehicle, attempt to place the pad as close to the ball joint as possible. In some instances, the lower control can be bent by improper placement of the lift pads under the vehicle. Damage to the vehicle could occur.

DIAGNOSING AND INSPECTING SUSPENSION AND STEERING PROBLEMS

At some time in their travels, everyone has seen obvious indications of poor suspension and steering components in vehicles on the road. The compact car with six passengers, the overloaded pickup, worn tires, and other indicators point to faults or potential faults. Each could eventually result in a visit to a shop. Most suspension problems can be prevented with some maintenance and timely service. The components discussed in this section are control arm bushings, ball joints, springs, and shock absorbers. Torsion bars are checked in the same manner as springs, and struts are checked in the same manner as shock absorbers. Initial diagnosis of automatic ride control (ARC) systems will also be discussed as well as electric and four-wheel steering (4WS) systems.

Driver Complaints

The typical driver's complaint on suspension systems involves noise, swaying, and sagging of a corner or end of the vehicle. Most noise complaints are stated as a heavy thump, grating, or grinding noise as the vehicle wheels moves over the road surface. This may indicate a weak spring that would allow the control arm to bottom out against the rebound bumper on the frame, a bad ball joint, a missing or worn control arm bushing, or another component failure (**Figure 9-1**). The driver may also complain that the vehicle seems to roll or lean excessively during a turning maneuver. This could be caused by a weak spring, a worn or missing sway bar bushing, or mountings that are worn or missing. A technician should test-drive the vehicle with the driver if possible. Test-drive results along with the following inspections will isolate the problem component within a minimum of time.

Control Arms

Most control arms do not suffer damage unless abused, but the bushings supporting the control arms do erode. The outer ends of the bushings are visible under the vehicle (**Figure 9-2**). The bushings should not be cracked, dry, or oil soaked. Pushing with a flat-ended rod or

Figure 9-1 The rebound bumper prevents the control arm from contacting the frame during hard shocks.

Figure 9-2 The general condition of the control arm bushings can be checked by inspecting the visible ends of the bushing.

screwdriver can check the bushings' flexibility. The bushing material should flex and return to its original condition. If the material is very soft or does not return, it should be replaced. Replace all the bushings on a control arm as a complete set. Worn bushings will allow the control arm to shift and may create a thud as the vehicle accelerates, brakes, or turns. Newer front-wheel-drive (FWD) vehicles have a lightweight lower control that can be bent by hard rearward shocks to the wheel or undercarriage.

Ball Joints

Ball joints attach the control arm to the steering knuckle and must not have excessive movement between the **ball joint stud** and the **steering knuckle**. Excessive movement is any movement more than specified. There are several means to check ball joint wear. The service manual will provide exact procedures for performing this inspection. Some general guides are listed here.

Ball joint stud on ball joints and steering linkages are tapered and are interference fitted into a tapered hole.

Steering knuckle: The part that contains the spindle or wheel hub which attaches to the steering components.

> ⚙ **SERVICE TIP** Some vehicles have what is called a "rubberized" ball or steering linkage joint. This type of construction allows movement between the stud ball and its socket to reduce road shock to the vehicle. Consult the service manual for the specific amount of movement allowed. Some service manuals cite "excessive movement" without defining what that means. Typically, when specific measurements are not provided and the technician feels the movement is excessive, it is best to replace the rubberized joint with a non-rubberized type. Before doing so, discuss the general technical issues and cost of changing to a non-rubberized joint with the customer. Technically, the only disadvantage to the technician and customer is the lack of "excessive movement" information and the fact that non-rubberized joint movement can be precisely measured. In fact, Ford, the primary user of rubberized joints, has never issued an exact movement specification for their rubberized joints and has discontinued their use. Both types have about the same life span, all things being equal.
>
> The cost of a rubberized versus a non-rubberized joint is about equivalent.

Some ball joints have an indicator that is visible when the joint is worn. The grease fitting is designed and fitted into a floating or moveable retainer. As the joint wears, the grease fitting can be moved by hand. In some vehicles, any movement is cause for replacement of the ball joint. Some ball joints have the fitting's retainer extending above the ball joint. If the grease fitting shoulder should sink and become flush with the ball joint, the joint must be replaced.

⚡ WARNING **Do not replace a ball joint relying on hand and visual inspection alone. The quick test is used to make an initial diagnosis. Always refer to the service manual for specific inspection and tolerances.**

Still other vehicles have ball joints that cannot be measured until the weight is removed from the joint. The weight of the vehicle must be on the control arm for a proper inspection. On a vehicle with the spring anchored on the lower control arm, place a jack under the control arm as close as possible to the ball joint (**Figure 9-3**). Operate the jack until the tire is an inch or two off the ground. Use a pry bar under the center of the tire and attempt to lift the tire (**Figure 9-4**). A good method is to place one hand on the top of the tire. This will give the technician a feel for any vertical movement of the wheel. If there is any movement, the ball joint needs to be checked further with a dial indicator or other procedures shown in the service manual (**Figure 9-5**). A typical ball joint measurement is outlined in **Photo Sequence 23—Typical Ball Joint Measurement.**

A short long arm (SLA) system with the spring on the upper control arm must have the upper ball joint checked in a similar manner. Systems that use torsion bars and struts also have specific inspection procedures to check the ball joints. Always check both wheels during the inspection.

Figure 9-3 The jack should lift directly under the ball joint or as near as possible.

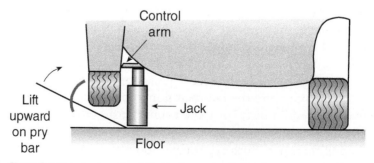

Figure 9-4 The wheel can be shifted by the pry bar if there is wear in the ball joint or the wheel bearing is loose. Before condemning the ball joint, ensure the wheel bearing is properly adjusted.

Figure 9-5 Dial indicator installed to measure horizontal joint movement.

PHOTO SEQUENCE 23
Typical Ball Joint Measurement

P23-1 Place a floor jack under a front, lower control arm, and raise the lower control arm until the front tire is approximately 4 inches (10 cm) off the floor.

P23-2 Lower the jack until the lower control arm is securely supported on a safety stand and remove the floor jack.

P23-3 Attach the magnetic base of a dial indicator for measuring ball joints to the safety stand.

P23-4 Position the dial indicator stem against the steering knuckle beside the ball joint stud.

P23-5 Preload the dial indicator stem approximately 0.250 inch (6.35 mm), and zero the dial indicator scale.

P23-6 Position a pry bar under the tire and lift on the tire-and-wheel assembly while a coworker observes the dial indicator. Record the ball joint vertical movement indicated on the dial indicator.

P23-7 Repeat step 6 several times to confirm an accurate dial indicator reading.

P23-8 Compare the ball joint vertical movement indicated on the dial indicator to the vehicle manufacturer's specifications for ball joint wear. If the ball joint vertical movement exceeds specifications, ball joint replacement is necessary.

Some ball joints have grease fittings that can indicate wear. Inspect these ball joints by inspecting the shoulder of the grease fitting for recession (**Figure 9-6**).

Springs and Torsion Bars

A worn or broken spring or torsion bar will not support its share of the vehicle's weight. A corner of the vehicle that is lower than the others indicates a weak or sagging spring. Some quick measurements can confirm the presence of a weak spring or torsion bar. Locate a point that can be found on each side of the vehicle. An example could be a permanent point on each front fender. The distance is measured from each point vertically to the floor and compared to each other (**Figure 9-7**). Some vehicles require measurements to be made for checking the **vehicle ride height** before a proper wheel alignment can be accomplished. **Figure 9-8** shows the measurement points on one vehicle. This diagnosis and test applies to all types of springs.

Vehicle ride height: The amount of space between the base of the vehicle and the road surface.

Figure 9-6 Ball joint wear indicator with grease fitting extending from ball joint surface.

Figure 9-7 Vehicle riding height measurement, torsion bar front suspension.

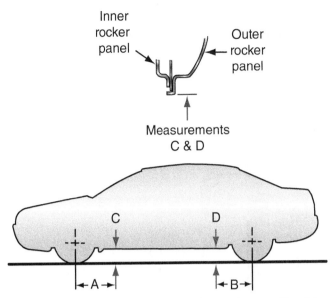

Figure 9-8 Vehicle ride height measurements are made at specific points under the vehicles as determined by the vehicle manufacturer.

Broken springs will cause the vehicle to drop in ride height, will not absorb shock, and will likely cause some noise. Any time the wheel with a broken spring encounters a dip or hump, the shock will be transmitted to the steering linkage and the vehicle frame. It will be readily noticeable by the passenger and particularly by the driver. A broken spring is replaced, not repaired. Some shops recommend that both springs on an axle be replaced together. That may be a valid recommendation in some cases, but it is not usually required by vehicle manufacturers and the makers of suspension components.

Shock Absorbers and Struts

Contrary to belief, shock absorbers and struts do not support the vehicle's weight. A weak spring does not necessarily mean the shock absorber is bad. However, a worn or broken spring may damage the shocks. Always check the shocks if a spring problem is found. Worn shocks and struts tend to make the vehicle **wave** or bounce as the wheels encounter road hazards. As the spring jounces and rebounds, the weak shock cannot control it. The end or corner of the vehicle will continue to move up and down until the spring's energy is exhausted.

To test a shock absorber or strut, forcibly press downward on the vehicle corner with the suspected shock or strut several times. Release and watch for the vertical movement of the vehicle. A good, functioning shock or strut should allow the typical spring to jounce and rebound no more than once before stopping the vehicle at its ride height. Struts and shocks, like springs, can only be replaced.

Shocks or struts can also be broken. The broken part is usually the rod end (**Figure 9-9**). It will also produce a banging or knocking noise as the wheel encounters road irregularities. Naturally, there will be no shock absorber action. A more common condition with older shock or struts is oil leakage. There may be a very light indication of oil on the outside of the housing. This is **seepage** and is a normal condition. However, any visible liquid oil or indications of heavy leaking is a sign that the seals are bad, and the shock or strut needs to be replaced (**Figures 9-10** and **9-11**). Some shops may recommend replacement of the shocks or struts on the same axle. With struts, it may be more desirable to replace both because front struts are part of the steering system also. It is the customer's choice, the overall condition of the vehicle, and the advice of the technician that determine the course of action.

> As the vehicle **waves** from poor shocks or struts, the alignment angle changes on the front wheels and can cause steering control problems.

> **Seepage** is the very small amount of fluid that seeps past the seal. Most seals are designed to allow some lubrication seepage around the seal's contact point with the shaft.

Typical breaking
point of rod

Shock absorber
or strut

Figure 9-9 The most common breaking point for a
shock absorber or strut rod is at the stepped-down
area near the top of the rod.

Oil film
OK

Any oil drop is
unacceptable

Figure 9-10 A leakage is present if liquid oil is present along the
outer shell of the shock absorber or strut housing.

Severe dents
or punctures

Bent or
broken rod

Figure 9-11 Damaged shock absorber inspection.

> **Isolators** are made of rubber or rubber-like material to reduce noise.

Additional malfunctions of the suspension may initially be thought to be steering problems. Weak springs, shocks, or struts could result in poor steering and ride complaints. Poor springs on the rear axle can result in a vehicle that wants to wander from side to side.

Bumping or grinding noise can indicate poor spring **isolators.** The same is true of shock and strut mountings. Damaged upper strut mountings and bearings could result in crushing or grinding noises as the wheels are steered or a bump is encountered. Worn or missing body bushings can result in a noise like that of worn shock bushings. A thorough investigation must be conducted to locate noises that seem to be suspension problems. Another fact to consider is that worn or broken suspension components can damage other suspension and steering components.

REPLACING SUSPENSION COMPONENTS

Front Shocks

Overall, shocks are very simple to replace. The wheel assembly may be removed for easier access. Remove the bolts or nuts that hold the shock in place. On some vehicles, the upper fastener is accessed through the engine compartment. Remove the top fastener on this type before raising the vehicle. The weight of the vehicle will help hold the rod while the nut is being removed. If the top nut and shock rod turn together, a special tool can be used to break off the nut (**Figure 9-12**), but in most cases, an impact wrench or air ratchet will remove the nut easily. Raise the vehicle and remove the lower fasteners. Remove the shock from the vehicle. Check to make sure all the old bushings are also removed. The bushings on the rod type are two-piece with metal retainers.

Figure 9-12 The end of the rod fits over the upper end of the shock absorber rod and its fastener. It is twisted back and forth until the rod breaks off.

⚙ **SERVICE TIP** Keeping the weight of the vehicle on the ground will help hold the shock's rod in as the upper fastener is removed. When installing a new shock, start the upper fastener on by hand and then lower the vehicle before completing the torque on the upper fasteners.

⚡ **WARNING Do not use pliers or other unpadded tools to hold the strut or shock absorber rod still while the nut is being installed. A damaged or marred rod can damage the upper seal.**

Gas-charged shocks hold the rod downward in the cylinder by a wire or lock. The rod must be released during installation. Inspect the mounting area and decide the best method of installing and releasing the rod. Many times, the best method is to fasten the lower end of the shock and then release the rod. The rod is not under a great deal of pressure, and extension is slow and can be controlled. The rod's tip is guided into place. Other than the automatic extension of the rod, the installation is the same as non-gas shocks.

> **Gas-charged shocks** are a hydraulic shock absorber with nitrogen gas installed to pressurize the hydraulic fluid for quicker and more responsive shock action.

⚙ **SERVICE TIP** On some vehicles, the rear shock absorbers extend into the very rear of the passenger compartment. It may be necessary to remove the rear seat back, package tray, or radio speakers to access the top fasteners. Ensure hands and tools are clean before working inside the vehicle's passenger compartment. Care must be taken when dealing with fragile speaker wire connections.

Install the rod's lower bushing retainer and bushing. Insert the rod into its mounting hole. Install the upper bushing, install the retainer, and loosely tighten the nut. Mount the lower end of the shock and install its bushings and fasteners. The shocks may have rod-type or plate-type mountings at each end or one of each (**Figure 9-13**). The rod-type mountings, where the nut screws directly onto the rod, always have bushings. The plate-type mountings usually do not have separate bushings. A special wrench is available that will hold the rod while the nut is being tightened (**Figure 9-14**). Torque all fasteners. Lower the vehicle, clean the work area, and complete the repair order.

Rear Shocks

Replacing the rear shocks is like replacing the front ones, and some rear struts can be replaced following the procedures for rear shocks. Do not remove rear struts if they are also used to mount coil springs. A service manual is required to remove this type of strut correctly and safely.

Figure 9-14 This special socket is made of two connected pieces. The bottom fits over the fastener and the top fits over the hex end of the rod. The top is then held while the bottom is turned by an open-end wrench. Space around the shock may prevent using this type of socket.

Figure 9-13 A shock absorber may have the types of mount shown or a combination of the two.

Classroom Manual
Pages 305

🔧 **Special Tools**

Jack stand, short or tall as needed

Gas-filled shocks have some nitrogen gas added to them to increase the shock's damping action

⚠ **Caution**

Removing shocks on some solid rear axles will release their coil springs and drop the axle. Failure to support this type of axle can cause damage or injury.

Before removing any fasteners, inspect the suspension to ensure that the spring will not come loose when the shock is disconnected. This is especially true of rigid axles with coil springs. The shock absorber helps hold the coil in place by preventing the axle from dropping too far. An axle with a leaf spring should also be supported. When the shock absorber is removed, the axle will drop, and the new shock absorber can be connected without raising the axle. The upper shock absorber fastener on some cars is accessible through the trunk.

If the upper mounting is in the trunk, raise the vehicle enough to gain access to both ends of the shock. The vehicle can be raised to a good working height if both mounts are under the vehicle. Remove the upper fastener and then the lower ones. Remove the shock and bushing. Most rear shocks have bushings pressed into the shock absorber mounting. Others use bushings like the ones on the front (**Figure 9-15**).

If a **gas-filled shock** is used, decide on the best method to remove the hold-down wire and install the shocks. Usually, the easiest method is to place the upper end in position, loosely tighten the fasteners, and cut the wire. The shock absorber rod can be guided to the lower mount as it extends. Mounting the lower end first and then guiding the rod upward into place is another method. It is usually easiest to guide the shock into place instead of extending it and then forcing it into place.

⚙ **SERVICE TIP** Before replacing some high-priced components, check the fuse for the air compressor and computer. Either one could cause the system to fail. Remember KISS ("Keep it simple, stupid."). If only one corner of the vehicle shows a possible suspension fault, then the problem is usually related to that wheel only. It could be an air spring, air valve, height sensor, or the airlines leading to the air spring. Assuming that the remainder of the system works, allow the compressor to switch on and listen for air leaks at the lowered corner. An air leak is indicative of a burst air spring, faulty air valve, or a burst or loose airline. A thorough visible inspection will pinpoint the malfunctioning component. If no air leak is heard, the air spring is not totally eliminated, but the fault is likely due to a height sensor or air valve. Shut down the system, find the airline connection to the air spring or valve and disconnect it. Start the compressor again and listen for air leaking from the line. Escaping air from the lines indicates that air is available, but for some reason it is not being allowed into the air spring. This narrows it down to the air valve, height sensor, or the electrical wiring to and from the computer. The service manual will provide electrical information for using a digital multimeter (DMM) to test each electrical component and the wiring in the system. Some of the newest ARC systems will illuminate a warning lamp in the instrument panel. This alerts the driver and may provide some assistance in diagnosing the system. Still other systems allow the technician to retrieve diagnostic trouble codes (DTCs) from the system with a scan tool. As usual, when in doubt consult the service manual. This is especially useful if only one function of an ARC is not functioning (e.g., system will not lower the vehicle at cruise speeds). In some instances, the ARC computer uses data from the powertrain control module (PCM) or break control module (BCM) to properly determine when and what function to activate. Although ARC systems and their individual components can be expensive, diagnosing the system is usually straightforward as long as there is an understanding of the system's operation and the service manual is used.

Figure 9-15 The shock absorber or strut's upper end will be isolated from the body or frame by rubber-type bushings fitted over the rod. A shock absorber may have the same type of bushing at its lower end.

Mount any necessary bushings and retainers. Install the shocks and torque the fasteners. Lower the vehicle, clean the work area, complete the repair order, and, if all repairs are complete, return the vehicle to the customer.

Automatic Ride Control

Diagnosis of the mechanical components of ARC systems is the same as for true mechanical suspension systems. One common fault visible on many older ARC-equipped vehicles is the complete lowering of the vehicle. This can be so bad that the steering wheel cannot be turned without the tire rubbing the inside of the wheel wells. On a rear-load leveling system, the same fault may be visible by the extreme lowering of the rear and the high angle of the vehicle's front end.

In almost all cases like this, it is a sign that the entire ride control system is malfunctioning. Normally all four air springs, or four air valves, or four (or three) height sensors will not fail together, but a common component such as a faulty compressor or computer will cause the system to cease functioning.

STEERING COMPONENT DIAGNOSIS

Many times, the first indication of a steering component failure is the customer buying new tires or complaining of poor steering. The technician can identify some problems by doing a quick inspection while performing an oil change or some other repair work. Each item of the steering system should be checked each time the vehicle has any work performed on the steering or suspension systems, including tire replacement or repair. Most of the parts can be easily checked without any special tools except for lifting equipment. Tires will be addressed in a later section.

Driver Complaints

Like the suspension system, the customer may complain of noise or improper movement of the vehicle that may indicate a steering component failure. A worn linkage socket or inner tie-rod may produce a popping or snapping noise as the steering wheel is turned. A loose or worn idler arm may be stated by the customer as a "wander" on a straight road or excessive play in the steering linkage. Excessive steering wheel play may be the result of worn gears in the steering gearbox, but this is more common with recirculating ball and nut than with rack-and-pinion systems.

A symptom that came to be known as "morning sickness" may be present on older rack-and-pinion steering systems. This is a result of poor power-steering boost after the vehicle has been shut down overnight. The system leaks pressure until the fluid warms up and the rack's piston seals swell. The driver must apply more force than usual to steer the vehicle during cold driveaway. The steering unit must be replaced or rebuilt to correct this problem. Because rack and steering units are rebuilt by aftermarket vendors, it is usually best to replace the steering unit with a rebuilt model instead of trying to rebuild it at the shop. The overall cost will be about the same, but the time will be much less. Like suspension problems, the technician may have to road test the vehicle to further isolate the failed component. After a road test, the technician should perform a good visual inspection that includes tire wear and a suspension inspection along with steering linkage inspection.

🔧 **SERVICE TIP** Before lifting a vehicle equipped with some form of ARC, turn the system off before lifting the vehicle. The on/off switch is found in the trunk of many vehicles (**Figure 9-16**).

Figure 9-16 On/Off switch.

Linkage

Before lifting the vehicle, check the linkage for **steering wheel free play.** With the engine off, move the steering wheel in each direction while watching the front wheels. Any movement of the steering wheel should result in an immediate corresponding movement of both wheels. If one wheel does not follow the steering wheel, there is probably a problem with the linkage to that wheel. This test is commonly called a **dry-park check**. The best way to perform a dry-park check is to have the vehicle parked on a drive-on lift or an alignment lift. Have an assistant sit in the driver's seat and raise the vehicle until you can comfortably stand under the front suspension. With the engine off, have your assistant move the steering wheel back and forth just off center. While your assistant is rocking the steering wheel, you should watch and listen to the steering linkage. Loose or damaged parts should be easily visible and will often make clicking or popping noises. By moving the steering wheel with the engine off and the weight on the front suspension, this will put stress on the steering linkage as the tires move back and forth. A loose or worn steering gear will result in erratic action of both wheels. If there is excessive free play between the steering wheel and the steering linkage, the steering gear or **coupler** must be replaced (**Figure 9-17**). Neither can be repaired even though the steering gear can be resealed.

A more accurate method of checking free play is to measure the amount of travel of one spot on the steering wheel. Slowly, move the steering wheel in one direction until all the free play is removed. Align a graduation on a rule or other linear measuring device with a spot on the wheel. Hold the measuring device in place and parallel to the wheel's direction of rotation (**Figure 9-18**). Turn the wheel in the opposite direction to the first movement until the free play is removed again. Note where spot on the wheel aligns on the measuring device. The distance between the two graduations is the amount of free play.

> **Steering wheel free play** is the amount of steering wheel movement before the steering gear or linkage moves.
>
> **Dry-park check:** Diagnostic procedure for checking loose steering components.

> The **coupler** is a flexible connection between the steering column and the gear or the upper and lower shafts.

Figure 9-17 The steering shaft coupling may be either a U-joint type or a rubber-type. A common symptom with the U-joint type is binding during a turn, whereas the rubber-type coupling tends to break loose or become chemically unstable.

Figure 9-18 Very few steering systems allow any free play at the steering wheel on modern vehicles. The free play can be measured with a ruler.

Make a check of the power-steering system by starting the engine and turning the wheels completely from side to side. The complete movement should be smooth and continuous. The engine should not stall at idle, nor should there be any belt squealing, pump noise, or jerks of the steering wheel. Replace or adjust the belt if it is worn or loose. Add power-steering fluid if there is pump noise. If the power-steering pump emits a moaning sound, check its fluid level. If the level is correct, the pump may need to be replaced.

Classroom Manual
pages 309–313

There are specific tests that can be used to diagnose the power-steering systems. One uses a beam-type torque wrench. Only an experienced technician should perform this test because the center pad of the steering wheel must be removed. The driver's airbag, if equipped, is part of the pad. Once the airbag is disarmed and the pad is removed, the torque wrench and socket are placed over the steering-wheel retaining nut (**Figure 9-19**). Lift the front wheels so they are free of the vehicle's weight. The steering wheel is turned using the torque wrench, and the amount of force is registered on the wrench. Another test is done with a pressure gauge set up for use with power-steering units. Although any pressure gauge reading over 300 psi can be used, finding the correct adapters may be a problem.

**WHEELS MUST BE FREE
OF VEHICLE WEIGHT**

Figure 9-19 The amount of torque it takes to move the steering can be measured with a dial-type torque wrench. It should be checked any time a new steering gear is installed, or new bearings are installed on the steering shaft.

Most power-steering test gauges come in sets with the correct adapters and can be used fairly easily. These tests would be performed by an "A" or "B" level technician; your job would be to determine if further testing is needed.

Most power-steering systems have a pressure switch. The switch sends a signal to the PCM to adjust idle speed during parking. A faulty switch may cause the engine to stall or stumble when parking. The engine may also have some problem in the fuel or air idle systems that may have caused it to stall during parking. The service manual lists the steps to test the switch and its circuits. However, the noted repairs do not have to be accomplished before the remainder of the inspection is performed.

The linkage must be checked for worn or broken connections or parts. A drive-on lift is best suited for the inspection. It will keep the weight of the vehicle on the wheels and apply normal operating tension on the steering and suspension systems. Raise the vehicle to a good working height. With the vehicle lifted, apply linear force against the connections of each of the sockets in the linkage (**Figure 9-20**). Apply the force back and forth in line with the socket stud. There should be absolutely no movement between the stud and the hole it fits through. Any movement of the stud within its mating hole requires the component to be replaced. Do not confuse this with the side-to-side movement of the ball within the socket. The ball should rotate from side to side, and some manufacturers allow for some vertical movement of the ball within the socket. Consult the service manual for specific tests.

The pitman arm located at the bottom of a recirculating ball and nut system is splined onto the sector shaft and held in place by a large nut and lock washer (**Figure 9-21**). There should be absolutely no movement between the pitman arm and shaft, and the nut should be completely tight. Here, any movement requires the replacement of the pitman arm, sector shaft, or both. If the sector shaft is the problem, it is recommended that the entire steering gear be replaced along with the pitman arm. It would be extremely unusual to find a problem at this connection. A nut that was not properly torqued is the only reasonable cause of slack between the arm and shaft. If there is a loose connection here, it advisable to closely inspect the entire system for similar loose fasteners or improperly installed components.

The idler arm, if equipped, must be checked for mounting and arm movement. The mounting bolts must be tight and the bracket flush against its mount, which is usually the frame rail (**Figure 9-22**). As the connection between the centerlink and idler arm is

> **⚠ Caution**
>
> Do not work on the steering column or steering wheel without disabling the airbag system. Severe injury or death can result if the airbag functions and the technician is working in the vicinity of the steering wheel or column. Consult the service manual for disarming procedures.

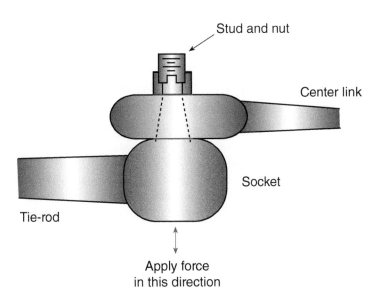

Figure 9-20 There should be no excessive movement between the stud and the tapered hole.

Figure 9-21 There should be no movement at all between the pitman arm and the sector shaft. In almost all cases, excessive steering movement can be traced to other steering components.

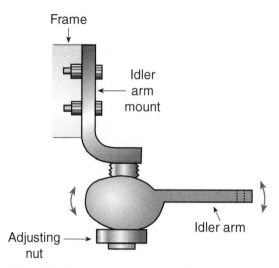

Figure 9-22 Some idler arms have an adjustment to remove excessive movement.

checked, notice the movement of the idler arm on its pivot. Some manufacturers allow for a limited amount of movement at this point; others allow no movement. A few idler arms can be adjusted to remove some movement. Always check the vehicle's service manual for specific procedures and specifications.

Have someone move the steering wheel side to side while watching the movement of the linkage (dry-park check). The entire system should move together. On a rack-and-pinion system, one common problem is a worn inner tie-rod socket. A short movement of the rack with no movement of the tie-rod indicates a worn socket. It can be verified by locking the steering wheel and trying to move the affected wheel by hand. If there is a socket problem, the wheel and tie-rod will move slightly. The same procedure can be applied to test the linkage on all types of steering systems. Remember, there is no movement allowed at the steering linkage connections and the steering should be smooth from one side to the other.

⚡ **WARNING** Do not apply heat to remove any steering linkage or fastener. Heat will change the temper of the component and could result in damage or injury.

REPLACING STEERING LINKAGE COMPONENTS

The power-steering belt is replaced by loosening the tensioner. The pump may be driven by a single serpentine belt or it may have a separate belt. The replacement of a serpentine belt was covered in Chapter 7.

A single belt, or belts, used to drive the power-steering pump usually employs the pump as the belt tensioner. To replace the power-steering belts, it may be necessary to remove other belts. It is recommended that all the belts be replaced at one time.

Some brackets have threaded adjusting bolts and brackets to adjust belt tension (**Figure 9-23**). The pump will have to be pushed or pulled on other types. Some pumps can be moved by hand whereas others must be forced with a bar of some type. In most cases, the bar will be a 1/2-inch or 3/8-inch breaker bar. The pump's bracket will have a square hole for the socket driver on the bar. Move the pump toward the engine until the belt is loose enough to be removed. Remove the belt and install the new one.

Adjusting
screw

Figure 9-24 This tool works like a press.

Figure 9-23 The screw-type adjuster can be found on many modern vehicles regardless of brand. The lock screw holds the pump in place after the belt is tensioned.

Position the bar to apply outward force to the pump. Do not pry on the reservoir. Pull the belt to the correct tension and hold in place. Tighten the easiest fastener to access. Release the bar and check the belt tension. If it is not adjusted properly, loosen the first fastener and tighten the belt again. When the tension is correct, torque all the fasteners.

Most steering linkage parts are replaced in similar methods by forcing the stud from its mounting hole. After lifting the vehicle, remove the cotter key, if used, and the nut from the stud. A special tool is then used to remove the stud from its mounting hole (**Figure 9-24**). Press the stud out and remove the steering linkage from the vehicle. In most cases, more than one connection will have to be removed to free the damaged components.

> SERVICE TIP Using a regular pry bar between the pump and a stationary component like the engine can damage the pump's reservoir.
> Use only the recommended tools to move the pump and tension the belt.

Tie-rod ends can be removed from the steering knuckle using the same tool. The end is then unscrewed from the tie-rod or adjusting sleeve. Count the number of turns as the end is screwed off. Install the replacement end by the same number of turns. This will approximately set the toe of the front wheels until an alignment is completed.

Installation is basically the reverse of removal. Before inserting the stud into its mating hole, ensure that the rubber boot is placed over the socket and a grease fitting is installed. Insert the stud and hand tighten the nut. Do the same with any other connection that needs to be done on this component.

The new part should come with a new nut and cotter key if used. (Do not reuse a locking nut or cotter key.) The nut will be a self-locking type or a **castellated nut** (**Figure 9-25**). The castellated nut is used with a cotter key. A standard torque wrench and socket cannot be used because of space and the arrangement of the components. If special torque wrenches

 Caution

Do not overtension the belt. The belt will wear quicker, and the pump's shaft bearings may be damaged.

Classroom Manual
Pages 313

 Special Tools

Ball joint or tie-rod
 end press
Impact wrench
Torque wrench

Special Tools

Inner tie-rod socket
 wrench
Ball joint or tie-rod
 end press

A **castellated nut** is so-named because of the turret-type projections on one side.

Figure 9-25 The slots between the projections allow for the insertion of a cotter key.

are not available, tighten the nut as much as possible using a standard box-end wrench. This is not really the proper procedure, but it is an accepted practice.

With a castellated nut, line up a slot in the nut with the cotter keyhole in the stud. Do not loosen the nut to align the hole or slot. Continue to tighten it a little at time until they are aligned. If the nut must be loosened, the complete connection has to be broken and the installation must be started from the beginning. The nut will have to be removed and the stud will have to be forced from the hole and reinstalled. This is due to the tapered shape of the stud and hole. Once they are forced or pressed into place, they must be locked at that point. Loosening the nut relieves force on the taper, and the two mating components will not have the correct fit.

On rack-and-pinion systems, the inner tie-rod and socket are removed in a little different manner. Disconnect the tie-rod end from the steering knuckle as previously stated. On the inner end, remove both boot clamps and slide the boot outward toward the end of the rod. It may be necessary to turn the steering wheel until the boot and socket are clearly visible. Inspect the socket connection. There should be a small hole with the end of a pin visible toward the inner end of the socket (rack end). The pin locks the socket to the rack and needs to be removed. A pin punch and small hammer will accomplish the task. Other systems may use a setscrew or other device to lock the socket. They must be removed before proceeding.

There is an air-equalizing tube that runs between the boots on each end of the rack (**Figure 9-26**). As a boot is removed, ensure that the tube remains in place in the other tube. When the boot is installed, ensure that the tube is properly positioned in each boot.

With the lock removed, locate two or more flat sides on the socket. There are special extended sockets available, or an open-end wrench can be used to remove or install the inner tie-rod. Remove the tie-rod end, lock nut, clamps, and boot from the inner tie-rod. Remove the inner tie-rod from the rack. Slide the boot, clamps, and screw nut onto the new tie-rod. Screw and torque the inner tie-rod onto the rack and install any locking device. Move the boot over the rack housing and install the inner clamp. Leave the outer

Classroom Manual
pages 312

Figure 9-26 The air-equalizing tube allows air trapped in the boots to move between the two as the rack is shifted left and right, collapsing and expanding the boots in the process. Loss or misplacement of this tube will result in grit and moisture entering the boots.

clamp loose for now because the toe must be checked and adjusted. Install the tie-rod end onto the tie-rod the same number of turns it took to remove it. Connect the end to the steering knuckle as previously stated. Hand tighten the lock nut until the toe is checked. The alignment can now be checked, and the repair completed.

TIRE WEAR DIAGNOSIS

Correctly interpreted, tire wear can point quickly to a steering or suspension malfunction. Abnormal tire wear is any one part of the tire that is worn more than a matching part of the same tire. It is primarily tread wear but may include damage to the sidewall or bead. Also, individual tires may show different types of wear.

A driver may complain of the vehicle **pulling** or **drifting** to one side on a straight road. This is usually a result of poor wheel alignment, and corrective actions can be taken at a shop. Usually, a vibration felt at certain speeds can be traced to unbalanced tires, but it could possibly be a driveline problem as well. If tire rotation and balance have been recently performed and a pulling to one side starts after the work, radial tire belt movement is usually the problem. Switching the two front tires to opposite sides may correct the pull, but at times the tires must be placed back in their original condition to eliminate the pull.

> **Pulling** usually refers to a sudden movement of the vehicle to one side if the steering wheel is released.

> **Drifting** usually refers to a slow movement of the vehicle toward one side if the steering wheel is released.

> ⚙ SERVICE TIP **Pulling** or **drifting** can be caused by poor alignment, dragging or grabbing brakes, road conditions on older two-lane roads, or bad tires. In the absence of obvious tire faults, do not condemn the tires until all related systems are checked.

An inspection of the tire's sidewall may show sloping ridges and valleys, and most radial tires will be affected to some extent. Normally, this is not a concern, but the driver may question the presence of the ridges. If the ridges seem to affect steering, ride, or durability, the tire needs to be replaced.

Tire Pressure Check and Tread Depth Check

Inspect the tire for the correct pressure. The correct tire pressure is always found on the tire placard (**Figure 9-27**). Tire pressure should be checked every month. Inform your customer to perform this service if they do not have tire pressure monitoring system (TPMS) sensors. When inspecting the tire tread wear, you must use a **tread depth gauge** (**Figure 9-28**). Always measure each tire in the middle and on both edges (inside, outside, and center). This will give you an early indication of any suspension, steering, or alignment problem. The legal minimum tread depth is 2/32 of an inch. A tire has wear indicators (**Figure 9-29**) that will be visible and flush with the tread when it is legally worn out. Most tires do not have good rain or snow traction below 4/32 of an inch. Most new passenger car tires start new with 12/32 to 13/32 of tread. Light truck tires have more.

Tire Pressure Wear

The tread should wear evenly around and across the tread area. The most common fault is incorrect tire pressure. An overinflated tire will ride on its center tread and not wear the other parts of the tire (**Figure 9-30**). An underinflated tire will ride on its edges with little wear to the inside area (**Figure 9-31**). Scuff marks on the sidewall may also be apparent on an underinflated tire. This results when the air pressure is not high enough to hold up the sidewall when cornering. The customer should be advised to check tire air pressure at least once a month or as the operator manual states. Underinflated tires also result in a loss of fuel mileage and poor steering control.

Thread depth gauge: A tool used to measure the tire thread left on the tire.

Figure 9-27 Wheel weights are removed and installed with special wheel weight pliers.

Figure 9-28 Tread depth gauge.

Wear indicator bars

Figure 9-29 Tire tread wear indicators.

Figure 9-30 Wear present on the center two or three treads is a sign of overinflation.

Figure 9-31 Even wear on each outside tread indicates a tire driven while underinflated.

Camber Wear

Camber wear is indicated by excessive wear on the edge of the tire with little wear on the center or opposite edge (**Figure 9-32**). Too much **positive camber** results in wear on the outer three or four treads, and excessive **negative camber** shows up as inner tread wear. The remainder of the tread will show even wear. Camber is corrected using an alignment machine. Some vehicles do not have camber adjustment, while others have adjustments on each wheel. If there is no adjustment mechanism, the suspension, frame, and body must be checked and damaged or worn components must be replaced.

Toe Wear

Toe wear will also show the most wear at the tire's edge, but it will be spread over most of the tread width. Toe wear is indicated by a sawtooth pattern across the tread (**Figure 9-33**). The pattern is obvious when the technician's hand, palm down, is drawn across the tread width. The edge of each tread will be sharpened as though it had been cut with a sharp object. The sharpened edges point in the direction of toe. If the edges point to the outside of the tire, the tire is toed-out too much. In this case, the greatest amount of wear will be on the inside treads, which is similar to camber wear, but camber wear will not create the sawtooth pattern.

Camber wear and toe wear can be confusing. Remember that toe wear will create a sawtooth pattern across all or most of the tread.

A positive camber is the inward tilt of the wheel at the top.

Negative camber is the outward tilt of the wheel.

Toe wear: Feather edge on a tire caused by the toe-in adjustment being out of specifications.

⚠ **Caution**

Before removing (or installing) a tire from a wheel equipped with an electronic tire pressure sensor, consult the service manual. There are specific steps in positioning the tire and wheel assembly on the tire machine. Failure to follow the specific steps for mounting the tire and wheel assembly on the tire changer could result in destruction of the sensor.

Wear on one side

Figure 9-32 Camber wear affects the outer tread(s) of the tire. The wear will be fairly even across the affected treads.

Feathered edge (sawtooth)

Figure 9-33 Toe wear leaves a sawtooth effect with the outer tire tread being the most worn and the wear on the individual treads decreasing toward the center of the tire tread area.

OUT OF BALANCE

LACK OF ROTATION

Figure 9-34 Separating out-of-balance wear from lack of rotation wear sometimes is difficult to do. The best solution is to balance and rotate the tires.

Wear Caused by Balance and Rotation

Balancing and rotating tires are both important for long tire life. They should be part of routine tire maintenance and be performed about every 5,000 miles. Failure to do so could result in poor riding and steering comfort.

Tires that are not kept in balance may develop **cupping** of the treads at irregular intervals around the tire (**Figure 9-34**). The cupping occurs when the heavy spot of the tire hits the road, causing a scuffing of the tire against the pavement. Tires that have not been rotated may also cup the tread. When determining if balance or rotation causes the wear, remove the tire assembly and check its balance on a tire balance machine. Even at this point the technician may not be able to completely eliminate rotation as the cause of the problem, so the best solution is to balance and rotate all tires on the vehicle.

Cupping wear caused by balance or rotation is not as clearly distinguishable as other types of tire wear.

WHEEL AND TIRE SERVICE

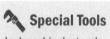 **Special Tools**

Jack and jack stands or lift

Impact wrench

Torque wrench or wheel torque wrench

Service manual

⚡ WARNING **The lug nuts on a wheel should never be tightened with a straight impact wrench. They must be torqued according to the manufacturer's specifications using a torque wrench, a torque stick, or a reliable adjustable impact wrench. Damage to the wheel, studs, hub, or brake disc could result.**

Some of the more common problems with wheels and tires can be traced to poor or no service. The driver or the technician may also cause a problem. Wheel and tire service is limited to replacement, proper torqueing, or repairing a hole in the tire.

Wheel Assembly Removal and Installation

⚡ WARNING **Apply extra care when working with chrome or magnesium wheels. Improper torqueing of the lug nuts could result in damage to the wheel.**

Almost everyone at some point in their driving career must remove and install a wheel. Most flat tires do not occur at a shop but on a dark, lonely road or in the middle of rush hour traffic. Drivers may carry some safety and repair equipment in their vehicles, but a torque wrench (available at the automotive shop) is the one thing they do not have. To change a tire, place a jack under the vehicle's lift point and raise it until the tire is clear of the ground or floor. If in a shop, a 1/2-inch impact wrench can be used to remove the lug nuts. Otherwise, the wrench that comes with the vehicle is used. Perform the necessary work on the wheel and tire or replace it with another assembly. Start the lug nuts on the studs and hand tighten them. Use a hand wrench to tighten the nuts as much as possible.

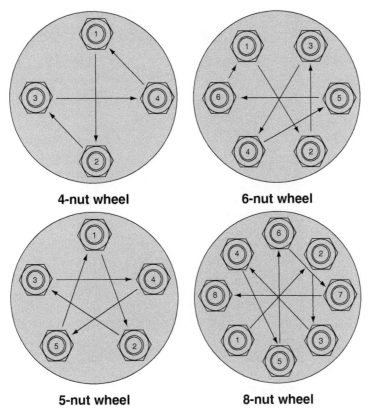

4-nut wheel   **6-nut wheel**

5-nut wheel   **8-nut wheel**

Figure 9-35 Lug nuts should be tightened in a crosshatch pattern similar to the pictured sequence.

If in a shop, use a torque wrench to tighten the nuts to specification. A cross pattern should be followed to torque or tighten the nuts so that the tightening process does not cock the rim on the hub (**Figure 9-35**). A cocked rim will prevent proper torqueing of some of the nuts and will be obvious once the vehicle is on the road.

> **SERVICE TIP** When working with a TPMS valve, remember that a standard wheel rotation requires that a "relearn" procedure be performed on the TPMS. The relearn procedure is conducted after the tire and wheel assemblies have been rotated and remounted to the vehicle. The procedure is specific to the manufacturer and must be followed exactly. Consult the service manual for the instructions and use of **special tools**. Some relearn procedures have to be completed with a special tool, while others can be reprogrammed by the driver controls. Some vehicles have smart sensors that will automatically adjust to their new position.

Special Tools
Tire changer
Service manual

Tire Replacement

There are several types of tire machines available. The following instructions are general guidelines for a tire changer with a side-bead breaker (**Figure 9-36**). The instructions for the tire changer must be followed. Remove the wheel assembly from the vehicle and release all the air from the tire. Select the proper mounting tools to install the wheel assembly on the tire changer. Chrome and magnesium wheels require special mounts to prevent the wheel from damage or marring. Place the deflated tire and wheel assembly onto the bead breaking portion of the machine. Ensure that the clamp does not lock over the rim but presses against the tire at a point near the rim. Operate the clamp until the tire bead breaks away from the rim. Turn the tire around and break the bead on the other side.

Figure 9-36 A tire changer with adapters for low-profile and other special tires.

 Caution

Before removing (or installing) a tire from a wheel with an electronic tire pressure sensor, consult the service manual. Failure to follow the specific steps for mounting the tire and wheel assembly onto the tire changer could result in destruction of the sensor.

Caution

Keep hands and fingers clear of the bead area as the tire is forced onto the wheel. Severe injury could result to hands or fingers if they get caught between the wheel and tire bead.

 Special Tools

Tire balancer
Wheel weight hammer or tool

⚡ **WARNING** Inspect the tire for evidence that a quick-seal agent has been used. Any kind of liquid or gas other than normal air exiting from the tire after the valve has been removed may be an indication that a chemical commonly known as "fix-a-flat" has been used. This gas or liquid may be explosive. Severe injury could result should a spark occur during the tire removal from the wheel.

AUTHOR'S NOTE The previous warning is given as a result of an accident in our town. A technician was removing a tire from a wheel when there was an explosion. The technician was seriously injured. Safety inspections later determined that the probable cause was that the customer had used a chemical for flat repair earlier that day and some of the chemical remained in the tire after deflation. A spark was generated from something (the exact source was never determined) that ignited this small amount of gas. Since about 1989, right around when this incident happened, manufacturers of tire chemical quick-seal kits have changed their propellant to an inert gas that is not flammable. However, exercising caution is always the best course of action.

Mount the wheel assembly onto the tire changer and lock it in place. Use the changer's bead guide to remove one side and then the other side from the rim. Before installing the tire, lube the beads with tire lubricant. Position one side of the tire on the rim and use the bead guide to install the tire. With the tire mounted, inflate it to the proper pressure. Some tires may require a band or air tool to initially seat the bead. Do not inflate the tire more than the maximum pressure noted on the sidewall. Once the tire is properly inflated, remove it from the changer. **Photo Sequence 24** shows the typical procedure for dismounting and mounting a tire on a wheel assembly.

PHOTO SEQUENCE 24
Typical Procedure for Dismounting and Mounting a Tire on a Wheel Assembly

P24-1 Dismounting the tire from the wheel begins with releasing the air, removing the valve stem core, and unseating the tire from its rim. The machine does the unseating. The technician merely guides the operating lever.

P24-2 Once both sides of the tire are unseated, place the tire and wheel onto the machine. Then depress the pedal that clamps the wheel to the tire machine.

P24-3 Lower the machine's arm into position on the tire-and-wheel assembly.

P24-4 Insert the tire iron between the upper bead of the tire and the wheel. Depress the pedal that causes the wheel to rotate. Do the same with the lower bead.

P24-5 After the tire is totally free from the rim, remove the tire.

P24-6 Prepare the wheel for the mounting of the tire by using a wire brush to remove all dirt and rust from the sealing surface. Apply rubber compound to the bead area of the tire.

P24-7 Place the tire onto the wheel and lower the arm into place. As the machine rotates the wheel, the arm will force the tire over the rim. After the tire is completely over the rim, install the air ring over the tire. Activate it to seat the tire against the wheel.

P24-8 Reinstall the valve stem core and inflate the tire to the recommended inflation.

DIAGNOSING AND SERVICING TPMS SENSORS AND TIRES WITH TPMS SENSORS

The first part of diagnosing and servicing a TPMS sensor is to inspect the warning light or warning message. During this inspection, make sure that the tires have the correct inflation by referencing the tire placard. Most systems are very accurate, and a customer complaint may just be that their tire is actually low on air, even though it does not appear to be. Inspect the sensors for obvious damage and missing parts.

TPMS relearn procedure Steps used to identity the proper location of the tire pressure sensors after a rotation or tire service.

After rotating the tires, installing a new sensor, or installing new tires, you must perform a **TPMS relearn procedure**. A TPMS sensor can be inspected further with a scan tool. Specific software must be used to access the TPMS data. When dismounting and mounting a tire with a TPMS sensor, you must take caution. First recognize the type of TPMS sensor used. If it is a valve-stem mounted sensor, then place the bead breaker of the tire machine 180 degrees away from it to minimize sensor damage. When removing the tire from the wheel, start with the arm of the tire machine 270 degrees away from the sensor. This will prevent sensor damage (**Figure 9-37**).

Wheel Balancing

⚡ WARNING **Before using any machinery, read and study the operating manual. Failure to do so could result in damage to the machine or injury to the operator.**

To properly balance a wheel assembly, it must be correctly mounted to the tire balancer (**Figure 9-38**). Various centering cones and fasteners are available on most tire balancers. Select a cone that fits approximately halfway through the wheel's large center hole. In most cases, all the old weights are removed before mounting the assembly onto the balancer. The following instructions may differ between balancer brands and models.

Place the cone onto the balancer's spindle. There should be a spring already in place on the spindle. If not, install the spring and then the cone. The spring keeps the cone

Figure 9-37 Proper tire and wheel position on a tire changer.

Figure 9-38 The circular ring around the tire bead must be centered on the wheel rim.

Locating ring molded in tire wall

Wheel

pressed into the wheel's center. Install the wheel assembly over the spindle and onto the cone. Install the protection devices for magnesium or aluminum wheels, if required, followed by the retainer nut. There are usually three measurements that must be entered into the balancer's computer. They **are rim diameter, rim width, and the distance from the edge of the rim to the balancer.** There is a measuring bar that slides out of the balancer and the end is placed against the rim. This measurement is automatically entered into the machine on some machines. With older machines, you must read the measurement and manually enter the number. This is the distance between the wheel assembly and the balancer's point of reference (**Figure 9-39**).

With the measurements entered, close the hood. Most balancers will automatically spin. The balancer will spin the wheel assembly and bring it to a halt. Some balancers have a roller that will simulate a road force on the assembly (**Figure 9-40**). The display or scale will indicate where weights are to be placed and the size of each weight. Usually, the wheel assembly is rotated until the point of the new weight is straight up. A light or the alignment of two or more lines shown on the balancer's display will indicate that point. There will be two points selected for weights: one for the inside edge of the wheel and another for the outside edge. They are usually not directly across each other. Install the proper weights, close the hood, and re-spin the tire. If everything is correct, the balancer will indicate zero (0) or the letters "OK" to indicate that the assembly is balanced. If the assembly is still out of balance, additional weights will be indicated. **Photo Sequence 25** shows the typical procedure for wheel balancing.

Figure 9-39 On-car wheel balancer.

Figure 9-40 The roller on the balancer senses force variation in the tire.

PHOTO SEQUENCE 25
Typical Procedure for Wheel Balancing

P25-1 Complete all the preliminary wheel balance checks.

P25-2 Follow the wheel balancer manufacturer's recommended procedure to mount the wheel-and-tire assembly securely on the electronic wheel balancer. Some wheel balancers have a centering check in the programming procedure. If the tire and wheel are not mounted properly, tire and wheel balance will be inaccurate.

P25-3 Use a pair of wheel weight pliers to remove all the wheel weights from the wheel rim.

P25-4 Enter the wheel diameter, width, and offset on the wheel balancer screen.

P25-5 Lower the safety hood over the wheel and tire on the wheel balancer.

P25-6 Activate the wheel balancer control to spin the wheel-and-tire assembly.

P25-7 Stop the wheel-and-tire assembly. Observe the wheel balancer display screen to determine the correct size and location of the required wheel weights.

P25-8 Install the correct size wheel weights in the location(s) indicated on the wheel balancer screen.

P25-9 Activate the wheel balancer control to spin the wheel again and observe the wheel balancer display to confirm that it indicates a balanced wheel-and-tire assembly.

P25-10 Stop the wheel, lift the safety hood, and remove the wheel-and-tire assembly from the balancer.

> ⚙️ **SERVICE TIP** At times, the wheel assembly may not balance without adding several ounces of weight at various points. This may be corrected by removing all the weights, deflating the tire, breaking the beads loose, and rotating the tire 90 degrees or 180 degrees on the wheel. The tire is reinflated and balanced. This procedure may cause heavy spots on the rim and tire to counteract each other, which will require less weight to balance the wheel assembly. Follow the wheel balancer's specific directions to achieve this. Larger tires typically require more weight.

Some customers with magnesium or chrome wheels desire that no weights be shown on the outside of the rim. Most new balancers have procedures for measuring and marking points for adhesive weight on the inside of the rims. The balancer must be set up differently. There are also special rim weights for chrome wheels. Follow the instruction manual for the special procedures.

Repairing a Tire

Tire repairs are a common procedure in automotive repair shops. However, the only repair that can be performed on a tire is to plug or patch a hole. This simple repair does have some drawbacks if not done correctly. The best method is to insert a plug patch, which is a combination of a plug and a patch (**Figure 9-41**). When inspecting a tire to repair, you must determine if the damage is in the repairable area (**Figure 9-42**). If the damage is outside of this area, you must install a new tire.

Once the hole is located, a plug can be inserted using a standard repair kit. Use the reamer in the kit to force the hole open. As the reamer is withdrawn from the hole, twist it in one direction. This forces the steel threads in the belt to align around the hole. Insert a plug into the plugging tool and coat the plug with vulcanizing cement. Push the plugging tool and plug through the hole. Do not twist the plugging tool as it is withdrawn. The plug should remain in the hole. Cut off any excessive plug material. Inflate the tire to its recommended pressure and check for leakage around the plug. The hole is now sealed and, in an emergency, the tire can be used. However, this is only a temporary seal and the following steps should be applied at the earliest time to complete the job.

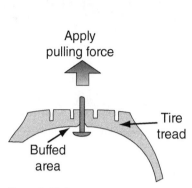

Figure 9-41 Plug installation procedure.

Figure 9-42 Repairable area on bias-ply and belted bias-ply tires.

SERVICE TIP A hole in the tire's sidewall should never be repaired. The tire must be replaced because any plugs or patches will work loose because of sidewall flexing.

SERVICE TIP Installing a plug alone is a temporary repair. The only correct repair is to patch the hole with an internal patch supported by a plug. Inform the customer of this before accepting the job. Plugging a tire can be done with the tire mounted on the vehicle. Before doing any repairs, locate the leak and determine if the tire can be patched. In most cases, any hole over 1/4 inch in diameter cannot be thoroughly sealed. There are plugs and patches to cover larger holes, but the results are not always satisfactory. Small leaks can be found by submersing the portions of the tire into a tire bath. A tire bath can be built by cutting a 50-gallon drum in half lengthwise and filling with water. Any hole will be evident by tracing the air bubbles.

Remove the tire from the wheel and trim the inside end of the previously installed plug. Install the tire on a bead-spreading tool if desired. The tire must be removed from the tool before the patch is installed. Buff on and around the plugged area with a fine buffing stone, but care must be exercised to prevent removing the tire's inner lining. The buffed area should be about twice as large as the patch being installed. Clean the inside of the tire of all buffed material and dust with a vacuum cleaner or a damp cloth. Remove the tire from the bead spreader if used. Apply a film of vulcanizing cement over the entire buffed area. Once the cement is dry, select a patch and remove the film covering the adhesive side of the patch. Do not touch the adhesive side. Locate the arrows on the patch and install the patch with the arrows pointing toward the tire's beads. Center the patch over the hole and press it firmly into place. Roll the patch in place with a corrugated tire stitcher moving from the center of the patch outward. Once the patch is rolled in, remove the plastic shield from the back of the patch. Mount the tire to the rim and inflate to the recommended pressure. Check for leaks around the sealed hole with a soapy solution.

Replacing a Damaged Wheel Stud

Many times, the wheel studs are damaged by overtightening the lug nut or cross-threading the nut onto the stud. Occasionally the stud will just be stripped or break because of old age, but this is very unusual. Replacing a stud can be easy on some wheel assemblies, whereas others require removal of the hub from the vehicle. The general procedure is the same in either case.

Remove the tire and wheel assembly and inspect the area on and around the stud being replaced. Determine if the stud can be replaced on the vehicle as it now sits or whether additional components require removal to fully access the stud. Once a sequence of activities is established, remove any component necessary to gain clearance for removing and installing the stud.

If the hub is still on the vehicle, one of the quickest ways is the use of washers and a lug nut. Select three or four washers with an inside diameter slightly larger than the knurled area of the stud. Insert the new stud in from the backside of the hub and slide the washers over it. Try to keep the washers centered on the stud as much as possible. Select an old lug nut without the chrome covering and screw it onto the stud, flat side first. Use a pry bar between two other studs to help keep the hub from rotating. While holding the hub stationary, use a box-end wrench or ratchet and socket to tighten the nut. This will draw the stud through the hole until the inside end of the stud is flat against the back of the hub.

> **⚙ SERVICE TIP** Very small holes will sometimes not leak except when the vehicle is operating, and the tire is flexing on the road surface. The customer may complain of having to add air every 2 or 3 weeks. The hole can sometimes be found by applying a soapy solution over the tire tread and moving the vehicle slowly while an assistant watches for air bubbles. This technique may require some extra time.

SETTING UP AN ALIGNMENT MACHINE

There are several brands and models of alignment machines in use today. They range from units that measure a white light (older models) to fully computerized units that do practically everything except turn the wrench. The following section covers a typical setup for a computerized system. Most alignment brands are very similar in their setup procedures, but you should always go through the manual or a training course before using it.

Switch on the alignment machine so that it warms up while the vehicle is positioned. Then use a ground guide mirror or another person to help you move the vehicle onto the lift. Keep the vehicle centered on the runway and stop when the front wheels are centered on the turntables. The ground guide should chock (i.e., block) the left rear wheel on the front and back prior to the driver getting out of the vehicle. With the wheel chock firmly in place, the driver can now exit the vehicle, leaving the transmission in neutral and the ignition key turned off. Remove one of the large sensors from the alignment rack and install it on the rear wheel. Install one of the smaller sensors on a front wheel. Each sensor is installed by clamping its wheel adapter over the outer rim of the wheel (or the inner rim on some vehicles). The head of the adapter should be straight up. Use the hand wheel to tighten the adapter onto the wheel and do not overtighten.

The alignment machine should be warmed up by now and show a graphic display screen. On the alignment machine, select the make, model, and year from the vehicle selection screen. At times, it may be necessary to input additional data such as ride height and number of drive wheels. The machine will prompt the technician for the necessary information. Note that the brand of car or truck must be selected on this screen (e.g., Chevrolet or Chevrolet truck). Also, note that full-size vans and some sport utility vehicles (SUVs) are listed as trucks under the appropriate brand.

⚙ SERVICE TIP If the hub is removed from the vehicle to gain clearance, it will be necessary to support it in some manner. A large vise will work or blocks of wood under the edges of the hub will provide a steady base. With the necessary clearance achieved, use a steel flat-nose punch of approximately the same diameter as the stud and a hammer to drive the stud out. There are two ways to install the new stud. If the hub is installed in a vise or supported by blocks, flip the hub over and drive in the stud with the hammer. Drive it in until the back of the stud is flush flat with the inside of the hub.

The next screen on the alignment machine is the sensor calibration screen. Each of the four sensors on the vehicle should be highlighted. If not, the sensor will be shown in red and the machine will not proceed. With all sensor signals received by the machine, a large green arrow will appear on-screen between the four wheels. This indicates that the vehicle is to be rolled backward until the arrow becomes a wide green line. There will also be a loud tone indicating the vehicle has been moved sufficiently for the machine to receive a calibration reading. The sensors will be calibrated at approximately 45 degrees to their installed position.

To properly move the vehicle for such a calibration reading, the technician should place the left hand on the front edge of the left rear tire without obstructing the sensor or receiver's line of sight. Move the rear chock about 20 inches away from and in line with the tire. Placing the right hand opposite the left, manually rotate the tire rearward until the arrow changes or the chock is reached. Do not rush the movement. Try to make the tire and vehicle move at a very steady slow rate until the arrow changes. If the arrow has not changed when the tire hits the chock, then the chock will need to be moved several more inches. It will take some practice over several alignments to get the distance and movement right the first time.

Once the arrow on the screen changes, hold the wheel (and vehicle) in place until the machine sounds a tone or changes color and the arrow changes back into an arrow pointing to the front of the vehicle display. Use the same procedures in the previous paragraph to roll the wheel forward until the arrow again changes into a straight line. Hold the vehicle until you hear the alignment machine's tone and then slide the rear chock back behind the wheel. Make sure it is firmly in place. The vehicle should be just about in the same position as when the calibration was started. The machine will automatically reset to the "Measure caster screen." Remove the locking pins from the turntables and slide plates.

The easiest way to measure caster on the Snap-on John Bean alignment machine is to sit in the driver seat of the vehicle, set the transmission to PARK, and apply the parking brake on firmly. On the alignment machine, there will be a center drawing of the steering wheel and the angle at which it is now positioned. On-screen, near the top of the graphic display, will be a small red dot. Turn the steering wheel slowly to the right while noting the red dot. When it reaches about 5 degrees of rotation, the ball will turn yellow. Continue guiding left until the ball turns green and the matching ball just under it also turns green. This will happen at about 10 degrees into a left turn. Hold the steering wheel in place until the upper ball turns red again. Steer to the right, past the center, until the upper and lower balls turn green, approximately 10 degrees right. Again, hold the steering wheel steady until the upper ball turns red and then steer left until the ball turns green over the green center ball. Adjust the steering wheel slightly until the degree reading in the center of the display screen shows zero-degree center. The caster has now been measured, recorded, and the machine will display the actual alignment angles for all four wheels. Turn the ignition key off and exit the vehicle.

Classroom Manual
pages 324–327

⚒ Special Tools
Alignment machine
Alignment lift

Most alignment machines have software to enter customer, vehicle, and other repair order information. They can also print out the data and include some diagnostic and repair information. The amount, style, and display methods differ on each brand and model. The machine's operating instructions will detail the exact procedures.

Alignment procedures taken after the setup will not be covered here, but for the reader's general information, the following steps are taken to complete an alignment:

- Sensors are compensated individually.
- Locking pins are removed from turntables and skid plates.
- The vehicle is lowered to lift.
- Caster measurement is taken.
- Rear caster, camber, and toe adjustments are made.
- Front caster, camber, and toe adjustments are made.
- Heads are removed and stored.
- The vehicle is removed from the lift and the turntable and skid plate pins are reinstalled.

Detailed information on suspension, steering, and alignment components and procedures can be found in **Today's Technician: Automotive Suspension and Steering Systems,** 5th Edition, Cengage Learning. Other sources include school, local, and Internet libraries. Suspension and steering parts manufacturers like MOOG and TRW have Internet sites with information on design and function of their products. **Photo Sequence 26** shows the typical procedure for setting up an alignment machine.

PHOTO SEQUENCE 26
Typical Procedure for Setting Up an Alignment Machine

P26-1 Display the ride height screen. Check the tire condition for each tire on the tire condition screen.

P26-2 Position the vehicle on the alignment rack.

P26-3 Make sure the front tires are positioned properly on the turntables.

P26-4 Position the rear wheels on the slip plates.

P26-5 Attach the wheel units.

P26-6 Select the vehicle make and model year.

P26-7 Check the items on the screen during the preliminary inspection.

P26-8 Display the wheel runout compensation screen.

P26-9 Display the turning angle screen and perform the turning angle check.

P26-10 Display the front and rear wheel alignment angle screen.

P26-11 Display the adjustment screen.

CASE STUDY

A technician was performing a routine tire rotation and balance along with other repairs. One tire proved difficult to balance and required excessive weights in the technician's opinion. The tire was deflated, broken down, and slid 180 degrees on the wheel. The amount of weight was reduced, and the tire finally balanced out. Due to the other repairs, the shop foreman tested the vehicle and it had a pronounced vibration similar to an unbalanced tire. The technician rechecked the difficult tire and found it to be more than 2 ounces off. The technician removed the tire from the rim. Inside were several large lumps of chassis grease and a new patch. Based on a conversation with the owner, it was determined that the tire had been patched several days earlier. The owner also stated that the vibration started after the tire was patched. The tire was a low-profile tire mounted onto a wide rim. It was deduced that the shop performing the patch could not seat the tire and had used chassis grease to fill the large gap between the tire and rim, so air could be held long enough for the tire to seat. Apparently, some of the grease got into the tire. The technician cleaned the interior of the tire and balanced it with no further problems.

ASE-STYLE REVIEW QUESTIONS

1. *Technician A* says the upper bearing on a shock absorber can cause a grinding noise during turns.

   Technician B says a broken coil spring may allow the frame to connect the bumper on the lower arm.

   Who is correct?

   A. A only
   B. B only
   C. Both A and B
   D. Neither A nor B

2. Coil springs are being discussed.

   Technician A says a broken spring may affect steering angles.

   Technician B says a weak spring on an SLA system may affect steering angles.

   Who is correct?

   A. A only
   B. B only
   C. Both A and B
   D. Neither A nor B

3. A rack-and-pinion system is being discussed.

   Technician A says a worn pinion gear may cause erratic movement between the steering wheel and wheels.

   Technician B says a worn inner tie-rod socket will cause both wheels to respond slowly to steering wheel movement.

   Who is correct?

   A. A only
   B. B only
   C. Both A and B
   D. Neither A nor B

4. The steering system is being discussed.

   Technician A says any movement in the socket or ball requires a replacement of that steering component.

   Technician B says any movement between the stud and its mating hole requires a replacement for that component.

   Who is correct?

   A. A only
   B. B only
   C. Both A and B
   D. Neither A nor B

5. A vehicle with a lowered corner is being diagnosed. This condition could be caused by any of the following EXCEPT:

   A. worn shock absorber.
   B. split air spring.
   C. broken leaf spring.
   D. broken torsion bar mount.

6. The LEAST likely cause of vehicle wandering is:

   A. worn recirculating ball nut.
   B. loose shock absorber mounts.
   C. broken rear springs.
   D. worn rack-and-pinion bushings (mounts).

7. Inspection of the steering linkage is being discussed.

   Technician A says sideways movement of the socket is normal.

   Technician B says force should be applied against the centerline of the socket stud to check for movement between the stud and hole.

   Who is correct?

   A. A only
   B. Both A and B
   C. B only
   D. Neither A nor B

8. A tire with abnormal wear is being discussed.

   Technician A says cupping of the tread could be a sign of poor maintenance.

   Technician B says balancing and rotation may help correct the problem.

   Who is correct?

   A. A only
   B. B only
   C. Both A and B
   D. Neither A nor B

9. *Technician A* says a sawtooth wear pattern across the tread and excessive wear on a tire's edge may indicate that both the toe and camber are incorrect.

   Technician B says excessive wear on the outer edge of a tire indicates the caster is too negative.

   Who is correct?

   A. A only
   B. B only
   C. Both A and B
   D. Neither A nor B

10. Wheel alignment is being discussed.

    Technician A says a cable is used to align the sensor with the wheel assembly.

    Technician B says that vehicle and customer data can be printed out by most computerized alignment machines.

    Who is correct?

    A. A only
    B. B only
    C. Both A and B
    D. Neither A nor B

ASE CHALLENGE QUESTIONS

1. Tires are being discussed.
   Technician A says that the contact patch will change due to different tire inflation.
   Technician B says the contact patch is where the tire supports the weight of the vehicle.
   Who is correct?

   A. A only
   B. B only
   C. Both A and B
   D. Neither A nor B

2. Tire pressure is being discussed.
   Technician A says that the proper tire pressure for each tire is on the side of the tire.
   Technician B says that the proper tire pressure can be found on the tire placard located on the drivers door frame.
   Who is correct?

   A. A only
   B. B only
   C. Both A and B
   D. Neither A nor B

3. Tire pressure monitor system is being discussed.
   Technician A says that the indirect system can identify the tire that is under inflated.
   Technician B says the direct system needs to be relearned after a tire rotation.
   Who is correct?

   A. A only
   B. B only
   C. Both A and B
   D. Neither A nor B

4. Two technicians are discussing coil springs.
   Technician A says that when the spring compresses it is called a jounce.
   Technician B says that when the spring extends it is called a rebound.
   Who is correct?

   A. A only
   B. B only
   C. Both A and B
   D. Neither A nor B

5. Suspension systems are being discussed.
   Technician A says that a dependent suspension is mainly used for light duty application.
   Technician B says that independent suspension can mainly be found on heavy duty trucks.
   Who is correct?

   A. A only
   B. B only
   C. Both A and B
   D. Neither A nor B

Name _____   Date _____

LUBRICATING AND INSPECTING SUSPENSION AND STEERING SYSTEMS

Upon completion and review of this job sheet, you should be able to lubricate the various components of the steering and suspension systems.

ASE Education Foundation Correlation

This job sheet addresses the following **MLR** tasks:

4.B.1 Inspect rack and pinion steering gear inner tie-rod ends (sockets) and bellows boots. (**P-1**)

4.B.7 Inspect pitman arm, relay (centerlink/intermediate) rod, idler arm, mountings, and steering linkage damper. (**P-1**)

4.B.8 Inspect tie-rod ends (sockets), tie-rod sleeves, and clamps. (**P-1**)

4.B.9 Inspect upper and lower control arms, bushings, and shafts. (**P-1**)

4.B.12 Inspect upper and lower ball joints (with or without wear indicators). (**P-1**)

This job sheet addresses the following **AST** task:

4.B.16 Inspect and replace pitman arm, relay (centerlink/intermediate) rod, idler arm and mountings, and steering linkage damper. (**P-1**)

This job sheet addresses the following **AST/MAST** tasks:

4.B.8 Inspect rack and pinion steering gear inner tie-rod ends (sockets) and bellows boots; replace as needed. (**P-1**)

4.B.17 Inspect, replace, and adjust tie-rod ends (sockets), tie-rod sleeves, and clamps. (**P-1**)

4.C.3 Inspect, remove and install upper and lower control arms, bushings, shafts, and rebound bumpers. (**P-3**)

4.C.5 Inspect, remove, and/or replace upper and/or lower ball joints (with or without wear indicators). (**P-2**)

This job sheet addresses the following **MAST** task:

4.B.16 Inspect, remove and/or replace pitman arm, relay (centerlink/intermediate) rod, idler arm, mountings, and steering linkage damper. (**P-2**)

We Support

ASE | Education Foundation

Tools and Materials

Service manual

Lift or floor jack with stands

Grease gun: powered or hand-operated

Goggles or safety glasses with side shields

Describe the vehicle being worked on:

Year _____ Make _____ Model _____

Vin# _____ Engine type and size _____

Procedure

                                                                        Task Completed

1. Raise the vehicle to a good working height.   ☐

Task Completed

2. Locate each grease fitting on the suspension.

Component	Fitting (Y/N)
Upper ball joint/Lower ball joint	_____
Steering knuckle (used on solid front axle without control arms; one top, one bottom; four total)	_____
Tie-rod ends	_____
Tie-rod or centerlink	_____
Idler arm mount	_____
Centerlink or idler	_____
Centerlink or pitman arm	_____
Front mount of rear leaf spring (rarely used, but there may be two)	_____
If the vehicle is RWD, check drive shaft U-joints, usually included as part of routine service (one per joint).	_____
Inspect other components for fittings.	
Are there additional fittings?	_____

If so, list. (Possible locations: trailer hitch, U-joint on rear-drive axle like some Corvettes; power takeoff and shaft; mounts on front leaf spring; U-joint-type coupling on steering shaft; other visible points where two or more components move or work together.)

_____

_____

3. Lube each fitting in turn, starting at one front wheel and moving across the vehicle to the other wheel. It may be necessary to turn the wheel left or right to access the fittings. ☐

4. Do not over lube. Stop lubing when a very small amount of grease is forced from the boot or a slight "pop" is heard. ☐

5. Lube any fittings at the rear of the vehicle. ☐

6. Lube any other fittings found under the vehicle. ☐

7. Lower the vehicle. ☐

8. Lube any suspension or steering components not accessible from under the vehicle. ☐

9. Check the power-steering fluid level. ☐

10. Clean the work area, complete the repair order, and return the vehicle to the customer. ☐

INSTRUCTOR CHECK POINT _____

Instructor's Response

_____

_____

_____

Name _____   Date _____

DISMOUNTING AND MOUNTING A TIRE

Upon completion and review of this job sheet, you will be able to dismount, inspect, and remount tire on a wheel equipped with a tire pressure sensor or without a sensor.

ASE Education Foundation Correlation

This job sheet addresses the following **MLR** tasks:

4.D.3   Dismount, inspect, and remount tire on wheel; balance wheel and tire assembly. (**P-1**)

4.D.4   Dismount, inspect, and remount tire on wheel equipped with tire pressure monitoring system sensor. (**P-1**)

This job sheet addresses the following **AST/MAST** tasks:

4.F.6   Dismount, inspect, and remount tire on wheel; balance wheel and tire assembly. (**P-1**)

4.F.7   Dismount, inspect, and remount tire on wheel equipped with tire pressure monitoring system sensor. (**P-1**)

We Support

ASE | **Education Foundation**

Tools and Materials

Goggles or safety glasses with side shields

Tire changer

1/2-inch drive impact wrench

Tire valve removal tool

1/2-inch drive torque wrench

 Caution

Treat the TPMS cap and valve with care. The cap is aluminum and the valve is nickel-plated to prevent corrosion. Failure to protect them may cause damage to the monitor.

 Caution

Use caution when the pry bar is inserted between the tire and guide or wheel while the turntable is rotating. Guide the bar until it can be lifted from the tire. Injury can occur if the pry bar jerks loose or becomes dislodged and hangs in the wheel assembly.

 Caution

Keep hands and fingers clear of the bead area as the tire is being forced onto the wheel. Serious injury could result if they get caught between the wheel and tire bead.

 Caution

Do not allow tire pressure to exceed the maximum pressure listed on the sidewalls. Serious injury or death could result from an exploding tire.

Describe the vehicle being worked on:

Year _____ Make _____ Model _____

VIN _____ Engine type and size _____

Procedure   **Task Completed**

1. Locate the tire data and lug nut torque.

   Tire size_____

   Tire type_____

   Recommended tire inflation_____

   Maximum pressure on sidewall_____

   Lug nut torque_____

2. Study all operating instructions for the tire changer.   ☐

3. Inspect the tire changer and operate each of the control valves. The valves should be   ☐
   spring loaded to the OFF position. If any valve does not return automatically to the
   OFF position immediately after being released, do not use the machine until repairs
   have been completed.

NOTE: Special instructions for a wheel with TPMS are listed after each appropriate step.

4. Lift the vehicle and remove the wheel assembly.   ☐

5. Remove the valve from the tire's valve stem. Place the TPMS cap and valve on a clean   ☐
   and dry surface.

6. Once the tire is completely deflated, remove all wheel weights.   ☐

⚠ **WARNING Do not place a magnesium or chrome rim onto the tire changer
without using special adapters. Damage to the wheel may result if protective
adapters are not used. Consult the machine's operating instructions.**

7. Place the tire in the bead breaker.   ☐

   A. If the breaker is on the side of the changer, move the tire until the clamp reaches
      the edge of the rim but is not directly on the rim. On a TPMS-equipped rim,
      position the bead breaker 90 degrees from the TPMS valve.

      1. Guide the clamp to the tire and operate the power valve.   ☐

      2. Maintain the valve in operating mode until the tire breaks from the bead.   ☐
         Release the valve.

      3. Turn the tire until the other side is under the clamp.   ☐

      4. Repeat steps 7.A.1 and 7.A.2 until the second side breaks loose.   ☐

   B. If the breaker is mounted on a turntable, place the tire on top of the turntable.

      1. Operate the power valve to lock the tire assembly to the platform.   ☐

      2. Position top and bottom clamps, as required, near the rim's edge.   ☐

      3. Operate the clamp-operating valve until the tire breaks loose from the rim on   ☐
         both sides.

Task Completed

4. If the tire is not on the turntable, place it there and clamp it in place using the changer's clamping device. ☐

5. Position the tire removal or installing guide and lock it in place. On a TPMS-equipped rim, position the guide so the pry bar can be inserted a little counterclockwise of the TPMS sensor when prying the tire over the dismounting guide. ☐

☐

6. Use the changer's pry bar to lift a section of the tire onto the guide. Do only one side of the tire at a time.

☐

7. Operate the turntable rotation valve until the pry bar can be removed and this side of the tire is separated from the rim. Turn off the rotation valve. On a TPMS-equipped rim, operate the turntable in a CLOCKWISE rotation only.

☐

8. Pull up on the tire near the guide. On a TPMS-equipped rim, position the guide so the pry bar can be inserted a little counterclockwise of the TPMS sensor when prying the tire over the dismounting guide.

☐

9. Insert the pry bar between the lower side of the tire and the wheel. Lift the tire over the guide.

☐

10. Operate the turntable rotation valve until the pry bar can be removed and this side of the tire is separated from the rim. On a TPMS-equipped rim, operate the turntable in a CLOCKWISE rotation only. Turn off the rotation valve.

☐

11. Move the guide to the side and remove the tire.

☐

INSTRUCTOR CHECK POINT _____

12. Lightly lubricate the bead area of the new tire with tire lubricant. ☐

13. Place the tire over the wheel. On a TPMS-equipped rim, position the mounting guide 180 degrees from the TPMS valve stem. Position the bead or transition at 45 degrees before (counterclockwise of) the valve. The transition area is the place where the mounted portion of the tire crosses from the rim's center up over the rim's lip to the nonmounted portion of the tire. ☐

14. Position the tire guide within the circle of the tire's beads and lock in place onto the rim. The tire should circle the guide. ☐

15. Start the front of the tire's lower bead down over the edge of the rim and over the guide. ☐

16. Operate the rotation valve and apply downward pressure on the tire as the guide forces one side of the tire onto the wheel. On a TPMS-equipped rim, operate the turntable in a CLOCKWISE rotation only. ☐

☐

17. Turn off the rotation valve.

18. Position the top of the tire over the edge of the rim and the guide. On a TPMS-equipped rim, position the mounting guide 180 degrees from the TPMS valve stem. Position the bead or transition at 45 degrees before (counterclockwise of) the valve. The transition area is the place where the mounted portion of the tire crosses from the rim's center up over the rim's lip to the nonmounted portion of the tire. ☐

19. Operate the rotation valve and apply downward pressure on the tire as the guide forces one side of the tire onto the wheel. On a TPMS-equipped rim, operate the turntable in a CLOCKWISE rotation only. ☐

☐

20. Turn off the rotation valve.

Task Completed

21. Connect the changer's tire inflation line to the tire's valve stem. ☐

22. Inflate the tire until the bead "pops" into place on the wheel. ☐

23. With the tire inflated, disconnect the airline and release the changer's locking clamps. ☐

24. Install the valve into the valve stem. ☐

25. Reconnect the airline to the valve stem and inflate it to specified pressure. ☐

26. Disconnect the airline and install the valve stem cap. On a TPMS-equipped rim, install the special aluminum cap and nickel-plated valve in place. Do not use caps made of any material other than aluminum. ☐

INSTRUCTOR CHECK POINT _____

NOTE: Wheel assembly should be balanced before completing step 27.

27. Install the wheel assembly onto the hub; install and torque lug nuts; and calibrate the TPMS if applicable. ☐

Instructor's Response

_____

_____

_____

Name _____   Date _____

BALANCING A WHEEL ASSEMBLY

Upon completion and review of this job sheet, you should be able to properly balance a wheel assembly.

ASE Education Foundation Correlation ——————————————————————

This job sheet addresses the following **MLR** tasks:

4.D.3   Dismount, inspect, and remount tire on wheel; balance wheel and tire assembly. **(P-1)**

4.D.4   Dismount, inspect, and remount tire on wheel equipped with tire pressure monitoring system sensor. **(P-1)**

This job sheet addresses the following **AST/MAST** tasks:

4.F.6   Dismount, inspect, and remount tire on wheel; balance wheel and tire assembly. **(P-1)**

4.F.7   Dismount, inspect, and remount tire on wheel equipped with tire pressure monitoring system sensor. **(P-1)**

We Support

ASE | Education Foundation

Tools and Materials

Goggles or safety glasses with side shields
Tire balancer
Wheel weight hammer or tool

Describe the vehicle being worked on:

Year _____ Make _____ Model _____

VIN _____ Engine type and size _____

Procedure   **Task Completed**

1. Study the balancer's operating instructions.   ☐

2. Select special adapters if the wheel is made of magnesium or chrome.   ☐

3. As required, select and install the centering cone onto the balancer's spindle.   ☐
    The cone may be installed on the outside of the wheel on some balancers or certain
    types of wheels.

4. Position the wheel assembly onto the cone or spindle as required.   ☐

5. Install the lock nut onto the spindle and screw it into place.   ☐

6. If the balancer is not powered, turn on the power switch.   ☐

7. Select the type of balancing procedures desired: static or dynamic. Some balancers   ☐
    have special procedures that can be used.

8. Measure the distance from the balancer to the edge of the wheel. Measure at the part   ☐
    of the rim where the weights are installed. Use the measuring device attached to the
    balancer.

9. Enter the measurement using the buttons or knobs on the balancer.   ☐

Task Completed

10. Enter the wheel's width into the balancer. ☐

11. Enter the wheel's diameter into the balancer. ☐

12. Ensure that all wheel weights are removed from the wheel. ☐

13. Close the hood and operate the balancer to spin the wheel assembly. ☐

14. After the wheel assembly has stopped, observe the reading. If it is not balanced, rotate the wheel by hand until the balancer's alignment marks are matched for the inside of the wheel. ☐

15. Attach the proper size weight, as determined by the balancer, to the inside of the wheel. ☐

16. With the balancer's marks aligned, install the weight onto the rim. Use the weight hammer. ☐

17. Rotate the wheel by hand until the marks are aligned for the outside of the wheel. ☐

18. Select and install the weight on the outside of the wheel. ☐

19. Close the hood and operate the balancer. ☐

   INSTRUCTOR CHECK POINT _____

20. If the wheel assembly is balanced, remove the wheel assembly from the balancer. ☐

21. If the wheel assembly is not balanced, repeat steps 13 through 19 as necessary to complete balancing. ☐

NOTE: If the wheel assembly cannot be balanced on the second attempt, recheck the measurements and the mounting of the wheel assembly on the balancer. The tire may have to be removed from the wheel and repositioned.

22. Once the tire is balanced, remove the wheel assembly from the balancer and install it on the vehicle. ☐

Instructor's Response

_____

_____

_____

Name _____   Date _____

REPAIR A TIRE

Upon completion and review of this job sheet, you should be able to inspect a tire and repair the tire.

ASE Education Foundation Correlation

This job sheet addresses the following **MLR** tasks:

4.D.5   Inspect tire and wheel assembly for air loss; perform necessary action. (**P-1**)

4.D.6   Repair tire following vehicle manufacturer approved procedure. (**P-1**)

This job sheet addresses the following **AST/MAST** tasks:

4.F.8   Inspect tire and wheel assembly for air loss; determine needed action. (**P-1**)

4.F.9   Repair tire following vehicle manufacturer approved procedure. (**P-1**)

We Support
ASE | Education Foundation

Tools and Materials

Goggles or safety glasses with side shields
Tire plug/patch
Plugging tool
Patching tool

⚠ **Caution**

Plugging alone is a temporary repair. The plug should be replaced or supported by an internal patch. Failure to patch a tire properly could result in a swift loss of tire air pressure and may cause damage or injury.

Describe the vehicle and tire being worked on:

Year _____ Make _____ Model _____

Vin# _____ Engine type and size _____

Tire size _____ Recommended tire pressure _____

Procedure

Task Completed

1. Remove the tire and wheel assembly from vehicle.   ☐

2. If necessary, inflate the tire to recommended pressure. Inspect the tire for damage and leaks. Record your findings for each area.

   a. Tire sidewall _____

   b. Tire tread area _____

   c. Tire bead_____

   d. Leaks. Where are they located on the tire? Is the hole small enough to be plugged and patched?

   _____

   _____

3. If there is no damage other than a repairable leak, is the hole readily identifiable?

   _____

4. If the leak is not easily located, place the tire assembly into the tire vat and force the lower rim under water. Check for bubbles. Rotate the tire assembly enough to place another portion of the tire under water. Continue until bubbles are seen and the leaking hole is located. Where is it located on the tire?

   _____

5. Insert a plug through the tines of the plugging tool. Apply vulcanizing cement to the plug.   ☐

Task Completed

6. Force the reamer through the hole until it is completely through the tire fabric. Rotate the reamer in one direction only to align the strands of the tire's steel belt in one direction.   ☐

7. Remove the reamer.   ☐

8. Using the plugging tool, push the plug through the hole until less than half of the plug is showing.   ☐

9. Remove the plugging tool and trim the plug even with the tread.   ☐

10. Inflate the tire to recommended pressure, if necessary. Place soapy water or other liquid around the plug and check for air bubbles. The plug is sufficient if no air bubbles are present. Record your findings.

_____

_____

11. If plugging only, advise the customer that the plug should be replaced with a patch as soon as possible.   ☐

To complete the repair, perform the following steps.

12. Remove the tire from the wheel.   ☐

13. Install a plug as directed in steps 5 through 11. This time also trim the plug flush with the tire's inner lining.   ☐

NOTE: A bead-spreading tool may be used during buffing, but the tire MUST be removed from the tool before applying the cement and patch.

14. Use a fine buffing stone to lightly clean the area around the plug. The cleaned area should be slightly larger than the patch to be applied and should not go through the lining.   ☐

15. Use a vacuum cleaner to clear the tire's interior of buffing dust.   ☐

NOTE: Remove the bead-spreading tool, if used.

16. Apply a film of vulcanizing cement over the entire buffed area.   ☐

17. Allow the cement to dry completely. Do not apply flame to quick-dry it.   ☐

18. Remove the backing from the adhesive side of the patch. Do not touch the adhesive side.   ☐

19. Position the patch so the bead arrows on the patch point to the tire beads.   ☐

20. Center the patch over the hole and press the patch into place.   ☐

21. Use a corrugated tire stitcher to roll the patch in place. Roll outward from the center with as much hand force as possible.   ☐

22. After completely rolling the patch, remove the plastic film from the top of the patch.   ☐

23. Mount the tire to the wheel and inflate to recommended pressure.   ☐

24. Check for leaks around the plug or patch. Record your findings.

_____

INSTRUCTOR CHECK POINT _____

25. Mount the wheel assembly on the vehicle.   ☐

Instructor's Response

_____

_____

_____

Name _____   Date _____

INSPECTING AND DIAGNOSING SUSPENSION CONCERNS

Upon completion of this activity, you will be able to properly diagnose the noises and ride height concerns associated with suspension systems.

ASE Education Foundation Correlation ————————————————————

This job sheet addresses the following **MLR** tasks:

4.B.10	Inspect and replace rebound bumpers. (**P-1**)
4.B.11	Inspect track bar, strut rods/radius arms, and related mounts and bushings. (**P-1**)
4.B.13	Inspect suspension system coil springs and spring insulators (silencers). (**P-1**)
4.B.14	Inspect suspension system torsion bars and mounts. (**P-1**)
4.B.15	Inspect and replace front stabilizer bar (sway bar) bushings, brackets, and links. (**P-1**)
4.B.16	Inspect, remove, and/or replace strut cartridge or assembly; inspect mounts and bushings. (**P-1**)
4.B.17	Inspect front strut bearing and mount. (**P-1**)
4.C.1	Perform prealignment inspection; measure vehicle ride height. (**P-1**)

This job sheet addresses the following **AST/MAST** tasks:

4.C.3	Inspect, remove and install upper and lower control arms, bushings, shafts, and rebound bumpers. (**P-3**)
4.C.4	Inspect, remove, and/or replace strut rods and bushings. (**P-3**)
4.C.7	Inspect, remove and install short and long arm suspension system coil springs and spring insulators. (**P-3**)
4.C.8	Inspect, remove and install torsion bars and mounts. (**P-3**)
4.C.9	Inspect, remove, and/or replace front/rear stabilizer bar (sway bar) bushings, brackets, and links. (**P-3**)
4.C.10	Inspect, remove, and/or replace strut cartridge or assembly, strut coil spring, insulators (silencers), and upper strut bearing mount. (**P-3**)
4.E.2	Perform prealignment inspection and measure vehicle ride height; determine needed action. (**P-1**)

We Support

ASE | **Education Foundation**

Tools and Materials

Goggles or safety glasses with side shields

Drive-on hoist

Basic hand tools

Note: Each Student Should Perform all of the tests. Take turns when inspecting the suspension. All students will be asked to identify components!

Describe the vehicle being worked on:

Year _____ Make _____ Model _____

VIN _____ Engine type and size _____

Procedure   **Task Completed**

1. Check for any **Technical service bulletins (TSB's)** that relate to this suspension system.

    List only **suspension** bulletins. (See instructor if more than four.)

    _____

    _____

    _____

    _____

2. **Perform a test-drive**. Fill in the chart below as best as possible.   ☐

3. **DO NOT LEAVE THE SCHOOL PARKING LOT!**

Type of Noise	Noise Heard	Driving Condition
Tire Noise	_____	_____
Clunk Noise over Bumps	_____	_____
Creaking or Whine Noise	_____	_____
Noise While Turning	_____	_____

4. During the test-drive, do you notice the vehicle pulling to one side or the other?

    _____

5. During the test-drive, do you notice the vehicle pulling to one side during braking?

    _____

6. Is the vehicle stable or does it wander?

    _____

7. Does the vehicle seem to bounce excessively?

    _____

8. Return to the shop. Adjust the tire air pressure to the manufacturer's specifications.

Tire pressure spec. FR. _____   **REAR** _____

9. Define the KEY TERM jounce.

    _____

10. Define the KEY TERM bounce.

    _____

11. Perform a **JOUNCE** test on all four corners of the vehicle. Does the vehicle bounce excessively on any one corner?

_____

12. Measure the ride height according to the manufacturer's specifications. **Print out and attach ride height procedure.**

Right Front	Left Front	Right Rear	Left Rear
_____	_____	_____	_____

Does the vehicle meet manufacturer's specifications?

_____

_____

13. Raise the vehicle on a **drive-on hoist.**   ☐

14. Describe the type of **FRONT** suspension on this vehicle.

_____

15. Describe the type of **REAR** suspension on this vehicle.

_____

16. Visually inspect the upper and lower control arm bushings. **JOUNCE** the vehicle up and down while you watch the bushings. Do the bushings have any movement in them?

_____

_____

17. Visually inspect the sway bar bushings. **JOUNCE** the vehicle up and down while you watch the bushings. Do the bushings have any movement in them? Any signs of cracking?

_____

18. Visually inspect the sway bar links and link bushing. Are there any concern? Any signs of cracking?

_____

19. If the vehicle has a strut system, lower the vehicle, so all the weight is on the tires. Open the hood. Check the upper strut bearing while jouncing the vehicle. Do you hear any noises? If so what?

_____

_____

20. Start the vehicle and turn the steering wheel from right stop to the left stop. Listen for any noise coming from the upper strut bearing. Did you hear any noises?

_____

_____

21. Turn off the engine. Re-hoist the vehicle (if necessary). Describe the ball joints and describe the type of load it is carrying.

_____

_____

22. Using service procedures, **review** the service procedure for testing/checking the ball joints. ☐

23. Be prepared to describe test procedure for checking ball joints to your instructor. ☐

24. Raise the tires off the hoist and rock each tire back and forth and up and down. If you feel looseness, have your partner inspect for movement in the steering and suspension linkage and joints while rocking tire. Repeat this procedure for all tires. Describe what you found.

_____

_____

_____

25. Inspect the rear suspension in the same manner you did for the front suspension. Inspect all components! Describe what you found.

_____

_____

26. Inspect shock absorbers or struts for bushing wear. Inspect the shock/strut for oil leaking. Describe what you found.

_____

_____

27. Inspect the coil or leaf springs and mounts for any cracks or damage. Describe what you found.

_____

_____

BE PREPARED TO IDENTIFY ALL OF THE SUSPENSION COMPONENTS BEFORE ASKING TO BE CHECKED OFF!

INSTRUCTOR CHECK POINT _____

28. Based on your inspection, what services would you recommend?

_____

_____

_____

_____

Instructor's Response

_____

_____

_____

Name _____   **Date** _____

SLA - BALL JOINT, CONTROL ARM, AND SPRING INSPECTION

Upon completion of this activity, you will be able to properly inspect the control arms, bushings, ball joints, steering knuckle, and spring and steering linkage.

ASE Education Foundation Correlation

This job sheet addresses the following **MLR** tasks:

4.B.10   Inspect and replace rebound bumpers. (**P-1**)

4.B.11   Inspect track bar, strut rods/radius arms, and related mounts and bushings. (**P-1**)

4.B.13   Inspect suspension system coil springs and spring insulators (silencers). (**P-1**)

4.B.14   Inspect suspension system torsion bars and mounts. (**P-1**)

4.B.15   Inspect and replace front stabilizer bar (sway bar) bushings, brackets, and links. (**P-1**)

4.B.16   Inspect, remove, and/or replace strut cartridge or assembly; inspect mounts and bushings. (**P-2**)

4.B.17   Inspect front strut bearing and mount. (**P-1**)

4.B.18   Inspect rear suspension system lateral links/arms (track bars), control (trailing) arms. (**P-1**)

4.B.19   Inspect rear suspension system leaf spring(s), spring insulators (silencers), shackles, brackets, bushings, center pins/bolts, and mounts. (**P-1**)

4.C.1   Perform prealignment inspection; measure vehicle ride height. (**P-1**)

This job sheet addresses the following **AST/MAST** tasks:

4.C.3   Inspect, remove and install upper and lower control arms, bushings, shafts, and rebound bumpers. (**P-3**)

4.C.4   Inspect, remove, and/or replace strut rods and bushings. (**P-3**)

4.C.7   Inspect, remove and install short and long arm suspension system coil springs and spring insulators. (**P-3**)

4.C.8   Inspect, remove and install torsion bars and mounts. (**P-3**)

4.C.9   Inspect, remove, and/or replace front/rear stabilizer bar (sway bar) bushings, brackets, and links. (**P-3**)

4.C.10   Inspect, remove, and/or replace strut cartridge or assembly, strut coil spring, insulators (silencers), and upper strut bearing mount. (**P-3**)

4.C.11   Inspect, remove, and/or replace track bar, strut rods/radius arms, and related mounts and bushings. (**P-3**)

4.C.12   Inspect rear suspension system leaf spring(s), spring insulators (silencers), shackles, brackets, bushings, center pins/bolts, and mounts. (**P-1**)

4.E.2   Perform prealignment inspection and measure vehicle ride height; determine needed action. (**P-1**)

We Support

ASE | Education Foundation

Tools and Materials

Goggles or safety glasses with side shields

Drive-on hoist

Basic hand tools

Service manual

Note: Each student should perform all of the tests. Take turns when inspecting the suspension. All students will be asked to identify compenents!

Describe the vehicle being worked on:

Year _____ Make _____ Model _____

VIN _____ Engine type and size _____

Procedure   **Task Completed**

1. Use a drive-on hoist.   ☐

2. **Perform a visual inspection on the complete suspension system**. (Front and Rear)

    Note any concerns.

    _____

    _____

    _____

    _____

    _____

    _____

    _____

3. Using service manuals, **review** the service procedure for testing the ball joints on this   ☐
    vehicle.

4. Position the jack on the desired area for testing ball joints. Use service procedure.

    Explain the procedure to test ball joint wear.

    Does the ball joint show signs of wear? Yes or no?

    Lower ball joint _____ specifications _____

    Upper ball joint _____ specifications _____

5. **Perform a visual inspection on the steering knuckles**. Do you see any physical damage to the steering knuckles?

    _____

6. **Inspect the control arms and the bushings for wear or damage**. Use a large pry bar to help inspect worn bushings. Inspect all 4 control arms. Did you find any concerns?

    _____

    _____

    _____

    _____

7. **Perform a dry-park steering check**. With the vehicle wheels on the drive-on hoist, have your partner sit in the driver seat, raise the vehicle to a comfortable height. With the engine off, have your partner rock the steering wheel back and forth. Listen and look for loose or worn steering components. Did you find any concerns? (Take turn inspecting the steering linkage!)

_____

_____

8. **Perform a steering linkage inspection**. With the front tires off the hoist, move the wheel up and down and from side to side while having YOU and your partner inspect for any loose joints. Do you feel any looseness on the right or left wheel? If so, explain what is causing the looseness?

_____

_____

9. **Inspect the sway bar and the sway bar links and bushings**. Do you see any broken links or worn bushings?

_____

10. **Inspect the coil springs/torsion bar**. Do you see any cracks or broken springs?

_____

11. Based on your inspection what repair or repairs would you recommend? Why?

_____

_____

_____

_____

_____

INSTRUCTOR CHECK POINT _____

12. Review the parts and labor prices for a ball joint replacement.

   Parts _____ Labor _____

13. What is the labor time to perform a wheel alignment?

_____

14. Describe in your own words, how to replace a front spring on this vehicle? (Use service manuals.)

_____

_____

_____

_____

Instructor's Response

_____

_____

_____

Name _____   **Date** _____

PERFORM A WHEEL ALIGNMENT

Upon completion of this activity, you will be able to properly perform an prealignment inspection and set up the alignment rack.

ASE Education Foundation Correlation

This job sheet addresses the following **MLR** task:

4.C.1   Perform prealignment inspection; measure vehicle ride height. (**P-1**)

This job sheet addresses the following **AST** task:

4.E.2   Perform prealignment inspection and measure vehicle ride height; determine needed action. (**P-1**)

This job sheet addresses the following **MAST** task:

4.E. 2   Perform prealignment inspection and measure vehicle ride height; perform necessary action. (**P-1**)

We Support

ASE | Education Foundation

Tools and Materials

Goggles or safety glasses with side shields

Drive-on hoist

Basic hand tools

Service manual

Describe the vehicle being worked on:

Year _____ Make _____ Model _____

VIN _____ Engine type and size _____

Procedure

Task Completed

1. Using service manuals, check for any alignment related Technical Service Bulletins (TSB's) or manufacturer alignment-related recalls.   ☐

2. **Perform a prealignment inspection**.

   Visual inspection: Tires, Tire sizes, Tire inflation, Wheels/Rims, Vehicle Ride Height, Suspension components: Shocks, Steering linkage, Suspension components. Note any concerns.

   _____

   _____

   _____

3. Perform a dry-park check. Note any concerns.

   _____

4. Set the tires to the correct pressure listed for this vehicle.

   Tire pressure spec. FR. _____   R. _____

5. Based on your prealignment inspection, should any suspension components be replaced before performing an alignment? Explain.

   _____

   _____

Task Completed

6. Prepare the alignment rack. Install the front radius plates onto the alignment rack.   ☐
Make sure holding pins are in place.

7. Drive the vehicle onto the alignment rack. Stop just before the radius plates. **DO**   ☐
NOT DRIVE ONTO THE RADIUS PLATES! Slowly hand roll the vehicle
forward.

8. Place the front tires in the center of the front radius plates. Chock the rear tires. Leave   ☐
the vehicle in neutral.

9. Turn the alignment machine on and get to the main screen.   ☐

10. Level the alignment rack.   ☐

11. Install the wheel sensors. Level, calibrate, and compensate the sensors.   ☐

12. Either using the software that is provided or the service manual, find the alignment
SPECIFICATIONS and write them down.

FRONT	Front Toe Specs. Combined	Front Camber left	Front Camber right	Front Caster left	Front Caster right
→	_____	_____	_____	_____	_____
REAR	Rear Toe Specs. Combined	Rear Camber left	Rear Camber right	Rear Caster left	Rear Caster right
→	_____	_____	_____	_____	_____
→	SAI Specs.	Trust Angle Actual	Thrust Angle Specs.	×	×
×	_____	_____	_____	×	×

13. Remove the locking pins from the radius plates.   ☐

14. Follow the alignment rack computer to perform the necessary steps to measure the   ☐
alignment angles.

15. Record the alignment angles and the specifications **LIVE READINGS**, in the chart
below.

Front Toe Actual	Front Toe Specs.	Front Camber Actual	Front Camber Specs.	Front Caster Actual	Front Caster Specs.
Rear Toe Actual	Rear Toe Specs.	Rear Camber Actual	Rear Camber Specs.	Rear Caster Actual	Rear Caster Specs.
SAI Actual	SAI **Specs.**	Trust Angle Actual	Thrust Angle Specs.	×	×
_____	_____	_____	_____	×	×

16. Based on your measurements and inspection, are the frame or any steering components bent?

_____

17. Do the measurements suggest that there could be tire wear?

_____

18. Describe the method used to adjust the front camber.

_____

_____

19. Describe the method used to adjust front caster.

_____

_____

20. Describe the method used to adjust the front toe.

_____

_____

21. Record your final **FRONT** alignment readings. **Live Readings!**

L. CASTER	R. CASTER	L. CAMBER	R. CAMBER	L. TOE	R. TOR
_____	_____	_____	_____	_____	_____

22. Record your final **REAR** alignment readings. **Live Readings**!

L. CASTER	R. CASTER	L. CAMBER	R. CAMBER	L. TOE	R. TOE
_____	_____	_____	_____	_____	_____

23. What adjustments are needed?

_____

_____

_____

_____

24. Describe the process of centering the steering wheel.

_____

_____

25. Why do you need to install a brake holding tool?

_____

Instructor's Response

_____

_____

_____

Name _____   Date _____

TPM SENSOR RELEARN

Upon completion of this activity, you will be able to relearn tire pressure sensors using the driver information center and using a scan tool.

ASE Education Foundation Correlation

This job sheet addresses the following **MLR** tasks:

4.D.7   Identify indirect and direct tire pressure monitoring systems (TPMS); calibrate system; verify operation of instrument panel lamps. (**P-1**)

4.D.8   Demonstrate knowledge of steps required to remove and replace sensors in a tire pressure monitoring system (TPMS) including relearn procedure. (**P-1**)

This job sheet addresses the following **AST/MAST** tasks:

4.F.10   Identify indirect and direct tire pressure monitoring system (TPMS); calibrate system; verify operation of instrument panel lamps (**P-1**)

4.F.11   Demonstrate knowledge of steps required to remove and replace sensors in a tire pressure monitoring system (TPMS) including relearn procedure. (**P-1**)

We Support

ASE | Education Foundation

Tools and Materials

Goggles or safety glasses with side shields

Drive-on hoist

Basic hand tools

Service manual

Scan tool

TPR tool

Describe the vehicle being worked on:

Year _____ Make _____ Model _____

VIN _____ Engine type and size _____

Procedure

Task Completed

1. Have your instructor assign a vehicle.   ☐

2. Using service manual procedure or owner's manual, locate and **review** a Tire Pressure Learn Procedure.   ☐

3. Using the Driver Information Center (DIC) on the dash, locate the tire pressure screen and record the actual tire pressures for all tires.

   LF _____

   RF _____

   RR _____

   LR _____

4. Locate the tire placard and record the tire pressure specifications.

    FR _____ **REAR** _____

5. Perform a TPMS sensor matching/relearn.   ☐

6. Set the parking brake.   ☐

7. Turn the ignition to ON/RUN with engine off.   ☐

8. With the up level DIC. Locate the tire pressure info page.   ☐

9. With the tire pressure info page displayed, press and hold the SET/RESET button on steering column. The horn should sound twice to signal the receiver is in the relearn mode.   ☐

10. Use the relearn tool to reset each tire. Start with the LF tire and activate the tool the horn should sound once. Move to the RF tire, then the RR tire and finish with the LR tire.   ☐

11. When the last tire is finished the horn should sound twice. Indicating the TPMS sensor matching/relearn process is no longer active.   ☐

Read TPMS Data

12. Perform a generic TPR tool test. Use a scan tool and TPR rest tool.   ☐

13. Power up the scan tool and use the **STAND-ALONE** function. Load vehicle information into the hand-held tool.   ☐

14. Engine off. Key turn off.   ☐

15. Scan each tire sensor one at a time. Record each tire sensors ID# and the tire pressure.

Tire location	I.D. Number	Tire Pressure
Left Front (LF)	_____	_____
Right Front (RF)	_____	_____
Right Rear (RR)	_____	_____
Left Rear (LR)	_____	_____

16. Do all the sensors show data? What does this tell you?

    _____

17. What type of TPMS does this vehicle have?

    _____

18. When should you perform a TPMS sensor relearn?

    _____

19. Describe two different types of TPMS systems.

_____

_____

20. Explain the procedure to replace a TPMS sensor.

_____

_____

Instructor's Response

_____

_____

_____

Name _____   Date _____

TIRE ROTATION

Upon completion of this activity, you will be able to remove and install tires to perform a tire rotation. Inspect tires and measure tread depth and adjust all tire to the proper inflation rate. Reset TPMS.

ASE Education Foundation Correlation

This job sheet addresses the following **MLR** tasks:

4.D.2   Rotate tires according to manufacturer's recommendations including vehicles equipped with tire pressure monitoring systems (TPMS). (**P-1**)

4.D.7   Identify indirect and direct tire pressure monitoring systems (TPMS); calibrate system; verify operation of instrument panel lamps. (**P-1**)

This job sheet addresses the following **AST/MAST** tasks:

4.F.3   Rotate tires according to manufacturer's recommendations including vehicles equipped with tire pressure monitoring systems (TPMS). (**P-1**)

4.F.10   Identify indirect and direct tire pressure monitoring systems (TPMS); calibrate system; verify operation of instrument panel lamps. (**P-1**)

We Support

ASE | Education Foundation

Tools and Materials

Goggles or safety glasses with side shields

Drive-on hoist

Basic hand tools

Service manual

Scan tool

TPR tool

Describe the vehicle being worked on:

Year _____ Make _____ Model _____

VIN _____ Engine type and size _____

Procedure

Task Completed

1. Have your instructor assign a vehicle for the activity.   ☐

2. Raise vehicle on hoist using proper lift points. Use service manual if you have any questions where to lift. Type of hoist? Drive-on or electric?

3. Identify tire size and brand. Inspect all four tires? Are they all the same size? Yes or no?

   Tire size _____ Brand _____
   If no, list different tires: Size _____ Brand _____

   _____

   _____

4. Check all wheels and tires for the following:

   a) Tire damage: tread and sidewall _____

   b) Tire wear: _____

   c) Valve stem damage: _____

    d) Tire Inflation: LF _____ RF _____ LR _____
        RR _____

    e) Lug nut condition: _____

    f) Wheel damage: cuts or gouges _____

    g) Hub cap or center cap damage: _____

5. Locate recommended tire rotation pattern, using service manuals. Draw a diagram of the pattern.

6. Remove all wheels and relocate to specified location. (Use the previous chart.) Note any issues of rust or corrosion between the wheel and the hub.

_____

_____

7. Using door placard or service manuals, write down recommended tire pressure.
    F _____ R _____ Set tire pressure to manufacturer specifications.

8. Using service manuals locate lug nut torque specification.
    Torque spec. _____

9. Using service manuals, draw a picture of the lug nut tighten pattern recommended.

10. Reinstall all tires and torque lug nuts to specifications. Describe proper method for torqueing lug nuts?

_____

_____

_____

_____

11. Reset TPMS if equipped. ☐

Instructor's Response

_____

_____

_____

Name _____   **Date** _____

POWER-STEERING FLUSH

Upon completion of this activity, you will be able to determine proper power-steering fluid type, inspect fluid condition and level, and perform a power-steering system flush. Also remove and replace a belt tensioner.

ASE Education Foundation Correlation

This job sheet addresses the following **MLR** tasks:

4.B.2	Inspect power-steering fluid level and condition. **(P-1)**
4.B.3	Flush, fill, and bleed power-steering system; use proper fluid type per manufacturer specification. **(P-2)**
4.B.4	Inspect for power-steering fluid leakage. **(P-1)**
4.B.5	Remove, inspect, replace, and/or adjust power-steering pump drive belt. **(P-1)**
4.B.6	Inspect and replace power-steering hoses and fittings. **(P-2)**
4.B.23	Describe the function of suspension and steering control systems and components (i.e. active suspension, and stability control). **(P-3)**

This job sheet addresses the following **AST/MAST** tasks:

4.B.9	Inspect power-steering fluid level and condition. **(P-1)**
4.B.10	Flush, fill, and bleed power-steering system; using proper fluid type per manufacturer specification. **(P-2)**
4.B.11	Inspect for power-steering fluid leakage; determine needed action. **(P-1)**
4.B.12	Remove, inspect, replace, and adjust power-steering pump drive belt. **(P-1)**
4.B.15	Inspect, remove and/or replace power-steering hoses and fittings. **(P-2)**

We Support

ASE | **Education Foundation**

Tools and Materials

Goggles or safety glasses with side shields

Drive-on hoist

Basic hand tools

Service manual

P/S flush machine

Belt tensioner tool

Recommended power-steering fluid

Describe the vehicle being worked on:

Year _____ Make _____ Model _____

VIN _____ Engine type and size _____

Procedure

Task Completed

1. Drive vehicle into a working area and install an exhaust hose.   ☐

2. With the engine running and the hood open, listen to the power-steering pump. Sometimes low fluid level can cause a noise. Do you hear any power-steering pump noises? If so, what do you hear?

_____

3. Turn off the engine and check the power-steering fluid level. What is the fluid level?

_____

4. Describe the condition of the fluid by smelling and touching it. Describe if it smells burnt or if it shows signs of dirt or metal in the fluid.

_____

_____

5. Explain what could possibly cause the power-steering fluid to become contaminated or aerated (foaming).

_____

6. Using service manuals, what type of power-steering fluid does the manufacturer suggest and at what service interval does it suggest replacement or inspection?

_____

_____

7. If the power-steering fluid level is low, add the correct type of power-steering fluid reservoir.   ☐

8. Start the engine and let it run for a minute. Do you hear any noise? Did the noise change?

_____

9. Turn off the engine and recheck the fluid level. What is the fluid level?

_____

10. Inspect the condition of the belt(s). Does it need replacement?

_____

11. Explain how a loose belt could cause a power-steering pump pressure problem.

_____

12. Inspect the belt tensioner. Remove the serpentine belt. ****NOTE BELT ROUTING BEFORE REMOVING BELT** Use service manuals to find belt routing if needed.**   ☐

13. With the belt removed, inspect the belt tensioner. Does the tensioner operate properly? _____ Does the pulley spin freely?

_____

14. With the belt removed, spin the power-steering pump pulley. Does the pulley spin freely?

_____

15. Using service manuals, **review** the service procedure for removing the belt tensioner. Remove the belt tensioner.   ☐

16. Reinstall the belt tensioner and the belt.   ☐

17. MAKE SURE THAT THE BELT IS ALIGNED PROPERLY BEFORE STARTING THE ENGINE. Start the engine and make sure belt and tensioner are operating properly.   ☐

Task Completed

18. With the engine off. Raise the vehicle and inspect for any power-steering fluid leaks. Inspect all hoses and lines, and connections, and pump housing. Record all power-steering fluid leaks.

_____

19. Replace the high-pressure power steering. (See instructor before removal) ☐

20. Based on your findings, what are your recommendations for any repairs?

_____

_____

21. Perform a power-steering flush using P/S flush machine. (See instructor or demo) ☐

22. Perform an air bleed procedure. With the engine running, and the front wheels slightly off the ground, remove the power-steering reservoir cap and leave off. Turn the steering wheel from full right stop to full left stop three times. DO NOT HOLD THE STEERING WHEEL AT THE FULL-TURN STOP FOR MORE THEN A FEW SECONDS. ☐

23. Turn off the engine, recheck the power-steering fluid level. Is the level at the full mark for a hot engine?

_____

24. Top off the fluid level if needed. ☐

25. Is the power-steering pump quiet? _____

26. When flushing and bleeding a power-steering system, why shouldn't the steering be held in the full-turn lock position for more than a few seconds?

_____

_____

27. Clean any fluid spilled off the engine/engine compartment, while performing a power flush. ☐

28. Clean your work area. ☐

29. Power-steering hoses wear from the inside out. True or false?

_____

_____

30. When a leaking hose is being replaced, is it a good idea to replace both hoses?

_____

31. Describe the function of a power-steering pressure switch.

_____

_____

Instructor's Response

CHAPTER 10
SERVICING BRAKE SYSTEMS

Upon completion and review of this chapter, you should understand and be able to:

- Describe the different types of automotive brake fluids and methods to test them.
- Diagnose brake systems and perform a basic inspection.
- Describe the safety precautions associated with ABS and airbag systems.
- Test, inspect, and replace brake power boosters.
- Test, inspect, and replace master cylinders.
- Inspect, clean, and replace disc brake pads.
- Inspect, clean, adjust, and replace drum brake shoes.
- Machine brake drums and rotors.
- Inspect, clean, and adjust parking brake cables.
- Perform the general replacement procedures for brake components.

Terms To Know

Burnishing	Grabbing brake	Pedal pulsation
C-clamp	Grease	Pressure bleeder
Cotter key	Hard pedal	Spongy pedal
Fading	Hot spots	Vacuum pump
Glazing	Locking brake	Wheel bearings
Glycol-based	On-vehicle brake lathe	

INTRODUCTION

The brake system provides safe, controllable slowing and stopping of the vehicle. Considering that most vehicles weigh between 1 and 2 tons and may be traveling at interstate highway speeds, major failure of a component can lead to a life-threatening or at least an unnerving situation. Although brakes are somewhat simple to repair, brake system service requires the technician to accurately install and test components to prevent a possible accident. The brake system is one of the most critical systems on a vehicle. Replacing and servicing bearings are also typical tasks for automotive technicians.

Classroom Manual
pages 332–334

This chapter will cover basic brake diagnosis and repairs. Because antilock brake systems (ABS) and traction control systems generally require the use of a scan tool and special procedures for bleeding and installing some components, repairs to those systems will not be discussed. For specific information on ABS and traction control systems, the technician may refer to *Today's Technician: Automotive Brake Systems*, 5th Edition, Delmar Cengage Learning, for detailed instruction on the two systems' operation, testing, and repair.

BRAKE FLUID INSPECTION AND SERVICING

Before beginning the diagnosis of a brake system, it would be wise to consult the service manual for the type of brake fluid to be used. There have been instances where adding a few ounces of fluid to top off the system have led to expensive repairs done at the shop's expense.

There are three basic compounds used for brake fluids: **glycol-based**, silicone-based, and petroleum-based. Installation of the wrong fluid into a system results in quick and complete deterioration of all the brake components that have a rubber seal. European vehicle manufacturers used a petroleum-based fluid in some of their older vehicles. Petroleum-based brake fluid has been discontinued for many years, but a classic vehicle technician may encounter it.

Silicone-based brake fluid is used on a few older vehicles (**Figure 10-1**). An inattentive technician may assume the brake fluid is glycol-based, top it off without checking, and cause serious problems in the brake system. Using the wrong type of fluid will cause deterioration of the sealing components and degrade the braking operation. Even if the mistake is caught before the vehicle leaves the shop, the entire system must be taken down and all rubber seals and flexible hoses must be replaced. This should be done at the shop's and technician's expense. Always consult the service manual.

Glycol-based brake fluids are used in almost every brake system sold. Their popularity and frequent use are the reasons why mistakes, like the one listed earlier, occur. This brake fluid performs well and is durable.

Brake fluids are classified by their resistance to boiling. The classes are DOT 3, DOT 4, and DOT 5 with the highest, 5, having the highest boiling point. DOT 3 is the most common brake fluid. If a system is being flushed, a lower-rated brake fluid should not replace a higher-rated one (DOT 3 *cannot* replace DOT 4), but a higher-rated brake fluid may replace lower-rated ones (DOT 4 may replace DOT 3). Consult the service manual for the proper type and classification.

The following precautionary statements should be as much a part of the technician's professionalism as the tools being used every day.

⚠ WARNING **Do not mix glycol-based and silicone-based brake fluids. The mixture could cause a loss of braking efficiency.**

Glycol is a member of a group of alcohols. It should not be confused with prefix "glyco," which is a combined form of glycerol, a sugar.

Figure 10-1 Silicone-based brake fluid cannot be mixed successfully with glycol-based fluids and never used in current ABS.

⚡ WARNING Protect the vehicle's paint surface from glycol-based brake fluids. This fluid can damage and remove the paint surface very quickly. If brake fluid gets on the vehicle's paint surface, wash the area immediately with cold, running water.

⚡ WARNING Store brake fluid in a sealed, clean container. Brake fluid will absorb water, and its boiling point will be lowered. Exposure to petroleum products such as grease and oil may cause deterioration of brake components when the contaminated fluid is installed.

⚡ WARNING Do not use petroleum-based (gasoline, kerosene, motor oil, transmission fluid) or mineral-oil-based products to clean brake components. These types of liquids will cause damage to the seals and rubber cups, and decrease braking efficiency.

⚠ **Caution**

Wear eye protection when dealing with brake fluid because it can cause permanent eye damage. If brake fluid gets in the eyes, flush with cold, running water and see a doctor immediately.

DIAGNOSING BRAKE SYSTEMS BY INSPECTION AND ROAD TESTING

Brake system failures can be classified as either hydraulic or mechanical failures. Both will create symptoms that can be readily connected to a component or assembly. Mechanical failures will be discussed by common symptoms and their probable causes.

Two conditions constitute hydraulic failure: external leaks and internal leaks. Although mechanical conditions will create the leaks, they will be addressed as hydraulic problems to clarify the operational symptoms. An external leak is defined as "brake fluid exiting the sealed system completely." An internal leak is defined as "a fluid leak within the sealed system where fluid does not exit the system."

Classroom Manual
pages 332–334

External Leaks

A driver may complain that, when the brakes were applied, the pedal dropped drastically, and it took much more distance to stop the vehicle. The most common cause of this problem is an external leak somewhere in the system. A complete leak where the fluid exits the sealed system results in a complete loss of fluid and pressure. The brake pedal will drop much lower than normal or go to the floorboard immediately upon brake pedal application. If a single-piston master cylinder is being used, the pedal will hit the floorboard and cause a complete loss of all brakes. Pumping the pedal in an effort to regain control only compounds the problem. On a dual-piston master cylinder, the pedal will drop much lower than normal, but the driver will have brakes on one of the **split systems** and can stop the car. Again, pumping the brake pedal will not improve braking. An external leak of this type is noticeable when performing a visual inspection. The brake fluid level will also drop in the reservoir.

If the master cylinder is leaking externally, brake fluid will be leaking between the rear of the master cylinder and power booster or firewall (**Figure 10-2**). Usually, an external master cylinder leak will not result in complete brake failure until the fluid in the reservoir is empty. An external leak at a line, wheel cylinder, or caliper will cause a total loss of brakes on that part of the split system. The leaking brake fluid will be visible on the inside of the wheel assembly and may be dripping from suspension or steering components. Small external leaks may be present but not readily visible. The leak is indicated when the reservoir must be topped off regularly. This type of leak is usually found at wheel cylinders during routine brake service. During inspection, the dust boots of the wheel cylinder are opened just enough to check for fluid inside the boots. The inside of the boots should be completely dry.

Split system Brake fluid is applied to the wheels on separate lines one is usually the front and rear.

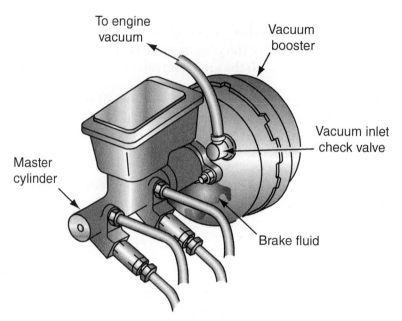

Figure 10-2 Missing paint and bare metal on the booster or firewall is an indication of an external master cylinder leak.

⚙ SERVICE TIP At times, a customer will complain of the brake warning light flashing on when the vehicle is turning a corner or stopping. This is caused by the low fluid level and may be an indication of a leak or normal wear and tear. As the disc pads wear, the caliper piston moves outward and pulls fluid from the reservoir. Before "topping off" the reservoir, check and measure the wear on the disc pads. They may be near the replacement point.

Internal Leaks

Drivers of older vehicles often complain of slow brake pedal drop while sitting at a stoplight. This situation is not as common on newer vehicles because of advances in machinery and materials, but it still occurs. The symptoms will usually occur as the car ages. Notice that the driver can stop the vehicle normally, but the braking action seems to lessen or disappear when the pedal is depressed for a time. Because the vehicle will stop with normal brake pedal movement, an external leak is eliminated. However, a pedal will drop when there is a loss of pressure, which points to an internal leak. An internal leak is almost always found in the master cylinder. Any other component that leaks will result in an external leak with the possible exception of ABS modulators. The modulators can leak internally, causing the same symptoms noted with a leaking master cylinder. However, because of regulations and new technology, this problem is usually found only on the isolation or dump solenoids on older rear wheel antilock (RWAL) systems.

An internal leak in the master cylinder allows fluid to flow from the compression chamber, past the seal or cup, and back into the reservoir (**Figure 10-3**). High pressure created during stopping may prevent the seals from leaking. However, as the driver reduces force to hold the vehicle in place, the pressure drops slightly. This pressure reduction is enough to relax the seals and allow fluid to seep. The fluid returns to the reservoir and the master cylinder pistons move forward, dropping the brake pedal. Eventually, the condition could cause the brake pedal to reach the floor if the brakes are applied long enough. The fluid level in the reservoir will remain the same. Usually, the cheapest, shortest, and best repair is the replacement of the master cylinder. There are rebuild kits for some master cylinders, but the additional labor cost may exceed the cost of a replacement component.

Figure 10-3 If the piston cup is corroded or damaged, pressurized fluid can escape back to the reservoir, causing the brake pedal to drop.

Grabbing or Locking

A **grabbing brake** occurs when one or more wheel brakes grab the rotor or drum and then quickly release it. The vehicle jerks as it slows. This situation could be caused by oil or brake fluid on the pads or shoes. **Hot spots** on the drum or rotor can also cause the problem as will worn or weak retaining hardware. Grabbing brakes are more common with drum brakes.

A **locking brake** is one that will cause the wheel to stop turning and slide when the brakes are applied normally. This is caused by the friction material on that pad or shoe engaging with more friction than the other wheels. This can be caused by oil or brake fluid on the friction material or a brake that is adjusted too tightly. Brake fluid in particular will soften the friction material, which tends not to slide on the drum or rotor. Shoes and pads that have been contaminated by brake fluid, grease, or oil must be replaced and the rotor or drum cleaned. A second cause could be the opposing brake's failure to work correctly. For instance, if the left front wheel consistently locks, the first thing to check is the right wheel brake to make sure it is working. Locking brakes are more common with drum brakes because leaking brake fluid flows directly on the shoes. A locking disc brake can usually be traced to a failure of the brake on the opposite wheel.

> **Grabbing brakes** are usually caused by oil or brake fluid on the friction material.

> **Hot spots** are formed by excessive heat on the weakest area of the disc or drum.

> **Locking brakes** may be caused by misadjusted shoes, blockage in the brake line, or slick tires/road surface.

Pulling to One Side

In this instance, a steady, consistent, smooth pull caused by brake failure on the opposite side of the pull is being discussed. The driver must hold the steering wheel to one side to counteract the pull. Correcting the problem can be accomplished by adjusting the brake shoes or cleaning the slide pins, assuming there is no mechanical damage or bad parts. A pull to the side is often caused by a problem on the front axle, but an overly tight rear brake may cause a steady drift (light pull) to that side during slow to moderate braking. Excessive position **camber** may also cause pulling to one side during braking, but the driver will probably report this as a steering symptom because it will be present without braking. Caster, camber, and toe are wheel alignment measurements that were discussed in Chapter 9.

> **Camber** Camber angle is the inward or outward tilt of the wheel from the top view.

Hose Damage

Another pulling condition can be caused by internal damage to a brake hose. The internal lining of the flexible hose can come apart, and every time the brake is applied, the torn lining moves and blocks the fluid flow (**Figure 10-4**). The result is good braking on one wheel and

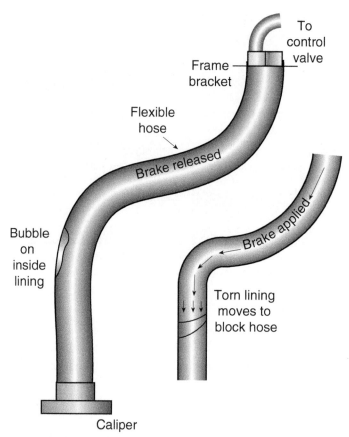

Figure 10-4 A torn or damaged interior lining can move and block the hose during brake application.

little or none on the damaged side. At times this same condition may exist, but the vehicle does just the opposite. The torn lining will allow fluid to flow to the brake component, but blocks its return when the brakes are released (**Figure 10-5**). This may cause a pull toward that brake during non-braking conditions. During the service, it will be noted that the linings for this wheel have worn much more than the other side, which is an indication that the caliper or wheel cylinder is stuck in the applied position. Usually, there are no external signs of the blockage. Over a period of time, the hose could become weak enough for a bulge to form during braking, but the driver will probably have the brakes checked before this happens. If a thorough inspection is completed without finding the cause of the locking brake, the technician should suspect the brake hose on the non-braking side. This hose can be removed, and the interior can be inspected by straightening the hose and looking at a light through the hole.

Fading

Fading occurs when the brakes are applied for long distances such as driving down a long grade. The brakes will appear normal at first, but braking action will be reduced, and the brake pedal feel will become firmer the longer the brakes are applied.

When the brakes are applied for long distances, they heat up and lose friction. As the friction between rotor or pad and drum or shoe decreases, the braking action decreases. This condition is known as brake **fade**.

The driver applies more force and the heating becomes more intense. The eventual result is a complete loss of braking and a very **hard pedal**. Releasing the brakes will allow them to cool, but damage to the brake friction components may have occurred. When a light-vehicle driver complains of a hard pedal but poor braking condition, the technician should check for a trailer hitch. Pulling a trailer with a light vehicle can cause a brake fade condition.

Fading is a condition where the brakes initially work but lose their stopping power because of heat or other conditions affecting the friction material. The most common cause of brake fade is heat, but it should be noted that water and hydraulic fluid can also cause brake fade.

Fade The reduction of stopping power usually due to extreme heat build up during extended high load braking.

A **hard-pedal** condition is when high driver input braking force does not sufficiently slow the vehicle.

Figure 10-5 The internal break in this hose may prevent the fluid from returning to the master cylinder, creating a constant brake application on this wheel.

Many times, signs are posted at the top of steep slopes warning drivers, particularly truck drivers, to select a lower gear before going down the hill (**Figure 10-6**). The vehicle's driveline and engine are used to slow the vehicle and reduce brake usage.

Hard Pedal but Little Braking

When a driver applies the brake and has a hard pedal, the vehicle may not slow as it should. Water between the rotor and pad or drum and shoes will cause this condition. The water will burn off quickly and the brakes will return to normal, but repairs are required if the same condition occurs every time the brakes are applied.

Figure 10-6 This sign indicates that a slower gear should be selected before beginning descent. This reduces the need to apply the brakes during descent.

In most cases, the condition stems from brakes that have been overheated. Failure to completely release the parking brake, repeated hard or long braking, pulling a trailer, misadjusted drum brakes, or a stuck caliper will cause the friction material to glaze from heat. The result is loss of friction between the pads or rotor and the shoe or drum. The pad or shoe slides over the machined surfaces without gripping. The primary cause is a driver who consistently waits until the last moment to apply the brakes. Hand sanding the friction material might provide a quick fix to remove the glaze, but the correct repair procedure is to replace the shoes or pad and machine or replace the drums or rotors.

An overheating or **glazing** on one wheel may cause the vehicle to pull to one side during braking. Like other braking faults, pad or shoe glazing may not be obvious on first inspection. A glazed pad or shoe normally produces a very shiny, hard look to the rotor or drum. Bluish-colored spots on the machined surface are usually a sign that the brakes have been overheated and hot spots have formed on the braking surface.

> **Glazing** is a condition when the friction material and rotor/drum develops a slick, nearly non-friction surface. This is usually caused by heat.

> ⚙ **SERVICE TIP** The wrong shoes or pads can cause a hard pedal. They can also create almost all of the brake symptoms discussed in this section. A hard pedal is defined as a pedal that will not depress past a certain point, but the vehicle does not slow as it should.

Pedal Pulsation and Vibration

A correctly machined drum and rotor, along with good, clean shoes and pads, will bring the vehicle to a smooth stop without vibration. During the life of the brake pad or shoes, heat attacks the working components during every braking action. This continuous heating and cooling, sometimes under poor weather and road conditions, causes the rotor or drum to warp or wear unevenly. Repeated hard braking increases the likelihood of damage. Disc brake systems are most susceptible to **pedal pulsation** and vibration during braking. Vibration on drum brake systems can be traced to hot spots on the drum, damaged shoes, or worn or damaged mounting components. For the most part, drum brakes do not produce noticeable pedal pulsation, but they do tend to grab.

On a disc brake rotor, braking surfaces that are not parallel will cause pedal pulsation. The rotor has two machined braking or friction surfaces that should be parallel to each other. If one surface has a high or low area compared to its companion, or if the rotor is warped, the brake pad and piston will be pushed in and out of the caliper during braking (**Figure 10-7**). This condition is known as rotor thickness variation or rotor parallelism. This action is transmitted back through the brake system, tries to kick up the brake pedal, and is repeated on every revolution of the rotor. In addition, there will be several points on the rotor that are not parallel or warped. The result is a light-to-heavy pulsation of the pedal and vibration of the vehicle. The damage could be severe enough to jerk the steering wheel back and forth as the two front brakes pulsate at different times. Another brake pulsation concerns that can cause the same symptom is called **rotor runout**. Rotor runout is a side-to-side movement of the rotor. This condition can be caused by a distorted rotor, hub flange distortion, or even a worn wheel bearing.

> **Pedal pulsation** is a rapid up-and-down movement of the brake pedal during brake application. The driver may complain that the vehicle shakes when coming to a stop.

> **Rotor runout** When a rotor deviates from its axial plane, also known as wobble.

> ⚙ **SERVICE TIP** Rotor runout can be caused by overtorqueing the wheel lug nuts, especially by technicians who like to use an impact gun to tighten lug nuts. The excessive torque can distort the hub surface and hat section of the rotor. Rotor runout can be checked by using a dial indicator.

Cooling vents

Machined friction surfaces

Figure 10-7 If the two-machined braking surfaces of a rotor are not parallel, the brake pedal will pulsate during brake application.

Customers who are not familiar with the action of ABS may complain that, at times, the brakes seem to vibrate and shudder during heavy braking, or the vehicle feels like it is coming apart during a panic stop. Loose gravel, mud, wet leaves, or anything that makes the road surface slick can cause the ABS to function under light to moderate braking. This condition can be referred to as a normal ABS "event." A brake inspection and customer education may be the only repair needed.

Spongy Pedals

A driver may complain that the vehicle stops quickly and quietly, but the pedal feels soft or bouncy. This is known as a **spongy pedal** and is caused by air in the brake system. As discussed in the section on hydraulic theory, liquids cannot be compressed. However, air can be compressed and used to perform certain types of work even though it does not transmit force well. When air enters the brake system, it forms an air pocket and compresses or expands every time the brakes are applied and released. What the driver feels through the pedal is the compression of the air pocket as the brake fluid tries to transmit the driver's force to the wheels. It will take more driver force to achieve the same braking action. The only solution is to locate and repair the air leak and bleed the entire brake system.

A **spongy pedal** is the result of air in the hydraulic system. It feels like stepping on an inflated balloon.

Grinding and Scraping Noises

Grinding and scraping noises are classic symptoms of completely worn out brake shoes and pads. The metal of the shoe web or pad mount comes into direct contact with the drum or rotor and causes the noise (**Figure 10-8**). The sound will become louder until it

New pad   **Worn pad**

Figure 10-8 As the linings wear, the wear indicator eventually hits the rotor and creates noise to warn the driver that new pads are needed.

occurs even when the brakes are not applied. It may also be caused by other problems such as a shoe-return spring breaking and dropping loose from its mounts. However, the probable cause is a lack of friction material or lining on the pad or shoe.

Another condition that can cause loud grinding or scraping noises is by installing the wrong type of pad or shoe. The friction material of the brake lining should always be suitable for the type of rotor used. With some composite rotors, a so-called "long-life pad" may cause very loud grinding noises during braking because metal particles or compounds in the pad friction material dig into the rotor surface and create a metal-to-metal noise. On a few vehicles, it may be necessary to install factory parts instead of aftermarket items to retain quiet braking.

Hygroscopic A fluid that absorbs moisture from the air.

🔧 **SERVICE TIP** Old brake fluid can cause an unsafe condition. Over time, the fluid is heated, cooled, and reheated. Brake fluid is **hygroscopic**, which means it absorbs moisture. This lowers the boiling point of the fluid, and as it is heated or cooled, some of the fluid will evaporate, causing gases to be formed with the system. Muddy-looking brake fluid is a sign that the fluid is old and contaminated, and that the system needs a complete flush and possibility new parts.

Classroom Manual pages 336–338

BRAKE BOOSTER DIAGNOSIS, INSPECTION, AND REPLACEMENT

⚡ **WARNING** Before moving the vehicle after a brake repair, pump the pedal several times to test the brake. Failure to do so may cause an accident with damage to vehicles, the facility, or people.

Figure 10-9 Bleed the booster with the key off, hold the pedal down, and switch key to START. The pedal should fall away when the engine starts.

 WARNING **Before working on the brakes of a vehicle with ABS, consult the service manual for precautions and procedures. Failure to follow procedures to protect ABS components during routine brake work could damage the components and cause expensive repairs.**

Master cylinders and boosters are easy to diagnose and repair. Listen to the operator's complaint, understand how the system works, and use the information presented on the last few sections to determine which component failed.

Brake Booster Operational Checks

To check the brake's power booster, turn off the engine and press down hard on the brake pedal several times to bleed (or exhaust) the booster's vacuum or hydraulic pressure. The pedal will rise with each stroke until it is firm and much farther from the floorboard than normal. Maintain force on the pedal and start the engine. When the engine starts, the pedal feel will become softer and the pedal will drop down as booster pressure builds.

Vacuum and Hydro-Boost

On vacuum and hydro-boost systems, bleed the boost pressure as previously discussed and start the engine (**Figure 10-9**). The pedal should immediately feel softer and drop lower. A firm, high, and steady pedal after engine start-up means there is no boost. On a hydro-boost system, attempt to turn the steering wheel. If power steering is present, the problem is in the valve mechanism mounted on the master cylinder. Consult the service manual to determine if the valve mechanism can be replaced or serviced. Many manufacturers require that the master cylinder and valve be replaced together. A lack of power steering indicates that the power-steering fluid may be low, the power-steering pump is inoperative, or the power-steering belt is missing or loose. Treat this condition as a steering component failure and perform the necessary repairs to that system.

A vacuum brake booster failure is caused by one of two things: booster failure or lack of vacuum. The first step is to disconnect the vacuum hose at the booster with the engine running (**Figure 10-10**). The engine will try to stall. Placing a thumb over the

 Caution

Do not work in or around the steering column without disarming the airbag systems. Serious injury could result from a deploying airbag.

Classroom Manual pages 336–338

 Caution

Do not work in or around the steering column without disarming the airbag systems. Serious injury could result from a deploying airbag.

To engine
vacuum

Vacuum
booster

Vacuum inlet
check valve

Master
cylinder

Figure 10-10 Vacuum can be measured or felt at the booster end of the vacuum hose.

A **vacuum pump** is a belt- or gear-driven pump to remove air from a system.

Special Tools

Service manual
Wheel blocks

end of the hose will steady the engine, and there should be a strong vacuum. A lack of vacuum usually means that the hose or check valve is closed or restricted. On a diesel engine, the **vacuum pump** and belt must be checked if there is no vacuum at the brake booster hose. **Photo Sequence 27** shows the typical procedure for vacuum booster testing.

PHOTO SEQUENCE 27
Typical Procedure for Vacuum Booster Testing

P27-1 With the engine idling, attach a vacuum gauge to an intake manifold port. Any reading below 14 in. Hg of vacuum may indicate an engine problem.

P27-2 Disconnect the vacuum hose that runs from the intake manifold to the booster and quickly place your thumb over it before the engine stalls. You should feel a strong vacuum.

P27-3 If you do not feel a strong vacuum in step 2, shut off the engine, remove the hose and see if it is collapsed, crimped, or clogged. Replace it if needed.

PHOTO SEQUENCE 27 (CONTINUED)

P27-4 To test the operation of the vacuum check valve, shut off the engine and wait 5 minutes. Apply the brakes. There should be power assist on at least one pedal stroke. If there is no power assist on the first application, the check valve is leaking.

P27-5 Remove the check valve from the booster.

P27-6 Test the check valve by blowing into the intake manifold end of the valve. There should be a complete blockage of airflow.

P27-7 Apply vacuum to the booster end of the valve. Vacuum should be blocked. If you do not get the results stated in steps 6 and 7, replace the check valve.

P27-8 Check the booster air control valve by performing a brake drag test. With the wheels of the vehicle raised off the floor, pump the brake pedal to exhaust residual vacuum from the booster.

P27-9 Turn the front wheels by hand and note the amount of drag that is present.

P27-10 Start the engine and allow it to run for 1 minute, then shut it off.

P27-11 Turn the front wheels by hand again. If drag has increased, this indicates that the booster control valve is faulty and is allowing air to enter the unit with the brakes unapplied. Replace or rebuild the booster.

Figure 10-11 The booster will have four to six fasteners through the firewall and must be removed from inside the vehicle.

VACUUM BOOSTER REPLACEMENT

If there is vacuum in the booster, consult the service manual for instructions on removing the booster and disarming the airbags. The general method of removing the booster is to disarm the airbag, disconnect the pushrod, disconnect and move the master cylinder, and remove the booster mounting fasteners (**Figure 10-11**). Do not disconnect the brake lines from the master cylinder. The master cylinder can be moved to the side without disconnecting the lines in most vehicles. Reverse the procedures to install the booster, vacuum hose, master cylinder, and the push rod. Check the service manual to determine if the push rod's length has to be measured and adjusted. The stoplight switch should be checked after installing the pushrod to the pedal.

Test the new booster by starting the engine and applying the brakes. The pedal should be firm but easy to apply. The engine should not stumble or stall while the brakes are on. Stop the engine and pump the pedal to exhaust the vacuum from the booster. The pedal should rise in height above the floor and become very firm or hard to push. Hold the pedal down while starting the engine. The pedal should fall away as the engine is started. The brakes should be easy to apply with a soft but not spongy feel.

Classroom Manual
pages 336–337

> ✹ **SERVICE TIP** An engine that operates poorly usually does not cause a total loss of vacuum in the brake booster. However, a tear in the booster's diaphragm can cause a good engine to stall and run roughly.

MASTER CYLINDER DIAGNOSIS, INSPECTION, AND REPLACEMENT

Special Tools

Wheel blocks
Vise
Master cylinder bleeder kit
Service manual

⚡ **WARNING** Handle master cylinders carefully to prevent spilling the fluid. Brake fluid is very corrosive and will cause paint and finish damage.

Master Cylinder Testing and Inspection

Check for cracks in the master cylinder housing, and look for drops of brake fluid around the master cylinder. A slight dampness in the area surrounding the master cylinder is normal. If the brake fluid is leaking, eventually the fluid level will be low enough and the warning lamp on the dash will illuminate. One of the best ways to check for leakage is to have a partner push on the brake pedal while you visually inspect the master cylinder and related components with a good lighting source.

Test-driving the vehicle to determine a bad master cylinder can be difficult. You should always inspect the vehicle with the wheels off, and on a hoist before going for this test-drive. Ensuring that the other mechanical components are not bad or causing issues will help determine and eliminate other components. If all other components check out to be good and you have a spongy brake pedal while braking, the master cylinder could have an internal leak.

Master Cylinder Replacement

The master cylinder creates and supplies the pressure for the brake system. An internal or external leak may affect both circuits of a dual system or just one circuit. In most cases, the master cylinder is easy to replace. This discussion will cover the replacement of a master cylinder, with vacuum boost, on a non-ABS system. Consult the service manual before attempting to replace any brake components on an ABS system.

⚠ **WARNING Some ABS components may be located on or near the master cylinder. Consult the service manual on removing the master cylinder if any ABS or other system components present a problem to master cylinder replacement. Damage to the ABS or other systems may occur if proper procedures are not followed.**

⚠ **WARNING Many ABS systems require special bleeding procedures. Do not remove a master cylinder on an ABS system before consulting the service manual. Damage to the brake system or loss of time could result if proper procedures are not followed.**

Move the vehicle into the service bay and block the wheels. Inspect the master cylinder mounting area. Remove the two or four metal brake lines attached to one side of the master cylinder body (**Figure 10-12**). An electrical connection for the fluid level sensor

Figure 10-12 Before loosening the brake fittings, place several cloths under them to catch any leaking fluid.

may be present. Two nuts should hold the master cylinder to the vacuum booster. Before loosening the nuts, place clean rags in the area under the master cylinder to capture any brake fluid that may leak. Remove as much fluid as possible from the reservoir.

Disconnect the electrical connection, if present. Select the proper line wrench and disconnect the brake lines. The flare nuts are loosened until the lines can be pulled out of the fitting. Bend the lines slightly so the master cylinder can be moved.

Select the correct wrench to remove the nuts. Generally, a 3/8-inch drive ratchet, socket, and a 3- or 6-inch extension are the only tools needed to remove the cylinder's mounting nuts. With all the nuts removed, slide the master cylinder toward the front of the vehicle until it clears the studs. Tilt the master cylinder slightly to one side with the line fittings up. This will reduce any leakage as the unit is moved.

Remove the reservoir cover and drain the remaining brake fluid into a suitable container. If necessary, remove the reservoir and install it on the new master cylinder using the new O-rings supplied (**Figure 10-13**). Place the old master cylinder in the box that the new one came in. It will be sent back to the parts vendor as a core for rebuilding. Secure a master cylinder bleeder kit or you can make one using two short pieces of brake line and two fittings.

Bench Bleeding a New Master Cylinder

Bench Bleeding
Removing air from a master cylinder on a workbench remove from the vehicle.

The kit basically consists of two plastic lines with two plastic male fittings (**Figure 10-14**). The threads on the fittings are cut to fit several types of master cylinder threads and will flex to hold the fittings to prevent any leakage during braking. Usually, there are three types of threads on the fitting, each one a little larger than the one before it. The fittings are screwed into the line ports on the master cylinder. They do not need to be very tight because there is not much pressure applied during **bench bleeding**. Screw the fitting in

Cylinder body

Grommets

Reservoir

Place reservoir top down on a hard, flat surface

Figure 10-13 Place the reservoir upside down, lubricate the new grommets, and press the reservoir straight down until it pops in place.

Figure 10-14 This bleeder kit can fit several types and makes of master cylinders.

by hand followed by only two or three turns with a wrench. Route the lines from the fittings back into the cylinder's reservoir, one per chamber. The lines should nearly touch the bottom of the chamber. The kit should include a small holding device to keep the lines in place during this procedure. Some master cylinders are packaged with a kit included, but most kits can be bought separately and kept on hand.

A bleeder kit can be made if necessary, but it is time consuming and can only be used for one type of master cylinder fitting thread. A brake line must be cut for two 6- to 8-inch pieces and bent to fit between the fitting and the bottom of the master cylinder's chamber. Select two male fittings that will slide over the line and screw easily into the master cylinder. The threads must be exact, because the fitting is either brass or steel and the wrong threads will cause problems. The fitting slides over the line, and one end of the line is flared. The line is then bent to fit into the master cylinder chamber. Generally, a technician does not make a master cylinder bleeder kit unless one type of vehicle is being maintained or a special-purpose vehicle requires special adapters. For the most part, it's not worth the trouble to build a bleeder kit for each vehicle that comes into the shop when an inexpensive kit can be bought at almost any parts vendor.

⚡ **WARNING** **Before adding brake fluid, consult the vehicle's service manual. Most manufacturers require a specific classification of brake fluid to be used.**

Mount the master cylinder in a vise with the forward end slightly elevated. Install the bleeder kit and remove the reservoir cover. Fill each chamber about half full with brake fluid. Use a punch or other rod to slowly apply force to the master cylinder's primary piston (**Figure 10-15**). Continue the stroke until both pistons bottom out. As the pistons are pushed in, air should escape from the bleeder tubes as evidenced by bubbles in the reservoir. Slowly release the force from the pistons and allow them to return to their resting position. Repeat the bleeding procedure until there are no bubbles visible in the reservoir during the pistons' inward movement. The rear or primary cylinder bore will probably bleed completely one or two strokes before the secondary chamber. During the bleeding process, always push both pistons to the bottom of their bore. Before removing the cylinder from the vise, close the two bleeder tubes with small paper clips and install the reservoir cover. The clips will prevent leaks and prevent air from entering as the cylinder is installed on the vehicle.

Place the master cylinder in position on the brake booster and torque the fasteners. Position the two brake lines near their fittings. Remove one of the brake bleeder fittings and quickly insert the correct brake line into the master cylinder. Hand tighten the fitting

Figure 10-15 A screwdriver can be used to move the master cylinder pistons during bench bleeding.

enough to prevent leaking. Do not tighten completely at this time. Install the second brake line in the same manner. Drain and store the bleeder kit. See **Photo Sequence 28** for details.

PHOTO SEQUENCE 28
Typical Procedure for Bench Bleeding the Master Cylinder

P28-1 Mount the master cylinder firmly in a vise, being careful not to apply excessive pressure to the casting. Position the master cylinder so the bore is horizontal.

P28-2 Connect short lengths of tubing to the outlet ports, making sure the connections are tight.

P28-3 Bend the tubing lines so that the ends are in each chamber of the master cylinder reservoir.

P28-4 Fill the reservoirs with fresh brake fluid until the level is above the ends of the tubes.

P28-5 Using a wooden dowel or the blunt end of a drift or punch, slowly push on the master cylinder's pistons until both are completely bottomed out in their bore.

P28-6 Watch for bubbles to appear at the tube ends immersed in the fluid. Slowly release the cylinder piston and allow it to return to its original position. On quick take-up master cylinders, wait 15 seconds before pushing in the piston again. On nonquick take-up units, repeat the stroke as soon as the piston returns to its original position. Slow piston return is normal on some master cylinders.

P28-7 Pump the cylinder piston until no bubbles appear in the fluid.

P28-8 Remove the tubes from the outlet port and plug the openings with temporary plugs or fingers. Keep the ports covered until the master cylinder is installed on the vehicle.

P28-9 Install the master cylinder on the vehicle. Attach the lines, but do not tighten the tube connections.

P28-10 Slowly depress the pedal several times to force out any air that might be trapped in the connections. Before releasing the pedal, tighten the nut slightly and loosen it before depressing the pedal each time. Soak up the fluid with a rag to avoid damaging the car finish.

P28-11 When there are no air bubbles in the fluid, tighten the connections to the manufacturer's specifications. Make sure the master cylinder reservoirs are adequately filled with brake fluid.

P28-12 After reinstalling the master cylinder, bleed the entire brake system on the vehicle.

It is best to have a second person assist during the final stage of master cylinder bleeding. The next step is to remove any air that may have entered the line or cylinder fitting during the change over from bleeder to brake line. Ensure that enough rags are placed under the master cylinder brake lines to catch any fluid. Ask the second person to press slowly on the brake pedal. Both brake lines will be loosened and retightened on each stroke of the brake pedal. Loosen the rear line slightly as the pedal is being applied. Watch this small leak for signs of air. If no air is present, tighten the fitting and loosen the front line. Ensure that both fittings are tight before the brake pedal is released. Repeat this step until no air is being expelled past either fitting. Generally, one or two strokes of the brake pedal are sufficient. Complete the installation by tightening the fittings and topping off the fluid level. It will be necessary to bleed the entire brake system. Before moving the vehicle from the service bay, apply the brakes and start the engine. If the vehicle has an automatic transmission, apply the brakes and put it in gear to see if the brakes hold the vehicle. On a manual transmission, release the brakes and engage the clutch just enough to move the vehicle slowly. Apply the brakes to ensure that they will stop and hold the vehicle.

Classroom Manual
pages 338–342

Brake Flushings

There are other facts that must be considered during master cylinder replacement. If the vehicle is more than 2 years old, or it has been more than 2 years since the last brake flushing, the old fluid should be flushed out with new fluid. This is done by bleeding each wheel until new fluid is visible. When a master cylinder fails because of age, it is probable that other components may require replacement. At the minimum, a thorough inspection of the brake system should be conducted.

DISC BRAKE PAD REPLACEMENT

 Special Tools

Lift or jack and jack stands

Impact wrench

Torque wrenches

C-clamp

Disc brake silencer

Disc brake lubricant

⚠ **WARNING** Before moving the vehicle after a brake repair, pump the pedal several times to test the brake and position the pads or shoes in place. Failure to do so may cause an accident that might damage the vehicle, the facility, or injure humans.

Figure 10-16 A typical C-clamp.

⚠ WARNING Before working on the brakes of a vehicle with ABS, consult the service manual for precautions and procedures. Failure to follow procedures to protect ABS components during routine brake work could damage them and require expensive repairs.

⚠ WARNING Failure to lower the reservoir brake fluid level during brake repairs may result in spillover during brake repair. Damage to the vehicle's finish can occur. The fluid will overflow as the caliper pistons are pushed back into their bores and the fluid behind them is forced back into the master cylinder.

A **C-clamp** is so-named because of its shape and function.

Disc brake pad replacement does not require any special tools except a **C-clamp** (**Figure 10-16**), although there are some specialty tools that will make the job go faster.

Before lifting the vehicle, inspect the fluid level in the reservoir. If the reservoir is more than half full, some of the fluid must be removed. A suction gun *should not* be used because the gun's rubber hose and piston seal will probably be damaged. A small, inexpensive syringe is the best choice. It can be discarded after several uses. Use cloth to capture any leaking fluid.

When replacing rear brake pads, you need to determine what type of caliper is used. If it has a regular piston like the front, then the service is the same. Some rear disk brake calipers have an integrated parking brake. The caliper piston will not retract with a C-clamp. You will need a special tool (**Figure 10-17**) to turn the piston—which will reset it back into the caliper housing to make room for the new pads.

Each brake pad installation will be a little different. It is noticing this difference and then knowing what to do about it that makes the difference between having a customer come back because you didn't clean or install some small part causing a noise and having a customer that will have many miles of trouble-free service.

On the back side of the brake pad, you must coat it with a noise suppression compound (**Figure 10-18**). Each technician has their own preference for what to use. Some brake pads have clips (**Figure 10-19**), and some calipers have support brackets that must be cleaned and lubricated (**Figure 10-20**). When servicing the brake pads, you should always clean (or replace if badly corroded) the caliper slides (**Figure 10-21**).

Rotate clockwise
until piston seats

$\frac{3}{8}$ in. rachet
and extension

Caliper

Piston

Drive
hole

Drive nibs
(2 or 4 per side)

Piston turning tool

Figure 10-17 One type of piston retracting tool.

Shim

Noise suppression
compound

**Typical disc pads
with a shim**

**Typical disc pads with
noise suppressant applied**

Figure 10-18 You can use noise suppression compound or a soft shim on the back of a brake pad to help reduce noise, but do not use both on the same pad.

⚡ WARNING Consult the service manual before pushing the caliper's piston into its bore. Some ABS systems require that the caliper's bleeder screw be opened, or they require other special procedures to prevent damage.

The next step is to compress the piston back into its bore. However, it is not desirable to push the old fluid back into the master cylinder. Instead, obtain a small container and piece of flexible hose about 12 to 18 inches long with a diameter that will allow it to fit tightly over the bleeder screw. Position the container in a secure place on the lift or have someone hold it. Route the hose from the bottom of the container and fit the other end

Figure 10-19 Many inboard pads have retainer springs that snap into the inner opening of the piston.

Figure 10-20 For a sliding caliper, lubricate the caliper ways, support spring, and the mating areas of the caliper housing as highlighted here.

Figure 10-21 Lubricate the inside of the caliper bushing as highlighted here.

over the bleeder screw (**Figure 10-22**). Use a C-clamp and a block of wood or an old brake pad and apply a small amount of force to the piston (**Figure 10-23**). Loosen the bleeder screw and continue using the C-clamp to compress the piston into its bore until the piston is fully seated close to the bleeder screw. This prevents almost any of the dirty fluid from returning to the reservoir.

Remove the fasteners holding the adapter to the steering knuckle. With the fasteners removed, the adapter, caliper, and pads will slide off the rotor (**Figure 10-24**). Be cautious at this point because many rotors can drop off the hub once the caliper is removed. Once the assembly is removed, hang it with mechanic wire or other strong wire so the caliper's weight does not hang on the flexible brake hose. Remove the pads, noting where they are located and what is used to hold them in place. Normally, antirattle clips keep the pads from clicking in the caliper when the brakes are applied and released.

Lube exterior of bushing and bolt with brake lubricant.

Hose to
fluid container

Bleed screw

Brake
fluid

Caliper

Flexible
bleeder hose

Rotor

Container

Figure 10-22 Route a hose from the bleeder screw to a capture container. Just before the piston bottoms out, close the bleeder screw and remove the hose and container.

Caliper

Screw of C-clamp
inside of piston

Figure 10-23 Use a C-clamp to compress the piston into its bore.

Caliper
support
(anchor plate)

Mounting
pin

Caliper
housing

Mounting
pin

Mounting
sleeve

Figure 10-24 Use care once the adapter fasteners are removed. The caliper and adapter can slide off and put unnecessary stress on the brake hose.

If there is brake fluid present on the outside of the caliper, the caliper must be rebuilt or replaced. Like the wheel cylinder, it is usually best and cheapest to replace the caliper, but there are rebuilding kits on the market.

Before installing the pads, the slide pins or guides on the caliper and adapter must be lubed. Inspect the caliper and adapter to determine the slide points. Some systems have slide pins only, whereas others slide the calipers along rails on the adapter. Once located, clean the slides and then lube with a disc brake lubricant.

Assemble the pads into the caliper, inboard pad first. Ensure that the retaining clips or springs are in place before fitting the assembly over the machined rotor. Install and torque the adapter bolts, and install the wheel assembly.

> ⚙⚙ **SERVICE TIP** Caliper adapter bolts and caliper bracket bolts should always be torqued to manufacturer's specifications. Overtorqueing or undertorqueing can cause severe injury.

When the vehicle is lowered to the floor and before the vehicle is moved, slowly pump the brake pedal several times to seat the pads against the rotor. The pedal should be firm and at a normal height above the floor before the vehicle is moved. Top off the brake fluid level as necessary and inspect for any overflow around the master cylinder.

BRAKE PAD BURNISHING

Burnishing is an operation used to heat and seat the brake pads to the rotor.

Whenever new brake pads are installed, they need a short period of controlled operation called a **burnishing** or "break-in" period. This period of time polishes the pads and mates them to the rotor finish. Road testing the vehicle after installing new pads performs this procedure and verifies that the brakes are working properly.

To do this, drive the vehicle at 30 mph (50 kph) and firmly but moderately apply the brakes to fully stop the vehicle. Repeat this five times within 20 seconds of driving time between each application. Then drive the car to highway speeds and apply the brakes moderately to slow the car to 20 mph (30 kph). Then let the brakes cool down for 10 minutes before letting the customer drive away. Tell the customer to avoid hard braking for the first 100 miles.

BRAKE ROTOR INSPECTION, MEASUREMENT, AND SERVICING

Rotors on modern light vehicles are thin and lightweight, when compared to the brake rotors of older vehicles. This design reduces vehicle cost and unspring weight; it also helps the vehicle's handling, adds to a smoother ride, and increases fuel economy.

Older-style, heavier, and thicker brake rotors (still found on heavier vehicles, light trucks, and SUVs) are often machined during brake pad replacement because they have more material on them and do not wear to the minimum legal thickness as quickly. You will find yourself replacing, instead of machining, a lot of brake rotors.

Brake rotors must be measured to see if they are causing some of the problems that the customer is experiencing. They must also be measured before a decision to refinish or replace them is made. Using a brake micrometer, measure the thickness of the rotor and compare it to the *machine to* and *discard* specifications. These specifications are found on the rotor itself (**Figure 10-25**) but are often not readable because of rust or dirt, so you need to refer to the service manual for specifications.

Figure 10-26 Check for thickness variations by measuring at 8 to 12 points around the rotor.

Figure 10-25 The discard dimension, or minimum thickness specification, is cast or stamped into all disc brake rotors.

With the brake rotor still on the vehicle, use the brake micrometer and measure the rotor at several spots (**Figure 10-26**). The difference in these measurements is the thickness variation. A thickness variation may be the cause of a brake pedal pulsation that the customer is concerned about. The thinnest dimension that you measure should be compared to the machine to and discard specifications.

With the rotor still mounted, reinstall and tighten the lug nuts and mount a dial indicator on the rotor surface. Rotate the rotor and measure the runout of the rotor surface (**Figure 10-27**). Compare this reading to the specification; too much runout can result in brake pedal pulsation. After removing the rotor, measure the runout of the hub; although it is rare and usually only caused by a vehicle accident, if the hub is bent it will cause even a new rotor to pulsate the brake pedal and measure out-of-round. Excessive rust will also cause an incorrect measurement. Make sure to clean the rust from the surface (**Figure 10-28**).

Figure 10-27 The rotor and hub runout is measured with a dial indicator mounted to some fixed point on the vehicle (upper right). The rotor thickness and parallelism can be measured with a standard outside micrometer, as shown on the lower left, but is more easily done with an electronic brake micrometer.

Figure 10-28 Clean the dirt and rust from the hub-mounting area of the hubless rotor.

Figure 10-29 The groove in this rotor surface is machined to aid in cooling and noise reduction.

Figure 10-30 This rotor has a shiny glazed surface.

After measuring the rotor, don't forget to visually inspect it. Look for grooves and a bluish glazed tint on the rotor surface. The glazing is caused by overheating of the rotor and can only be machined or resurfaced (**Figures 10-29** and **10-30**).

BRAKE ROTOR REMOVAL AND RESURFACING

To remove the rotor, lift the vehicle to a good working height, set the locks, and remove the tire and wheel assembly. Inspect the disc brake area to determine the type of rotor or hub assembly and the method used to attach the caliper. Some systems require the adapter to be removed before the rotor is free. The rotor and hub may be one- or two-piece. If equipped, the wheel bearings and seal have to be removed on a one-piece assembly before the rotor can be machined. For this example, the rotor is separate, and the adapter must be removed. **Photo Sequence 29** shows the typical procedure for replacing the brake pads.

PHOTO SEQUENCE 29
Typical Procedure for Replacing Brake Pads

P29-1 Begin front brake pad replacement by removing brake fluid from the master cylinder reservoir or reservoirs servicing the disc brakes. Use a siphon to remove about two-thirds of the fluid.

P29-2 Raise the vehicle on the hoist, making certain it is positioned correctly. Remove the wheel assemblies.

P29-3 Inspect the brake assembly, including the caliper, brake lines and hoses, and rotors. Look for signs of fluid leaks, broken or cracked lines or hoses and a damaged rotor. Correct any problems found before replacing the pads.

PHOTO SEQUENCE 29 (CONTINUED)

P29-4 Loosen and remove the caliper-to-knuckle bolts.

P29-5 Lift and rotate the caliper assembly up and off the rotor.

P29-6 Remove the old pads from the caliper assembly.

P29-7 To reduce the chance of damaging the caliper, suspend the caliper assembly from the underbody using a strong piece of wire.

P29-8 Check the condition of the caliper locating pin insulators and sleeves. Make sure all slide or guide hardware is lubricated with brake lubricate before assembly.

P29-9 Install a hose from the bleeder screw to the container. Open the bleeder screw.

P29-10 Place a block of wood over caliper piston and install a C-clamp over the wood and caliper. Tighten the C-clamp to force the piston back into its bore.

P29-11 Remove the C-clamp and check the piston boot, then install new locating pin insulators and sleeves if needed.

P29-12 Install the new pads in the caliper. Then set the caliper with its new pads onto the rotor. Check the assembly for proper positioning.

P29-13 Install and torque the caliper mounting bolts. Install the tire and wheel assembly and tighten to torque specifications. Then press slowly on the brake pedal several times to set the brakes.

An **on-vehicle brake lathe** is designed to machine a brake rotor on the vehicle.

Most rotors can be removed and turned on the brake lathe (**Figure 10-31**), but there are portable lathes that can machine the rotor on the vehicle (**Figure 10-32**). They were originally sold to service one-piece hubs or rotors on front-wheel-drive (FWD) vehicles where the drive axle and the wheel bearing were press-fitted to the hub. This lathe requires some special setup procedures, but these can be learned quickly. The **on-vehicle brake lathe** can be used for machining other rotor setups if the rotor, machining head, and hub can be clamped tightly together with the lug nuts. Machining a rotor or drum should be done by an experienced technician following the lathe operating instructions and the procedures set forth in the vehicle's service manual. **Photo Sequence 30** shows the typical procedure for mounting a brake rotor on the bench lathe.

Figure 10-31 AAMCO brake lathes like the one shown are found in most brake repair shops and are preferred by many older technicians.

Figure 10-32 This on-car lathe has its own drive motor.

PHOTO SEQUENCE 30
Typical Procedure for Mounting a Brake Rotor on the Bench Lathe

P30-1 Clean the inside and outside of the rotor where the lathe's centering cone and clamps will be in contact with the rotor. Select a centering cone that fits into the rotor's center hole without protruding too much in either direction.

P30-2 Slide the inner clamp onto the arbor followed by the centering cone tension spring.

P30-3 Slide the cone onto the arbor followed by the rotor. The rotor is mounted in the same way as on the vehicle.

P30-4 Push on the rotor to compress the spring, and then hold the rotor in place as the outer clamp is fitted onto the arbor. The two clamps, inner and outer, must be of same size.

P30-5 Keep the pressure on the install parts as the spacers and bushing are installed on the rotor.

P30-6 Hold all parts in place as the arbor nut is installed and tightened to lathe manufacturer specifications.

P30-7 Install the vibration damper. The damper may be a flexible strap or spring that fits the outer circumference of the rotor, or two pad-like devices that are forced against each side of the rotors similar to installed disc brake pads.

P30-8 The rotor is now properly mounted and ready for the final check before machining.

DRUM BRAKE SHOE AND WHEEL CYLINDER INSPECTION, MEASURING, AND REPLACEMENT

Classroom Manual
pages 347–352

Drum Brakes

Drum brake shoe replacement is a fairly straightforward process of removing the old and installing the new. However, drum brake shoes are held in place with return springs, retainers, parking brake cable, and adjusters, most of which can easily be installed incorrectly (**Figure 10-33**). As always, the service manual contains instructions for the repair. A quick method of learning drum brake assemblies is to lay out all parts in the order in which they were removed and **NEVER** remove both sets of brakes at the same time. Another quick method is to use your cell phone's camera and take pictures during the disassembly process. You can use the pictures to check the proper assembly during reassembly. Many new technicians lose time because they pull both sides down at the same time and do not remember how the drum brakes were assembled. Keeping one set intact provides a model, or mirror, to reassemble the other side. Because the drum may have to be machined, no time is lost by keeping one assembly intact. It is suggested that this practice be followed during most drum brake repairs.

Removing and installing drum brakes may also expose the technician to brake dust. Older brake materials contained asbestos. The brakes should be cleaned with a low-pressure aquatic cleaner to prevent dust (**Photo Sequence 31**). Another method is to use a

Special Tools

Brake spring pliers
Brake spring hold-down tools
Drum brake micrometer

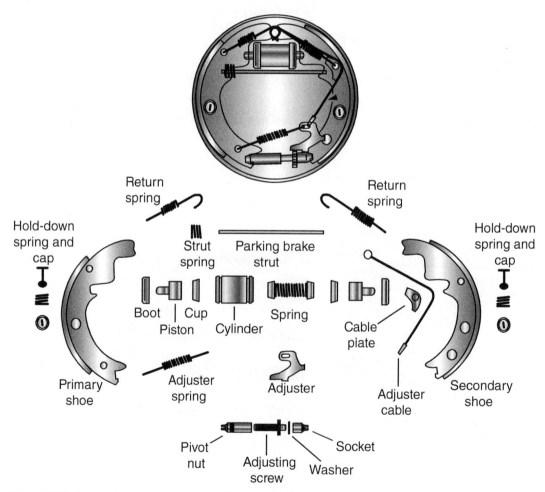

Figure 10-33 Care must be taken when reassembling a drum brake.

PHOTO SEQUENCE 31
Typical Procedures for Wet-Cleaning Drum Brakes

P31-1 After removing the wheel assembly, position the cleaner so the runoff waste is captured in the cleaner.

P31-2 Use a wet brush to remove any brake dust from the outside of the drum. This brush is not used to clean mud or accumulated dirt from the drum. If mud is present, use a dry brush to remove it.

P31-3 Remove the drum and use a wet brush to clean the inside of the drum. Once clean, set the drum to the side.

P31-4 Use a wet brush to wash the brake components free of dust. Ensure the waste material or liquid is draining into the cleaner's tub.

P31-5 Once the brake has been disassembled, wash the individual parts separately, holding them so the waste drains into the tub. Lay the parts on rags or a table for drying.

P31-6 After cleaning the various individual parts, complete the cleaning by washing down the backing plate.

P31-7 After cleaning is complete, check the cleaner's waste filter and replace it if necessary. The full filter must be treated as hazardous waste and disposed of accordingly.

Figure 10-34 This negative-pressure enclosure is used to contain brake dust during cleaning.

specially made high-efficiency particulate air (HEPA) vacuum cleaner (**Figure 10-34**). Although there are several ways to clean the brake area and not spread brake dust, there is only one right way. The correct method is to use a washer or vacuum-type collector built for this purpose. One of the most common units is a portable, heated, low-pressure aquatic washer. This unit has a catch container that filters out the dust and shavings from the washer fluid and captures it in a holding device. The waste filter is treated as hazardous waste when it is full. There are different brands and models available at nominal cost to the shop, which are much cheaper in terms of dealing with the EPA, OSHA, and, more importantly, employee safety in general.

BRAKE SHOE, DRUM, AND RELATED COMPONENTS INSPECTION AND MEASUREMENT

The first procedure for inspecting and measuring the brake shoes is removing the wheel and drum. Sometimes the drum will just come right off, sometimes you need to lightly hit it with a hammer to break up the rust on the hub, and sometimes you need to adjust the parking brake cable to ensure there is enough shoe clearance for drum removal (**Figure 10-35**). Be

Figure 10-35 Loosen the parking cables if necessary to ensure enough shoe clearance for drum removal.

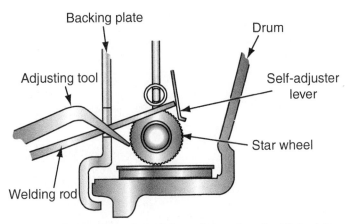

Figure 10-36 Use a heavy piece of wire, such as a piece of welding rod, to move the adjusting lever away from the star wheel. Then rotate the star wheel to release brake adjustment for drum removal.

Figure 10-37 These shoes have been adjusted to fit a badly worn drum, and the drum will jam on the lining if you try to remove it without backing off the brake adjustment.

Figure 10-38 Use a depth micrometer to measure lining thickness precisely.

careful when removing the drum; sometimes adjustment of the parking brake cable is necessary and other times you must open the adjustment cover on the back side of the drum and adjust the shoes themselves (**Figure 10-36**). This may be because the drum is badly worn, and the shoe is jammed into the groove (**Figure 10-37**).

Once you have the drum off and the brake shoes are cleaned, you can inspect the components. Use a depth micrometer (**Figure 10-38**) to measure the lining thickness. Inspect the shoes for cracks, glazing, and unequal wear (**Figure 10-39**). Pull the wheel cylinder boot slightly away and check for brake fluid leakage (**Figure 10-40**). Inspect the brake springs for cracks. Usually, when you install new brake shoes you will install new hardware, which will include new springs.

Next, remove the brake shoes. Follow the steps in **Photo Sequences P32-1** and **P32-8** for drum brake disassembly and reassembly. After removing the shoes, inspect the backing

Figure 10-39 Inspect the linings for unequal wear, as well as for thickness, cracks, fluid or grease contamination, and other damage. Unequal wear is most common at the indicated points.

Figure 10-40 Pull the boot slightly away from the cylinder and check for liquid or fluid within the boot.

PHOTO SEQUENCE 32
Typical Procedure for Disassembling Brake Drum Shoes

P32-1 Remove the top return spring from the anchor and shoe. Remember which shoe's return spring is on top. It must be reassembled in the same position.

P32-2 Remove the retaining spring from the secondary shoe. Do not remove the primary shoe retaining spring at this time.

P32-3 Rotate the secondary shoe forward and down. Catch the self-adjuster as the shoe releases the tension.

P32-4 Remove the parking brake strut and spring from the primary shoe. Withdraw it to the front. Remember which end holds the spring and which end faces forward.

P32-5 Remove the retaining spring on the primary shoe. Do not allow the shoe and self-adjuster mechanism to drop.

P32-6 There are two means to disconnect the parking brake from the primary shoe. One way is to remove the clip from the parking brake lever pin where it passes through the shoe. Do not lose the waved washer behind the shoe. The other way is to disconnect the cable from the lever as shown. The lever can be removed from the shoe later.

PHOTO SEQUENCE 32 (CONTINUED)

P32-7 Clean and lubricate the shoe support pads now. This is a common area that may be forgotten.

P32-8 Lay out the brake components in roughly the same position as they were on the vehicle will help in repositioning them correctly during assembly.

plate and support pads (**Figure 10-41**). If needed, you should replace the wheel cylinder based on your inspection (**Figure 10-42**). Use caution, as the brake line going to the wheel cylinder is often rusty and may break if you are not careful.

In general terms, the following steps for shoe removal and reinstallation may be followed and are illustrated in **Photo Sequences 32** and **33**.

1. Remove both return springs.
2. Remove the retaining spring on one shoe and remove that shoe from the backing plate ensuring that the parking brake and self-adjuster components are held in place and then removed.
3. Remove the second shoe and disconnect the parking brake mechanism from the shoe to which it is attached.
4. Compare the new parts to the old ones and reverse the procedure to install.

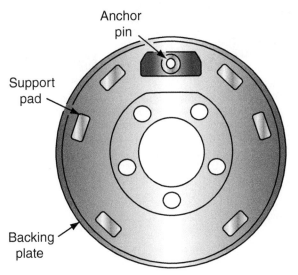

Anchor pin

Support pad

Backing plate

Figure 10-41 Failure to lube the shoe or backing plate rub points will result in noise during brake application.

Figure 10-42 Two common ways to mount wheel cylinders to backing plates are with small bolts or cap screws (A) or with a spring-type retainer ring (B).

PHOTO SEQUENCE 33
Typical Procedure for Reassembling Brake Drum Shoes

P33-1 Compare the new parts with the old to ensure that the correct ones are available.

P33-2 Depending on how the parking brake was disconnected, there are two means to connect it: slide the lever pin with the washer installed through the primary shoe and install a new retaining clip, or connect the cable to the lever as shown.

P33-3 Hold the self-adjuster mechanism in place on the primary shoe as the retaining spring is attached.

PHOTO SEQUENCE 33 (CONTINUED)

P33-4 Pull the top of the secondary shoe forward and fit the parking brake strut between the two shoes. The large slot in the strut goes to the rear and the spring to the front.

P33-5 Pull the lower ends of the shoes apart and fit the self-adjuster between the two shoes. After releasing the shoes, check to make sure that the self-adjuster is correctly linked to the shoes and that the star wheel is to the rear.

P33-6 Hook the cable to the pawl and push up on the pawl. Route the cable over the cam. After releasing the pawl, ensure that the cable stayed in place on the cam.

P33-7 Install the pawl and its spring(s).

P33-8 Pull the lower ends of the shoes apart and fit the self-adjuster between the two shoes. After releasing the shoes, check to make sure that the self-adjuster is correctly linked to the shoes and that the star wheel is to the rear.

P33-9 Hook the cable to the pawl and push up on the pawl. Route the cable over the cam. After releasing the pawl, ensure that the cable stayed in place on the cam.

P33-10 Check each component of the assembled drum brake for proper positioning and anchoring.

Figure 10-43 Measure the inside diameter of the drum with the brake drum micrometer and tighten the locknut.

Figure 10-44 Flip the caliper over and place it over the widest point of the brake shoes. Adjust the brake shoes until they touch the caliper pads.

5. Preset the shoe adjustment by measuring the drum and setting the brake shoes to that measurement (**Figures 10-43** and **10-44**).
6. After installing the drum, adjust the shoes by moving the adjuster with the drum on. Spin the drum until it catches on the brake shoes only slightly.

Parking Brakes

Parking brake repairs normally consist of cable replacements and adjustments or replacement of the ratchet locks on the pedal or lever (**Figure 10-45**). Most repairs are straightforward procedures laid out in the service manual. However, in most systems, some of the repairs require removal whereas some require partial disassembly of the rear brakes.

Once the new brake parts have been installed, the rear brakes must be adjusted before the parking brake is adjusted. The parking brake should lock solidly about halfway along the pedal or lever's travel. Some manufacturers require that the parking braking locks after a certain number of clicks are heard or felt as the parking brake is applied. During the installation of any parking brake cable, ensure that the cable and sheaths are not clamped shut by fasteners. An inspection of the parking brake cable is done by moving it back and forth and making sure that the drum shoe moves as well. Some brake cables (especially those on automatic transmission vehicles) are rusty and will not move easily; they may break or get stuck.

The parking brake warning light switch may need adjustment after work is performed on the parking brake lever or pedal mechanism (**Figure 10-46**). The light should come on as soon as the parking pedal or lever is moved. It should not go off until the brake is completely released. After repairs are complete, spin the tires with the wheels off the ground to make sure the parking brake is not causing the shoes or pads to drag.

> ⚙ SERVICE TIP The RED brake warning light can be stay illuminated because of three reasons: low brake fluid level in the master cylinder, the parking brake being applied, or low fluid pressure in the hydraulic system. The light will normally come on during a bulb check when the ignition turned on. The light should turn off when the engine is started.

Rear Disc Brakes

On vehicles with rear discs, the procedure for replacing the pads becomes a little more complicated. The rear disc may also be used as the parking brake or the disc rotor may have a small, internal drum brake whose sole function is for parking (**Figure 10-47**). In

Figure 10-45 The shaded points are the typical wear points on the parking brake actuation cable. They are also common lubrication points.

Figure 10-46 Most parking brake warning light switches must be adjusted if the pedal is removed for any reason.

Parking brake shoes

Rear axle

Rotor with internal parking brake drum

Figure 10-47 This parking brake is located within a small drum built into the back of a rear disc brake rotor.

Figure 10-48 It is not obvious in this picture, but this tool has six types of adapters to fit almost any rear-disc caliper piston. (Courtesy of MATCO Tool Company)

either case, compressing the piston back into the bore requires special tools and procedures. Normally, the piston is screwed into the bore with a special wrench (**Figure 10-48**). At times, even the correct procedures may not work properly, and the technician must spend time analyzing the problem. In addition, ABS present special situations that must be dealt with before a proper repair can be completed. Always follow manufacturer's procedures when working on ABS brake systems.

Brake Bleeding

⚠ **Caution**

The heat from a torch can damage the rubber components inside the caliper or wheel cylinder. All components should be inspected for proper operation after using the heat method.

The brake system must be bled if the system was exposed to air at any time. This includes disconnecting the lines from the master cylinder, disc calipers, or wheel cylinders and external leaks at any point in the system. Bleeding the brakes completely is similar to flushing or replacing the brake fluid. Use caution when opening a bleeder valve, bleeder valves are tapered and can become wedged or rusted in the bottom of the caliper causing the bleeder valve to break or shear off when attempting to remove.

⚙ **SERVICE TIP** Shocking or heating the bleeder screw is what will help break loose the bleeder. If you just turn the bleeder screw it often results in breaking off the bleeder screw. Some methods to try are: Hit and Tap with penetrant oil, Heat and Tap, or Wax method.

A **pressure bleeder** is a small holding tank with a diaphragm between a compressed air chamber and a fluid chamber.

The key to brake bleeding is to locate the bleeder screws (**Figures 10-49** and **10-50**) and the bleeding sequence from the service manual (**Figure 10-51**). Bleeding can be done in several ways. Special suction equipment can be used as well as a **pressure bleeder** or two persons working together (**Photo Sequence 34**). In most systems, the wheels are bled starting with the one farthest from the master cylinder. An example of

Figure 10-49 Bleeder screws, such as this one on a disc brake caliper, are also installed on wheel cylinders and some master cylinders and valves.

Figure 10-50 Bleeder screws are located at the highest points on calipers and wheel cylinders.

a typical sequence is right rear, left rear, right front, and left front (**Figure 10-51**). Depending on the repairs accomplished, the master cylinder and some control valves may have to be bled before bleeding the wheels. Most ABS brakes require special bleeding procedures. Improper bleeding of the ABS may not lead to damage, but it could cost in labor time and frustration on the part of the customer, the technician, and the supervisor.

PHOTO SEQUENCE 34
Typical Procedure for Manually Bleeding a Disc Brake Caliper

P34-1 Be sure the master cylinder reservoir is filled with clean brake fluid. Recheck it often to replace the fluid lost during the bleeding process.

P34-2 Attach a bleeder hose to the bleeder screw.

P34-3 Place the other end of the hose in a capture container partially filled with brake fluid. Be sure that the free end of the hose is submerged in brake fluid. This helps to show air bubbles as they come out of the system and prevents air from being accidentally sucked into the system through the bleeder screw.

P34-4 Have an assistant apply moderate (40–50 pounds), steady pressure on the brake pedal and hold it down.

P34-5 Open the bleeder screw.

P34-6 Observe the fluid coming from the submerged end of the hose. At the start, you should see air bubbles.

P34-7 When the fluid is clear and free of air bubbles, close the bleeder screw.

P34-8 Have your assistant release the brake pedal. Wait 15 seconds and repeat steps 2 and 3 until no bubbles are seen when the bleeder screw is opened. Close the bleeder screw at that wheel and move to the next wheel in the bleeding sequence. Bleed all four wheels in the same manner. Check the master cylinder fluid level often.

P34-9 When the entire system has been bled, start the engine and allow it to idle.

PHOTO SEQUENCE 34 (CONTINUED)

P34-10 Check the pedal for sponginess.

P34-11 Check the brake warning lamp for an indication of unbalanced pressure. Repeat the bleeding procedure to correct the problem.

P34-12 Top off the master cylinder reservoir to the proper fill level.

Right rear

1

Left rear

3

Right front

4

Left front

2

Figure 10-51 Recommended bleeding sequence for a diagonally split brake system.

Figure 10-52 shows a typical pressure bleeder. To use it, connect the adaptor to the master cylinder reservoir (**Figure 10-53**), and pressurize it to the specifications outlined in the equipment's owner's manual. Do not depress the brake pedal while performing this procedure, as the fluid is already pressurized. Once pressurized, open the bleeders one at a time (and in proper sequence) and bleed the air out of the system.

Figure 10-53 Many different adapters are needed to properly pressure bleed the various master cylinder configurations.

Figure 10-52 The lower chamber is compressed air; the upper chamber holds brake fluid. The chambers are separated by a flexible, air-tight diaphragm.

Figure 10-54 A brake flushing machine.

Another brake bleeder device is shown in **Figure 10-54**. It can be connected to all four wheels at once as well as to the master cylinder. A reservoir on the device holds enough brake fluid to completely flush the system. Fluid is pumped into the master cylinder's reservoir, through the brake system, and then returned to the brake bleeder device. Each wheel hose has a clear section near the wheel's bleeder screw. As new fluid is observed in these "windows," the bleeder screw is then closed for that wheel. A complete flush, using this device or a similar one, takes less than 10 minutes, even when performed by a newly trained technician or student.

BRAKE LINES AND HOSES INSPECTION AND SERVICE

All manufacturers include brake line and hose inspection in their maintenance schedules. Rigid hydraulic lines are made of double-wall, welded steel tubing coated to resist corrosion. Brake hoses are flexible and have multiple layers to withstand high pressures. Brake hoses will wear out due to movement and time. Brake lines will need to be replaced due to excessive rust that eats away the lines. Sometimes a hard line will have to be replaced because it was weak and brittle and the part it was connected to (like the wheel cylinder) was removed and moved the line enough to break it.

Start your inspection by looking for kinks in the lines and other damage due to rust and obvious leaks. Brake hoses can be inspected for cracks, leaks, and broken hold-down clips (**Figure 10-55**). If you must remove a brake hose from a caliper, you have to install a new brake banjo washer. This washer is made of copper and creates a seal as you install it. This is a one-time-use component, so you must discard it and install a new one when removing a hose or caliper (**Figure 10-56**).

Brake hoses can be damaged on the inside due to a lack of fluid maintenance. To diagnose this, you need to have a coworker pump the brakes while feeling the hose for swelling or bulging as pressure is applied internally (**Figure 10-57**). Ask the coworker to quickly apply and release the brakes and then quickly spin the wheel. If the brake at any wheel seems to drag after pressure is released, that brake hose may be restricted internally.

> ⚙ **SERVICE TIP** When you diagnose a vehicle as having one bad wheel bearing, remember that the other side likely has the same amount of wear. It isn't necessary to replace both bearings at the same time. If the other side is not making noise and is not loose, then it is not bad. Notify the customer that the other wheel bearing has the same amount of service on it. Sometimes he or she will want you to install two-wheel bearings during the same visit. But if the customer does not want to remove a part that isn't bad yet, then at least he or she will be prepared and aware when the other side does need replacement.

Figure 10-55 The locking clip is tapped into place with a small hammer.

Figure 10-56 Note the sealing washers on this banjo fitting. They should be replaced anytime the fitting is loosened.

Figure 10-57 Possible internal defects in a brake hose.

INSPECTING WHEEL BEARINGS BEFORE REMOVAL

Inspection of a wheel bearing is usually part of a periodic inspection. There are three methods of inspecting a wheel bearing. Raising the vehicle and shaking the tire while everything is still installed and looking for movement because of a loose bearing is the most common method. Removing the bearing and disassembling it is a less preferred method and is only possible with a non-sealed bearing. Test-driving the vehicle to listen for a bad bearing is also common.

A sealed bearing assembly can be disassembled for inspection, but you will have to purchase a new bearing assembly after disassembly. Older style non-sealed wheel bearings are easy to inspect but must still be disassembled to inspect, and you must install a new seal to put it back together.

To inspect a wheel bearing during a test-drive, you must drive the vehicle in an open parking lot. Always use caution when test-driving a vehicle. Turn the vehicle at low speeds and listen carefully for a growling or humming noise that occurs only when turning. After the test-drive return to the shop where you can lift it up and inspect the bearings.

When wheel bearings and hubs are a sealed (integral) assembly, the bearing endplay should be measured with a dial indicator stem mounted against the hub. If the endplay exceeds manufacturer specifications, the hub bearing assembly must be replaced. A typical specification is 0.005 inch (0.127 mm). Some manufacturers require replacement when there is any movement. **Photo Sequence 35** shows a typical procedure for measuring front wheel hub bearing endplay.

REMOVING A NON-SEALED WHEEL BEARING AND SEAL

We will discuss the removal of a wheel bearing and seal from a non-driving wheel in this section. (Later, we discuss their installation.) Lift the vehicle and remove the wheel. Use a screwdriver or the appropriate special tool to remove the dust cover (**Figure 10-58**). Under

Typical Procedure for Measuring Front Wheel Hub Endplay—Integral, Sealed Wheel Bearing Hub Assemblies

P35-1 The vehicle should be properly positioned on a lift and raised to a comfortable working height for performing this measurement.

P35-2 Remove the wheel cover and dust cap.

P35-3 Attach a magnetic dial indicator base securely to the inside of the fender at the lower edge of the wheel opening. Position the dial indicator stem against the vertical wheel surface as close as possible to the top wheel stud and preload the dial indicator stem.

P35-4 Zero the dial indicator pointer.

P35-5 Grasp the top of the tire with both hands. Push and pull on the top of the tire without rotating the tire and note the dial indicator readings with the tire pushed inward and the tire pulled outward. The difference between the two readings is the wheel hub endplay. Repeat this procedure twice to verify the endplay reading.

P35-6 Maximum wheel bearing endplay should be 0.005 inch (0.127 mm). If the endplay measurement is not correct, wheel bearing hub replacement is necessary.

P35-7 Remove the dial indicator and install the dust cap and wheel cover.

Figure 10-58 These tools can remove most automotive seals and dust covers.

A **cotter key** is a piece of metal doubled back on itself. One use is to keep a nut from loosening.

⚠ **Caution**

Before lifting the vehicle, ensure that the lift pads are correctly placed under the vehicle's lift points. Consult the service manual for the location of the lift points. Failure to properly place the hoist lift pads and arms could result in damage and injury.

 Special Tools

Dust cover removal tool
Cotter pin remover

Classroom Manual pages 352–353

the dust cover, a nut with a **cotter key** will be visible. Remove the key and nut and place to the side. A flat, keyed washer is removed next. Many times, the outer bearing will slide out with the washer. Shaking the hub will dislodge the bearing and washer for an easier grip.

With the outer bearing, washer, nut, and cotter key removed, the hub can be removed. However, read the next short section on how the inner bearing and seal can be removed quickly.

Thread the nut back onto the spindle about two or three threads. Grasp the edges of the hub and pull outward and downward (**Figure 10-59**). The inner bearing will hang on the nut and force the seal out. Force is not needed to perform this task. If the nut will not fit through the hub, or the seal and bearing do not come out, remove the hub and place it on the table with the seal facing upward.

A seal remover or large flat-tipped screwdriver can then be used to remove the seal. Place the tool or screwdriver across the hole in the seal (**Figure 10-60**). Hook the end of the tool or screwdriver under the inner edge of the seal and use the hub as a pivot to apply force to the seal. The seal should pop out easy. The bearing can now be removed. A picture of a typical non-sealed bearing setup and related components is outlined in **Figure 10-61**.

Before beginning work on the bearings, clean the inside of the hub completely. Dry it with compressed air and set it aside.

Figure 10-59 Screw the nut slightly onto the spindle and pull outward and downward on the hub assembly.

Figure 10-60 A screwdriver can be used to remove many automotive seals.

Figure 10-61 Components of a non-sealed wheel bearing (this setup is found on both brake drums and rotors).

> ⚙️ **SERVICE TIP** The terms inner and outer do not necessarily refer to the position of the races to the bearing. Inner, with regard to bearings, races, and seals, generally means the part that is closer to the center line of the vehicle. Outer refers to the part furthest from the centerline. For example, an outer bearing, seal, and race will be visible after the wheel and the hub's dust (or grease) cap is removed. The inner ones will not be visible until the hub is removed from the vehicle.

INSPECTING AND REPACKING A WHEEL BEARING

During a typical brake job, it is usual for the technician to service the **wheel bearings** on non-driving wheels. During the service, the bearings should be inspected. The inspection discussed next applies to every bearing but directly highlights ball and roller bearings. Engine bearing failures are usually made obvious by sound and engine operation.

⚠️ **Caution**
Always use protective gloves and face protection when cleaning bearings. Skin and eyes can be injured by the cleaning solution.

🔧 **Special Tools**
Seal remover
Blow gun
Parts washer

Wheel bearings A wheel bearing, also known as a hub assembly is a component that helps decrease friction when the wheel is turning by the use of ball bearing.

Classroom Manual
pages 356–357

Spalling, from the German, means "to break or split off."

Brinelling, from the French, is a type of bearing failure where dents appear due to the roller "hammering" against the race.

⚠ **Caution**

Do not let the bearing spin while drying it with compressed air. The bearing may fly apart, thereby causing injury, or it could catch skin between the bearing and cage or race.

🔧 **Special Tools**

Parts washer
Blow gun
Catch pan or vat

Inspection

Bearings fail because of a lack of proper lubrication, overloading, or improper installation. In most cases, the design of the total component prevents overloading. Few bearing failures result directly from improper installation, thereby leaving lubrication as the main cause of failure.

Most shops have a standard fee for replacing the brakes on one axle. The fee includes repacking the wheel bearing and a new seal.

Bearing failure is indicated by a noise from the bearing area. At times, it is almost impossible to isolate the problem bearing. Axle bearings can be confused with wheel or differential bearings. A slow test-drive around the lot may find the end of the vehicle from which the noise originates. Turning in tight circles may pinpoint the side of the car with the most noise. An experienced technician should perform the test-drive if one is needed.

Bearing failure causes two types of damage. When pieces break off the bearing, the damage is called **spalling** (**Figure 10-62**). Dents in the bearing or race are signs of **brinelling** (**Figure 10-63**). Either can cause more damage if not replaced. A dry bearing can weld itself to the races and the components by friction-produced heat.

Remove the suspected bearing and clean it thoroughly with a parts cleaner. Ensure that all of the old lubricant is removed. The bearing can be air dried with compressed air. This will help remove the cleaning solution and any particles trapped in the bearing. However, when using compressed air to dry a bearing, there are some critical things that must be done. Hand protection must be used to protect the skin from the cleaner, compressed air, and any particles dislodged by the air (**Figure 10-64**). Eye and face protection should always be worn during the cleaning process. The bearing must be held in a manner to prevent any of the rollers or balls from spinning. A free-spinning bearing can disintegrate, thereby causing injuries. If at all possible, use a regulator to reduce air pressure. This will lessen the possibility of injury due to spinning bearings and loose particles within the bearing.

Dry the bearing over a vat or pan to catch the cleaner and particles blown off by the air (**Figure 10-65**). The waste will probably have to be treated as hazardous waste and stored accordingly. With the bearing cleaned, clean the races. Most can be cleaned while still installed in the component.

Figure 10-62 Note how the metal is flaking from this race (left) and bearing.

Figure 10-63 Denting of the bearing metal is caused by the bearing moving up and down within its races. This may be a result of improper installation.

Figure 10-64 Protective gloves can be used by automotive technicians, and they can still retain a good sense of touch.

Figure 10-65 Always clean the bearing over a catch container or parts washer to capture the old grease and debris you have removed.

With all of the bearing components clean and dry, inspect each ball or roller for dents, nicks, **burned areas,** or any other damage to the machined surfaces (**Figure 10-66**). Inspect the cage for any bending or breakage. Inspect the races as much as possible. Most races can be inspected while installing a component. Also, inspect the component into which the bearing and races are fitted. If a race has been removed from

A **burned area** will be bluish in color.

Tapered Roller Bearing Diagnosis

Consider the following factors when diagnosing bearing condition:
1. General condition of all parts during disassembly and inspection.
2. Classify the failure with the aid of the illustrations.
3. Determine the cause.
4. Make all repairs following recommended procedures.

Abrasive Step Wear

Pattern on roller ends caused by fine abrasives.
Clean all parts and housing, check seals and bearings and replace if leaking, rough, or noisy.

Galling

Metal smears on roller ends due to overheating, lubricant failure, or overload. Replace bearing, check seals and check for proper lubrication.

Bent Cage

Cage damaged due to improper handling or tool usage.
Replace bearing.

Abrasive Roller Wear

Pattern on races and rollers caused by fine abrasives.
Clean all parts and housings, check seals and bearings and replace if leaking, rough, or noisy.

Etching

Bearing surfaces appear gray or grayish black in color with related etching away of material usually at roller spacing.
Replace bearings, check seals, and check for proper lubrication.

Bent Cage

Cage damaged due to improper handling or tool usage.
Replace bearing.

Indentations

Surface depressions on race and rollers caused by hard particles of foreign material.
Clean all parts and housings. Check seals and replace bearings if rough or noisy.

Good Bearing

Misalignment

Outer race misalignment due to foreign object.
Clean related parts and replace bearing. Make sure races are properly sealed.

Figure 10-66 Typical bearing and race problems and their probable causes.

the hub or **shaft,** carefully inspect the area where the race fits (**Figure 10-67**). This area must be as clean and undamaged as other parts of the bearing. Any type of damage to any part requires replacement of the complete bearing, including the races.

If the shaft or hub is damaged, consult the owner. Replacing or machining some shafts and hubs can be an expensive job. Although there is not much choice concerning the repairs needed in this situation, the owner is the one paying and has the final word based on the technician's recommendation.

Repacking the Bearing

For a bearing to be completely lubricated, each roller or ball must be greased before installation on the vehicle. *Repacking* is usually the term used to indicate the procedure for pre-lubricating the bearing.

Several types of tools can be used to repack bearings. Some operate by using compressed air whereas others are hand operated (**Figure 10-68**). The tools that have been

The **shaft** with the type bearings discussed here is called a *spindle*. The shaft (or spindle) is part of a steering knuckle, which is replaced as a unit.

 Special Tool
Bearing packer

Classroom Manual
pages 357–359

Figure 10-67 The spindle or shaft must be as clean and as clear of damage as the bearings and races.

Figure 10-68 Typical hand-operated bearing packer.

around the longest and are sometimes the fastest and most accurate are, quite simply, the technician's hands. That is the procedure we will discuss here.

Before attempting to repack the bearing, ensure that it is clean and dry. Parts cleaner left on the bearing will break down the grease and lubrication will be reduced. Also ensure that the correct grease is being used. It should be multipurpose or a G classification. After repacking, make sure there is a clean strip or ribbon of grease extending upward and past the ends of each roller (**Figure 10-69**). Even with the best cleaning, a little dirty grease may sometimes be trapped within the bearing. The new grease will force it out.

With the grease on hand and the bearing clean and serviceable, scoop some **grease** and place it in the palm of one hand. The amount of grease depends on the technician's hand, of course, but it should be about the size of half a Mounds candy bar. Holding the bearing with the smaller cone (end) up, press a portion of the outer edge of the bearing down into the grease from the top of the mound (**Figure 10-70**). When the bearing edge

A rough cloth will remove most of the grease from a technician's hands after the bearing has been repacked.

Figure 10-69 Until ribbons of grease appear above the rollers, the bearing is not properly packed.

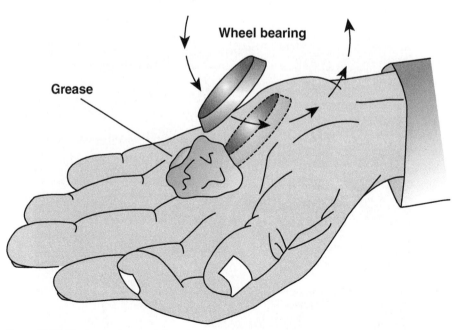

Figure 10-70 The bearing is moved down through the grease and rotated toward the heel of the palm.

touches the palm, press just slightly harder and drag that edge across the palm toward the wrist. It does not have to be dragged far. Repeat this step until clean grease ribbons are visible at the top end of the rollers. Rotate the bearing until the next rollers are aligned to be placed in the grease. Repeat the pressing, dragging, and rotating until all of the bearing rollers have clean grease between them. More grease may be needed to replenish what is in the palm. Before placing the bearing on a clean cloth or rag, spread a light coat of grease around the outside of the bearing. This will provide lubrication during the first several rotations of the bearing within its supported component. Repeat the procedures for any other roller bearing.

Ball bearings can be repacked in a similar manner, but it may take a little longer. Most automotive ball bearing assemblies are not used in areas where they have to be repacked. Many ball bearing assemblies are pre-lubed and sealed at the factory. Some ball bearings are exposed to the lubricant within the component. However, new ball bearings used in these applications should be pre-lubed before installation. The best way is to drip the bearing in the same lubricant used in the component. A new bearing will run dry and be damaged if not pre-lubed in some manner. If the technician works on large vehicles, it is common practice to pre-lube the large bearings even if it is an old one being reinstalled.

Repacking a bearing with packing tools is fairly easy and much cleaner because each tool does what can be done by hand. The bearing is placed in a cone-shaped holder and another cone is fitted over it (refer to **Figure 10-68**). The grease is forced through the bearing rollers. Ensure clean ribbons of grease protrude past the rollers and some grease is applied to the outside by hand.

REMOVING AND INSTALLING A RACE IN A HUB

A race is removed from a hub or shaft with a hammer, punch, or a puller. On the hub we are working with in this chapter, the races can be removed with a hammer and pin punch. Set the hub on a solid surface such as a metal workbench. If the race for the inner bearing is being removed, some type of spacers may have to be placed under the hub so the race has room to slide out. Two short pieces of 2 x 4 lumber work fine without damaging the hub. The punch will enter the hub from the opposite side of the race being removed.

A typical hub with two bearings and one seal will be used for this purpose. **Photo Sequence 36** shows a typical procedure for removing a bearing race. The procedures can be followed on other applications except that some bearings may need to be pressed out and in. This is best done by an experienced technician.

With the hub cleaned and in position, observe downward through the opening. A very small edge of the bottom race or two notches in the hub directly behind the race will be visible (**Figure 10-71**). In many cases, both are visible. The notches provide a place to rest the end of the punch against the race.

Set the punch in place and strike it sharply with the hammer (**Figure 10-72**). Move to the other notch and repeat. Do not strike the punch very hard unless absolutely necessary. Most races will pop out without too much trouble. It will be necessary to move the punch from notch to notch to keep the race fairly even as it is forced out. This is a simple task, and there is no need to rush. Rushing usually causes a mistake with the hammer and results in a sore hand or finger. Repeat the procedure for the other bearing race.

Installing the race is pretty much the opposite of the removal. A bearing and seal driver should always be used to install a race. The driver set comes with different-sized drivers (**Figure 10-73**). Select a driver that fits snugly in the bearing side of the race and will slide inside the hub opening.

 Special Tools

8-ounce ball peen hammer

6-inch punch

Spacer material

Classroom Manual
page 359

 Caution

Always wear gloves and use a chisel or punch holder when removing or installing races and seals with a driving tool. A misaimed swing could result in a serious hand or finger injury.

Caution

Always wear gloves and use a chisel or punch holder when removing or installing races and seals with a driving tool. A misaimed swing could result in a serious hand or finger injury.

PHOTO SEQUENCE 36
Removing and Installing a Bearing Race on a Non-Sealed Wheel Bearing

P36-1 Before attempting to remove the bearing race, clean the hub thoroughly.

P36-2 Flip the hub so the race to be removed is down. It may be necessary to place short 2×4 strips under the hub for clearance.

P36-3 With the driving end of the punch held against the inner edge of the race, use the hammer to drive the bearing race out.

P36-4 The space where the bearing was installed is plainly visible.

P36-5 Center the new race over the hub's cavity and fit the driver into the race.

P36-6 Use the hammer to drive the bearing race into the hub. Continue to drive until the race bottoms against its stop inside the hub.

PHOTO SEQUENCE 36 (CONTINUED)

P36-7 Typically when a bearing race is fully installed, there will be a space between the outer edge of the race and the outer edge of the hub.

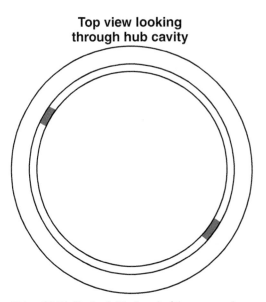

Top view looking through hub cavity

Figure 10-71 The backside (inner) of the race can be seen through the two notches at the lower edge of the hub's opening.

Figure 10-72 A hammer and pin punch are used to drive the race from the hub.

Set the race over the opening with the larger opening upward. Place the driver into the race and tap slightly with a hammer (**Figure 10-74**). The race and driver must be completely aligned with the opening. If the race is off just a little, it will not enter the opening or will be cocked in the cavity. Once the race is started, harder strikes can be made with the hammer until the race bottoms out. This is noticeable because of the sound made when the race hits the bottom. The sound will be dull compared to the sounds made when the race is being driven in. Remove the tool and inspect the race for nicks or scratches. There should be no damage if the proper driver is used. Flip the hub over and install the other race.

 Special Tools

8-ounce hammer
Bearing and seal
   driver

Classroom Manual
page 357

Figure 10-73 A bearing and seal driver set.

Figure 10-74 Proper selection and positioning of the correct driver prevents damage to the race.

SERVICE TIP In the past, the hub grease cap was packed with grease. Many technicians, and most vehicle owners, figured that if a little bit is good, then a lot must be very good. Before you fill that hub cavity with grease, think about what happens inside that hub after several hours of driving at interstate speeds (actually it doesn't take that long). The grease dissolves into a near liquid, increasing pressure within the hub. The seal is not designed to retain that pressure nor the semiliquid; grease is forced out of the hub. The problem starts when the vehicle stops for a period of time and what remains of the grease solidifies. At the next start-up, the now-dry bearings will have little or no lubrication. In a short period of driving, the damage is done.

INSTALLING A WHEEL BEARING AND SEAL

With the race installation completed, the inner bearing and seal can be installed. Put some loose grease into the hub. It should fill the inner portion of the hub's opening but not overfill it (**Figure 10-75**). Excessive grease here will cause heat buildup, which will melt the grease on the bearing, thereby causing damage to the bearings and seal. Place the hub with the inner bearing race facing up. This is normally the larger race of the two.

Set the greased inner bearing into its race. It should drop into the race far enough to be almost flush with the race. If it is not flush, it is probably in backward. The bearing should not block the seal.

With the bearing in place, align the wheel seal over the opening. Remember to place the seal's lip inward or toward the lubricant (**Figure 10-76**). The same driver used to install the race can be used to install the seal, but normally a hammer is used. Hold the seal in place by hand and tap slightly around the edges of the seal until it is started evenly. Use the hammer to tap the seal the rest of the way into the cavity. If the seal has a flange that overlaps the hub, the sound will change as the flange mates with the hub. Many seals do not have a flange. They are driven in until the outer edge of the seal is even or flush with the outer edge of the hub.

There are many ways used to adjust the wheel bearing tension. The only right way is to follow service manual instructions. The other ways rely on touch, feel, and guessing.

Reinstall the hub and slide the outer bearing over the spindle and into the hub. Screw the nut on while turning the hub. This helps center the bearings on the spindle and in the hub. Consult the service manual for the torqueing specification on the nut. Many times, the nut is tightened to a torque and then backed off a certain amount. Overtightening may result in bearing failure because of excessive friction and heat. Undertightening will also damage the bearing and cause the wheel assembly to be loose on the spindle.

With the nut torqued, select a new cotter key and install it. Place some grease into the dust cup, about 1/3 full, and install it into the hub.

Special Tool
Torque wrench

Classroom Manual
pages 356–357

REPLACING A SEALED HUB BEARING ASSEMBLY

Sealed wheel bearing removal and replacement will vary from model to model. Most removal procedures are similar or follow a similar pattern. In this section, you will learn the typical procedures needed to remove and replace a bolted-in sealed hub bearing

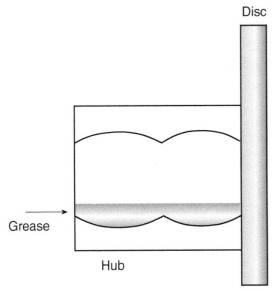

Figure 10-75 Only fill the center of the hub anywhere from a third to half full of grease. Too much could ruin the bearing.

Figure 10-76 Installing a seal with a seal driver.

assembly. The procedures for replacement of a pressed in sealed bearing are similar, with the exception of a press needed to move the bearing out. In practice, always remember to follow the manufacturer's guidelines for replacement.

If the bearing is on a driving axle, the axle shaft (or CV shaft) must be removed from the hub unit first. This usually includes using an air tool to remove the axle nut and a hammer and brass punch to move the shaft inward. In some cases, a special hub puller can be used (**Figure 10-77**). In most cases, you will only need to remove the outer axle from the hub. Some vehicles will require removal of the strut-to-knuckle bolts (**Figure 10-78**), the ball joint (**Figure 10-79**), and the outer tie rod end to remove the shaft. After removing the shaft, remove the bolts of the hub assembly from the knuckle (**Figure 10-80**).

At this point, the hub bearing assembly should just come off, but it won't because it is a close and tight fit and there is usually corrosion. You will have to use a chisel and hammer, and sometimes heat, to get the assembly out. Once it is out, clean the hub surface of any rust or corrosion before reinstalling the new one. Reverse the procedure to install.

Figure 10-77 Removal of the outer axle joint from the front wheel hub.

Figure 10-78 Mark eccentric strut bolt before removal.

Figure 10-79 Removal of ball joint to steering knuckle clamp bolt.

Figure 10-80 Remove the bolts attaching the hub to the steering knuckle.

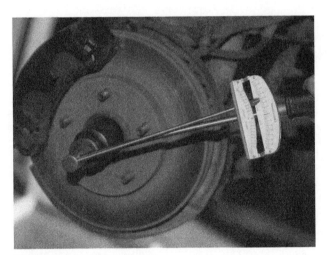

Figure 10-81 Hub nut torqueing.

Figure 10-82 Nut lock and cotter pin installation.

Make sure to torque all fasteners and the wheel nut to specifications (**Figure 10-81**). It helps to have someone holding the brake or wedging the tire against the ground (just like when torqueing lug nuts, do not let the vehicle all the way down to the ground when torqueing the nut). Make sure to reinstall the cotter pin (**Figure 10-82**) and tighten the lug nuts in the proper order (**Figure 10-83**).

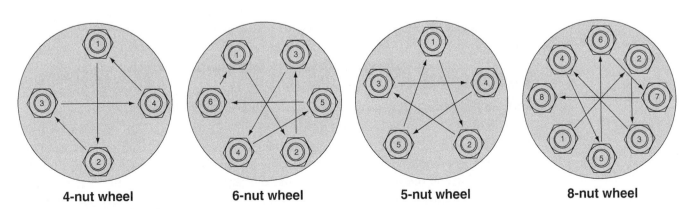

4-nut wheel **6-nut wheel** **5-nut wheel** **8-nut wheel**

Figure 10-83 Wheel nut tightening sequence.

Figure 10-84 This is a typical type of warning for ABS systems. All manufacturers have cautions or warnings printed at the front of their manuals and in each section to which the caution or warning applies.

Classroom Manual
pages 359–364

A sealed bearing that is a press fit must be removed with the knuckle (Figure 10-80). This is a more advanced technique that you will learn later. You can reference the book *Today's Technician: Automotive Suspension and Steering*, 6th edition, for more information.

ABS DIAGNOSIS AND SERVICE

Repairs to service brakes should not be started until the service manual has been consulted concerning cautions or warnings on the ABS system (**Figure 10-84**). Improper procedures or tools could damage wheel-mounted ABS components. In addition, some ABS systems have specific steps for bleeding or flushing the brakes. Although most "routine" repairs to the service brakes can be performed without fear of ABS damage, it is always too late to check the service manual after the damage is done.

Generally speaking, most shops will assign an experienced technician to diagnose and service an ABS system, but many of the simple steps can be performed by a new technician with adequate supervision. The first is the retrieval of diagnostic trouble codes (DTCs).

When a customer complains that the ABS light (or other warning light) is on, the first tool used is the scan tool. All ABS DTCs can be retrieved with the vehicle's manufacturer-specific tool, and most can also be retrieved with an aftermarket scan tool like the ones sold by Snap-on, MasterTech, OTC, and others. In most instances, it is just a matter of identifying the vehicle to the tool and plugging the data link cable between the tool and vehicle. On-board diagnosis (OBD) I vehicles usually have separate data link connector and scan tool for ABS communication. Some of those vehicles required the DTC to be retrieved manually, and they will not communicate with the scan tool. OBD II vehicles have a more complex system for monitoring vehicle electronics. In most cases, but not all, the ABS computer can be accessed using the same OBD II Diagnostic Link Connector (DLC) and can sometimes be accessed with a generic (Global) OBD II scan tool.

IMPORTANT NOTES ABOUT SERVICING BRAKES ON HYBRID ELECTRIC VEHICLES

When servicing the brakes on a hybrid electric vehicle (HEV), make sure that you read all service precautions first. Some vehicles have an electronically controlled accumulator that pressurizes the brakes. This is unique to some hybrids because the engine will shut off when braking and the vehicle will mainly use the regenerative electric motor to slow the vehicle down. The brakes on these vehicles will typically last a little longer because they are not under the same braking stresses and conditions as non-hybrid vehicles. Follow the manufacturer's direction to depressurize or disable the system before servicing the brakes. Many times, you will perform a brake inspection on a HEV and find that the brakes are not worn down. You should still inspect the caliper slides and moving parts. Those parts must be lubricated periodically as well, mostly from lack of use.

CASE STUDY

A student brought his car to our lab stating his brakes would not stop the car properly and he had added some brake fluid "a while back." A quick visual inspection showed the disc brake pads were completely worn out, to the point there was no lining left on any pad. If the brake fluid is low, obviously check for leaks but don't forget to check lining wear. This is particularly true of disc brake systems.

ASE-STYLE REVIEW QUESTIONS

1. Drum brake repairs are being discussed.
   Technician A says to keep one side intact while the other is disassembled to save some time.
   Technician B says wheel grease can be used to lube the rub points.
   Who is correct?
   A. A only
   B. B only
   C. Both A and B
   D. Neither A nor B

2. Pedal pulsation may be caused by any of the following EXCEPT:
   A. Warped rotors.
   B. Out-of-round drums.
   C. Broken or weak return springs.
   D. Damaged wheel bearings.

3. All of the following can be used to check a wheel speed sensor EXCEPT:
   A. Voltmeter.
   B. Test light.
   C. DMM on voltage.
   D. DMM on Resistance.

4. The driver complains of the vehicle that front disc is pulling to the left during braking.
   Technician A says the left front brake is the cause.
   Technician B says the right front pads may be glazed from heat.
   Who is correct?
   A. A only
   B. B only
   C. Both A and B
   D. Neither A nor B

5. Diagnosing a disc or drum brake system is being discussed.
   Technician A says a bad hose on the rear brakes may cause the front wheels to lock too quickly.
   Technician B says pedal pulsation may be caused by incorrect pressure between the front and rear brakes.
   Who is correct?
   A. A only
   B. B only
   C. Both A and B
   D. Neither A nor B

6. Drum brake diagnosis is being discussed.
   Technician A says a grinding noise may be caused by broken hardware.
   Technician B says a firm pedal with little braking action could be caused by a parking brake misadjustment.
   Who is correct?
   A. A only
   B. B only
   C. Both A and B
   D. Neither A nor B

7. Rotor and drum machining is being discussed.
   Technician A says a drum can be machined on the vehicle.
   Technician B says a disc rotor must be removed from the vehicle for machining.
   Who is correct?
   A. A only
   B. B only
   C. Both A and B
   D. Neither A nor B

8. The customer complains of a slight pull to the left during braking. The technician finds the right inner pad worn out and the right outer pad with almost no wear.

   Technician A says the inner pad wear could be caused by dirty caliper slide pins.

   Technician B says the left pull could be a result of the right side brake failure.

   Who is correct?

   A. A only   C. Both A and B

   B. B only   D. Neither A nor B

9. The vehicle has very little brake boost and the engine stumbles severely during braking.

   Technician A says the hydro-boost power valve may be the problem.

   Technician B says the diaphragm in the booster is probably split.

   Who is correct?

   A. A only   C. Both A and B

   B. B only   D. Neither A nor B

10. A vehicle pulls slightly to one side during normal or slow braking.

    Technician A says a right rear drum brake shoe that is too tight could be the cause.

    Technician B says brake shoes soaked with brake fluid are the probable cause.

    Who is correct?

    A. A only   C. Both A and B

    B. B only   D. Neither A nor B

ASE CHALLENGE QUESTIONS

1. Repacking a bearing is being discussed.

   Technician A says the grease should be an L classification.

   Technician B says the grease must extend through the rollers.

   Who is correct?

   A. A only   C. Both A and B

   B. B only   D. Neither A nor B

2. Laying an RTV bead is discussed.

   Technician A says RTV should cover the entire sealing surface.

   Technician B says to remove the tube from the work until the complete bead is laid.

   Who is correct?

   A. A only   C. Both A and B

   B. B only   D. Neither A nor B

3. All of the following are used to clean a bearing EXCEPT:

   A. compressed air.

   B. hand, eye, and face protection.

   C. gasoline or kerosene.

   D. parts washer.

4. Removing a wheel bearing is being discussed.

   Technician A says the seal must be removed before the outer bearing can be removed.

   Technician B says the nut or a screwdriver can be used to gain access to the outer bearing.

   Who is correct?

   A. A only   C. Both A and B

   B. B only   D. Neither A nor B

5. Installing a wheel bearing is being discussed.

   Technician A says the driving tool can be used to install the seal behind the inner bearing.

   Technician B says the outer bearing race may be installed with the same tool used to install the inner race.

   Who is correct?

   A. A only   C. Both A and B

   B. B only   D. Neither A nor B

Name _____   Date _____

REPLACING A MASTER CYLINDER

Upon completion and review of this job sheet, you should be able to remove, bench bleed, and install a master cylinder.

ASE Education Foundation Correlation ————————————————————————

This job sheet addresses the following **MLR** tasks:

5.B.2	Check master cylinder for external leaks and proper operation. (**P-1**)
5.B.3	Inspect brake lines, flexible hoses, and fittings for leaks, dents, kinks, rust, cracks, bulging, wear, loose fittings/supports. (**P-1**)
5.B.4	Select, handle, store, and fill brake fluids to proper level; use proper fluid type per manufacturer specification. (**P-1**)
5.B.5	Identify components of hydraulic brake warning light system. (**P-3**)
5.B.6	Bleed and/or flush brake system. (**P-1**)
5.B.7	Test brake fluid for contamination. (**P-1**)

This job sheet addresses the following **AST/MAST** tasks:

5.B.2	Check master cylinder for internal/external leaks and proper operation; determine needed action. (**P-1**)
5.B.6	Inspect brake lines, flexible hoses, and fittings for leaks, dents, kinks, rust, cracks, bulging, and wear; check for loose fittings and supports; determine needed action. (**P-1**)
5.B.9	Select, handle, store, and fill brake fluids to proper level; use proper fluid type per manufacturer specification. (**P-1**)
5.B.11	Identify components of hydraulic brake warning light system. (**P-2**)
5.B.12	Bleed and/or flush brake system. (**P-1**)
5.B.13	Test brake fluid for contamination. (**P-1**)

We Support
ASE | Education Foundation

Tools and Materials

Goggles or safety glasses with side shields

Vise

Oil drain pan and rags

Recommended brake fluid

Torque wrench

Line wrenches

Fender cover

Service manual

Master cylinder bleeder kit

⚠ **Caution**

Wear eye protection when dealing with brake fluid. Brake fluid can cause permanent eye damage. If brake fluid gets in the eyes, flush them with cold running water and see a doctor immediately.

⚠ **Caution**

Do not work in or around the steering column without disarming the airbag systems. Serious injury could result from a deploying airbag.

Describe the vehicle being worked on:

Year _____ Make _____ Model _____

VIN _____ Engine type and size _____

Procedure

1. Determine the following information from the service manual.

   Master cylinder fasteners torque _____ Brake boost?
   _____ Yes _____ No

   If yes, what type? _____

   Brake fluid type and classification _____

   ABS cautions _____

   _____

⚡ **WARNING** Do not mix glycol-based and silicone-based brake fluids. The mixture could cause a loss of braking efficiency.

⚡ **WARNING** Before working on the brakes of a vehicle with ABS, consult the service manual for precautions and procedures. Failure to follow procedures to protect ABS components during routine brake work could damage the components and cause expensive repairs.

2. Place thick wipe cloth under the master cylinder brake line connections.   ☐

3. Disconnect the fluid level sensor if equipped and remove fluid from the reservoir.   ☐

4. Loosen and remove the brake lines.   ☐

5. Remove the master cylinder fasteners. Remove the master cylinder.   ☐

6. Drain remaining fluid into an oil drain pan and dispose of the old fluid.   ☐

7. Place the master cylinder body in a vise.   ☐

8. Use a small pry bar to remove the reservoir from the master cylinder. Be careful not to crack the reservoir.   ☐

9. Lube the new O-rings with brake fluid before installing the reservoir on the new master cylinder. The new master cylinder should have new O-rings.   ☐

10. Install the reservoir to the master cylinder.   ☐

11. Install the bleeder kit.   ☐

⚡ **WARNING** Do not mix glycol-based and silicone-based brake fluids. The mixture could cause a loss of braking efficiency.

12. Fill the reservoir chambers about half full of approved brake fluid.   ☐

13. If necessary, reposition the master cylinder in the vise so the front (closed) end is higher than the rear.   ☐

14. Use a punch or cross-tip screwdriver to push in the secondary piston. Push it on enough to bottom both pistons.   ☐

Task Completed

15. Observe the reservoir for bubbles emerging from the bleeder hoses. ☐

16. Repeat steps 14 and 15 until there is no air coming out of the brake. Ensure that the line ends are kept below the level of the fluid. ☐

INSTRUCTOR CHECK POINT _____

17. Close the lines with small paper clips. ☐

18. Install the reservoir cover. ☐

19. Install the master cylinder to the brake booster or firewall. ☐

20. Install and torque the master cylinder's fasteners. ☐

⚡ **WARNING** **Protect the vehicle's paint surface from glycol-based brake fluids. This type of brake fluid can damage and remove the finish quickly. If brake fluid gets on the vehicle's finish, wash immediately with cold, running water.**

21. Remove the bleeder lines and attach the vehicle's brake line to the master cylinder. Do one line at a time and do not completely tighten the fittings. ☐

NOTE: Step 22 should be done on a single stroke of the brake pedal.

22. Ask a second person to slowly depress the pedal while the secondary line fitting is loosened slightly to allow brake fluid to escape. Observe the fluid for air. Loosely tighten the fitting. Repeat with the secondary line. Ensure that both fittings are tightened before the pedal is released. ☐

23. Release the brake pedal and repeat step 22 as necessary to ensure that the lines are clear of air. If the procedure is not completed within two or three strokes of the pedal, the master cylinder is not bled properly and should be removed for bench bleeding. See steps 2 through 7 and 11 through 23. ☐

24. Test the system by applying the brakes with and without the engine running. The pedal should be firm and there should be no spongy feel. If the pedal is spongy, the complete brake system will have to be bled. It is suggested that the system should be bled any time the system is exposed to air. ☐

25. Remove the wipe cloth and ensure that the area is clean of any brake fluid. ☐

⚡ **WARNING** **Before moving the vehicle after a brake repair, pump the pedal several times to test the brake. Failure to do so may cause an accident with damage to vehicles, the facility, or it may cause personal injury.**

26. Test the brakes before moving the vehicle from the bay. ☐

27. Clean the area and complete the repair order. ☐

Instructor's Response

_____

_____

_____

Name _____   Date _____

REPLACING DISC BRAKE PADS

Upon completion and review of this job sheet, you should be able to replace the brake pads on a front wheel disc brake and inspect the brake components.

ASE Education Foundation Correlation

This job sheet addresses the following **MLR** tasks:

5.D.1   Remove and clean caliper assembly; inspect for leaks and damage/wear; determine necessary action. (**P-1**)

5.D.2   Clean and inspect caliper mounting and slides/pins for proper operation, wear, and damage; determine necessary action. (**P-1**)

5.D.3   Remove, inspect, and replace pads and retaining hardware; determine necessary action. (**P-1**)

5.D.4   Lubricate and reinstall caliper, pads, and related hardware; seat pads and inspect for leaks. (**P-1**)

5.D.9   Retract and readjust caliper piston on an integral parking brake system. (**P-2**)

5.D.10   Check brake pad wear indicator; determine necessary action. (**P-1**)

This job sheet addresses the following **AST/MAST** tasks:

5.D.2   Remove and clean caliper assembly; inspect for leaks and damage/wear to caliper housing; determine needed action. (**P-1**)

5.D.3   Inspect caliper mounting and slides/pins for proper operation, wear, and damage; determine needed action. (**P-1**)

5.D.4   Remove, inspect, and/or replace brake pads and retaining hardware; determine needed action. (**P-1**)

5.D.5   Lubricate and reinstall caliper, brake pads, and related hardware; seat brake pads; inspect for leaks. (**P-1**)

5.D.10   Retract and readjust caliper piston on an integrated parking brake system. (**P-2**)

5.D.11   Check brake pad wear indicator; determine needed action. (**P-1**)

We Support
ASE | Education Foundation

Tools and Materials

Goggles or safety glasses with side shields

Lift or jack and jack stands

Impact wrench and socket

C-clamp

Disc brake silencer

Wire or cord

Torque wrench

Describe the vehicle being worked on:

Year _____ Make _____ Model _____

VIN _____ Engine type and size _____

Procedures

1. Determine the following information.

   Vehicle Make _____ Model _____ Year _____

   Caliper and caliper adapter torque _____

Task Completed

Brake boost? _____ Yes _____ No If yes, type?   ☐

Brake fluid type and classification _____

ABS _____ Yes _____ No If yes, type? _____

ABS cautions _____

⚠ **WARNING** Before working on the brakes of a vehicle with an ABS system, consult the service manual for precautions and procedures. Failure to follow procedures to protect ABS components during routine brake work could damage the components and cause expensive repairs.

2. Inspect and adjust the fluid level in the master cylinder so the reservoir is about half full.   ☐

3. Lift the vehicle and remove the wheel assembly.   ☐

4. Inspect the brake caliper mounting area for ABS components, caliper and adapter, and general condition of the steering, suspension, and brake components on each side of the vehicle. Record your observations.   ☐

⚠ **WARNING** Do not use petroleum (gasoline, kerosene, motor oil, transmission fluid) or mineral-oil-based products to clean brake components. This type of liquid will cause damage to the seals and rubber cups and it decreases braking efficiency.

5. Position a catch basin and spray down the braking components with a low-pressure washer.   ☐

6. Remove the caliper mounting fasteners.   ☐

7. Slide the caliper from the rotor. Use a large, flat screwdriver, if necessary, to pry the pads far enough to clear any ridge.   ☐

8. Use wire or cord to support caliper weight from a vehicle component.   ☐

9. Remove the pads and antirattle clips from the caliper or adapter.   ☐

10. Remove the caliper adapter and rotor.   ☐

11. Inspect the rotor for damage. Measure the thickness of the rotor and compare it to specification. If the rotor requires machining, consult with an experienced technician. Record your measurements and recommendation.   ☐

12. Inspect the caliper piston boot, slide pins, and slide areas on the adapter and caliper. Record your observations.

_____

13. Clean the slide pins or areas. Lube the pins or areas with disc brake lubricant.   ☐

14. Coat the metal portion of each of the new pads with disc brake silencer. Allow them to set for about 5 minutes.   ☐

INSTRUCTOR CHECK POINT _____

Task Completed

15. Connect a flexible hose to the bleeder screw and route to a capture container. ☐

16. Loosen the bleeder screw and use a C-clamp to compress the caliper piston into its bore. Close the bleeder screw just before the piston bottoms out. ☐

17. Remove the hose and capture container. ☐

18. Install the inner pad into the caliper ensuring the pad mates and snaps into the piston. Install the antirattle springs. Install the outer pad. ☐

19. Install the rotor onto the hub. Install and torque the adapter. ☐

20. If necessary, install the slide pins into the rotor. ☐

21. Slide the caliper or adapter with pads installed over the rotor and align the fastener holes. ☐

22. Install and torque the caliper fasteners. ☐

23. Install the wheel assembly and torque the lug nuts. ☐

24. Repeat steps 2 through 22 for the opposite wheel. ☐

25. Lower the vehicle when both wheel assemblies have been installed. ☐

26. Press the brake pedal several times to seat the pads to the rotor. ☐

⚠ WARNING **Before adding brake fluid, consult the vehicle's service manual. Many manufacturers require a specific classification of brake fluid to be used.**

27. Check the brake fluid level and top off as needed. ☐

⚠ WARNING **Before moving the vehicle after a brake repair, pump the pedal several times to test the brake. Failure to do so may cause an accident with damage to vehicles and the facility, or it may cause personal injury.**

28. Perform a brake test to ensure that the brakes will stop and hold the vehicle. Do this test before moving the vehicle from the bay. ☐

29. Clean the area and complete the work order. ☐

Instructor's Response

_____

_____

_____

Name _____   Date _____

DIAGNOSING DRUM BRAKE HYDRAULIC PROBLEMS

Upon completion and review of this job sheet, you should be able to diagnose poor stopping, noise, pulling, grabbing, dragging, or pedal pulsation concerns.

ASE Education Foundation Correlation ——————————————————

This job sheet addresses the following **MLR** task:

5.C.4   Inspect wheel cylinders for leaks and proper operation; remove and replace as needed. **P-2**

This job sheet addresses the following **AST/MAST** tasks:

5.C.1   Diagnose poor stopping, noise, vibration, pulling, grabbing, dragging, or pedal pulsation concerns; determine needed action. **(P-1)**

5.C.5   Inspect wheel cylinders for leaks and proper operation; remove and replace as needed. **(P-2)**

We Support

ASE | Education Foundation

Tools and Materials

Goggles or safety glasses with side shields

Basic hand tools

Describe the vehicle being worked on:

Year _____ Make _____ Model _____

VIN _____ Engine type and size _____

Procedure   **Task Completed**

1. Determine the following information from the service manual.

    Lug nut torque _____

    Brake boost _____ Yes _____ No If yes, type?

    Brake fluid type and classification _____

    ABS Cautions _____

2. Begin the inspection of the drum brake hydraulic system by checking the fluid level in   ☐
    the master cylinder reservoir.

3. Check for leaks between the master cylinder and the power booster (bulkhead),   ☐
    between the reservoir and master cylinder, and the master cylinder outlet lines.

4. Lift the vehicle to gain access to the undercarriage.   ☐

5. Inspect for indication of fluid leakage down the inner side of each tire and wheel   ☐
    assembly.

6. Inspect for leakage in and around the wheel cylinder area of each wheel.   ☐

7. Inspect for leakage or damage on each line and hose with emphasis at each   ☐
    connection.

8. Road test the vehicle. As you apply the brake pedal, check for excessive travel and
    sponginess. What did you find?

    _____

9. If the vehicle pulls or grabs to one side, raise the vehicle and remove the tire and wheel assembly. ☐

10. Remove the drum and inspect the brakes. Is there grease, fluid, or oil on the linings? Is the wheel cylinder firmly mounted without leaks? Are the wheel cylinder pistons seized in place?

_____

_____

11. If the problem is not found, consult with your instructor. ☐

    INSTRUCTOR CHECK POINT _____

Instructor's Response

_____

_____

_____

Name _____   Date _____

REPLACE DRUM BRAKE SHOES

Upon completion and review of this job sheet, you should be able to remove, inspect, install, and adjust drum brakes.

ASE Education Foundation Correlation

This job sheet addresses the following **MLR** tasks:

5.A.1   Install wheel and torque lug nuts. **(P-1)**

5.C.1   Remove, clean, inspect, and measure brake drum diameter; determine serviceability. **(P-1)**

5.C.3   Remove, clean, and inspect, and/or replace brake shoes, springs, pins, clips, levers, adjusters/self-adjusters, other related brake hardware, and backing support plates; lubricate and reassemble. **(P-1)**

5.C.5   Preadjust brake shoes and parking brake; install brake drums or drum/hub assemblies and wheel bearings; make final checks and adjustments. **(P-1)**

This job sheet addresses the following **AST/MAST** tasks:

5.A.4   Install wheel and torque lug nuts. **(P-1)**

5.C.1   Diagnose poor stopping, noise, vibration, pulling, grabbing, dragging or pedal pulsation concerns; determine needed action. **(P-1)**

5.C.2   Remove, clean, inspect, and measure brake drum diameter; determine needed action. **(P-1)**

5.C.4   Remove, clean, inspect, and/or replace brake shoes, springs, pins, clips, levers, adjusters/self-adjusters, other related brake hardware, and backing support plates; lubricate and reassemble. **(P-1)**

5.C.6   Preadjust brake shoes and parking brake; install brake drums or drum/hub assemblies and wheel bearings; perform final checks and adjustments. **(P-1)**

We Support
ASE | Education Foundation

Tools and Materials

Goggles or safety glasses with side shields

Basic hand tools

Drum brake tools

Lift or jack and jack stands

Brake washer or vacuum cleaner

Brake lubricant

Describe the vehicle being worked on:

Year _____   Make _____   Model _____

VIN _____   Engine type and size _____

Procedure

1. Determine the following information from the service manual.

    Vehicle Make _____ Model _____ Year

    Is this a duo-servo or leading-trailing drum brake? _____

    Lug nut torque _____

Task Completed

Brake boost _____ Yes _____

No If yes, type? _____

Brake fluid type and classification _____

ABS Cautions _____

2. Begin the repair of the drum brake hydraulic system by checking the fluid level in the master cylinder reservoir. ☐

3. Lift the vehicle to gain access to the undercarriage. ☐

4. Inspect for indication of fluid leakage down the inner side of each wheel and wheel assembly. ☐

5. Remove the wheel assemblies and drums. Inspect the brakes. Is there grease, fluid, or oil on the linings? Are the wheel cylinders firmly mounted without leaks? Are the wheel cylinder pistons free to move?

_____

6. Use the proper cleaning procedures and equipment to clean the brake area including the backing plate. ☐

🛠️ **SERVICE TIP** Lay the brake components out in a manner similar to the way they are mounted on the backing plate.

7. Use the brake return spring tool to remove the shoes' return springs. ☐

8. Use the brake retaining spring tool to remove one shoe's retaining spring. It can be either shoe. ☐

9. On a leading-trailing system, remove the self-adjuster spring from the leading-trailing system. ☐

10. On a leading-trailing system, rotate one shoe away from the bottom anchor and disconnect the self-adjuster and parking brake strut at the top. ☐

    OR

11. On a duo-servo system, rotate one shoe away from the wheel cylinder link and disconnect the self-adjuster mechanism at the bottom and the parking brake strut at the top. Remove the second shoe. ☐

12. Disconnect the parking brake cable from the lever or the lever from the shoe. ☐

13. Thoroughly clean and lubricate the six shoe rubbing pads on the backing plate and the thread portion of the self-adjuster. ☐

    INSTRUCTOR CHECK POINT _____

14. Pull each wheel cylinder boot partially open and check for fluid inside the boot. Record your observations and recommendations.

_____

15. If necessary, install the parking brake cable to the lever or install the lever to the rear shoe or both. ☐

Task Completed

> ⚙⚙ **SERVICE TIP** It does not matter which shoe goes on first. This job sheet will list the steps for installing the rear shoe on the right rear wheel. However, a right-handed person may find it easier to install the front shoe first.

16. Fit the top of the rear shoe web into the wheel cylinder link and install the retaining spring. ☐

17. Install the return spring for the rear shoe. ☐

18. On a leading-trailing system, install the self-adjuster and parking strut onto the rear shoe. On a duo-servo system, install the parking brake strut onto the rear shoe. ☐

19. Align and fit the top of the front shoe web to the parking brake strut and self-adjuster, if necessary, and install the retaining spring, return spring, and self-adjuster spring (leading-trailing only). ☐

20. On a leading-trailing system, fit the bottom of the shoes into place on the anchor. ☐

21. On a duo-servo system, fit the self-adjuster spring into place on the shoes. ☐

22. If not already done, screw the self-adjuster to its minimum length. Place the front end of the self-adjuster into place on the bottom of the front shoe. Pull the bottom of the shoes apart and snap the rear end of the self-adjuster into place on the rear shoe (duo-servo only). ☐
☐

23. Inspect the assembly for proper installation with emphasis on the springs and self-adjuster. ☐

24. Measure the inside of the drum and preadjust the shoes to match that measurement. Record the drum measurement.

_____

25. Slide the drum over the brake assembly. If it does not slide on easily, adjust the shoes in until the drum slides on with some drag. If it slides on too easy, adjust the shoes out. The drum should turn with a slight drag. ☐

    INSTRUCTOR CHECK POINT _____

26. With the drum installed, apply the brakes to center the shoes. Use a brake adjusting tool to make the final adjustment. ☐

27. Install the wheel assembly and torque the lug nuts. Lower the vehicle and top off the brake fluid as needed. ☐

28. Test the brake system by applying the brake pedal several times without and with the engine operating. Record your observations.

_____

_____

_____

Instructor's Response

_____

_____

_____

⚠ **Caution**

Before applying the brakes, ensure the brakes on any other wheels are installed and the drum(s) are installed.

Name _____   Date _____

TEST AND REPLACE WHEEL SPEED SENSOR

Upon completion and review of this job sheet, you should be able to replace a wheel speed sensor for the ABS.

ASE Education Foundation Correlation

This job sheet addresses the following **AST/MAST** task:

5.G.1   Identify and inspect electronic brake control system components (ABS, TCS, ESC); determine needed action. **(P-1)**

This job sheet addresses the following **MAST** tasks:

5.G.4   Diagnose electronic brake control system electronic control(s) and components by retrieving diagnostic trouble codes, and/or using recommended test equipment; determine needed action. **(P-2)**

5.G.7   Test, diagnose, and service electronic brake control system speed sensors (digital and analog), toothed ring (tone wheel), and circuits using a graphing multimeter (GMM)/digital storage oscilloscope (DSO) (includes output signal, resistance, shorts to voltage/ground, and frequency data). **(P-2)**

We Support

ASE | Education Foundation

Tools and Materials

Goggles or safety glasses with side shields

Basic hand tools

Lift or jack and jack stands

GMM or DSO

Service manual

Describe the vehicle being worked on:

Year _____   Make _____   Model _____

VIN _____   Engine type and size _____

Procedure

Task Completed

1. Determine the following information from the service manual.

   What are the voltage or resistance specifications for the speed sensor?

   _____

   General replacement procedures including fastener torque:

   _____

ABS Cautions

2. Use the scan tool to retrieve and record any DTCs.   ☐

3. Lift and support the vehicle to gain access to the sensor electrical connection. Unplug the sensor connection. What color are the conductors?

   _____

Analog Meter

4. Select the AC volt position with the rotary switch.   ☐

5. Connect the meter leads to the terminals in the sensor plug. ☐

6. Rotate the wheel while observing the action of the meter's needle. Record the needle action or inaction and make your suggestions on the sensor's serviceability.

_____

NOTE: If there was no action, double-check with the meter set in the DC volt position.

Digital Multimeter

7. Select the resistance or ohm position on the meter with the rotary switch. ☐

8. Connect the meter's leads to the terminal in the sensor's plug. Record the resistance reading. Make your suggestion on the sensor's serviceability.

_____

_____

9. Follow the service manual's instruction to gain access to the sensor and its fastener. ☐

10. Clean the area around the sensor and then remove the fastener. ☐

11. Remove the sensor. ☐

12. Install the new sensor following the service manual instructions. Install and torque the fastener. ☐

13. Reconnect the sensor harness connection and any components removed for access. ☐

14. Lower the vehicle, and, if not already done, clear any DTCs from the electronic control brake module (ECBM). ☐

15. Start the engine. With the engine idling, pump the brake pedal several times. Did the ABS warning light stay off? If not, follow the scan tool's instructions and retrieve the DTC(s) and determine the necessary action(s).

_____

_____

INSTRUCTOR CHECK POINT _____

Instructor's Response

_____

_____

_____

Name _____   Date _____

CALCULATING THE PRESSURES AND FORCES WITHIN A HYDRAULIC CIRCUIT

Upon completion and review of this job sheet, you should be able to calculate the values of a hydraulic circuit and make suggestions on possible faults or repairs.

ASE Education Foundation Correlation ⎯⎯⎯⎯⎯⎯⎯⎯⎯⎯⎯⎯⎯⎯⎯⎯⎯⎯⎯⎯⎯⎯⎯⎯⎯

This job sheet addresses the following **AST/MAST** task:

5.B.1   Diagnose pressure concerns in the brake system using hydraulic principles (Pascal's law). **(P-1)**

We Support

ASE | Education Foundation

Tools and Materials

Calculator

Procedure

1. The following information is provided by the service manual and initial diagnosis on a nonworking circuit.

   Four-wheel disc brakes with a diagonal split system; master cylinder piston (2) area 0.5 square inch each; caliper piston (8) area 1.5 inch each.

   For the student's information: A diagonal split system has each master cylinder piston supplying equal fluid/pressure to each side of the split. One split includes the LF and RR, the other split includes the RF and LR.

   Assume you have done performance and inspection checks of the brake system and have determined that the system is not supplying the necessary forces to safety slow the vehicle. There are no oblivious leaks.

   What are some of the possible faults for this circuit? What would be your next step to locate the fault(s) if any?

   _____

   _____

2. The following information was obtained from making hydraulic measurements and observations.

   The brake pedal is applied with 50 ft.-lb. of force. A pressure gauge inserted into the brake system at each of the calipers indicate the pressures are different when the brakes are applied and held on.

   LF 20 psi; RF 25 psi; LR 25; RR 20

   Using Pascal law, F PA, calculate the theoretical pressures that should be present at each wheel. What are they?

   LF_____ RF_____ LR_____ RR_____

   $F = PA$, $P = F/A$, $A = F/P$

   $P = F$ (brake pedal input force) / A (area of the master cylinder piston) hydraulic pressure (psi) throughout system.

3. Comparing your calculation to the measured pressure, which component(s) is/are at fault in this system?

_____

_____

4. What is your suggested repair?

_____

_____

_____

_____

Instructor's Response

_____

_____

_____

Name_____   Date_____

REMOVING, REPACKING, AND INSTALLING WHEEL BEARINGS

Upon completion and review of this job sheet, you should be able to remove, repack, and install a wheel bearing and seal.

ASE Education Foundation Correlation

This job sheet addresses the following **MLR** task:

5.F.1   Remove, clean, inspect, repack, and install wheel bearings; replace seals; install hub and adjust bearings. **(P-1)**

This job sheet addresses the following **AST/MAST** task:

5.F.2   Remove, clean, inspect, repack, and install wheel bearings; replace seals; install hub and adjust bearings. **(P-2)**

We Support
ASE | Education Foundation

Tools and Materials

Goggles or safety glasses with side shields

Seal remover

Bearing packer

8-ounce ball peen hammer

Seal driver

Hub with inner bearing and seal installed

Blocking materials

Catch basin

> ⚠ **Caution**
>
> Use eye, face, and hand protection during parts cleaning. Injury or skin irritation may result if protective clothing is not worn.

> ⚠ **Caution**
>
> Do not allow the bearing to spin when using compressed air as a drying agent. Serious injury could result if the bearing comes apart.

Describe the vehicle being worked on:

Year _____ Make _____ Model _____

VIN _____ Engine type and size _____

Procedure

Task Completed

1. Place the hub with the outside facing upward. If necessary, use blocks so that clearance is available between the hub and bench.   ☐

2. Use the seal remover to force the seal from its cavity.   ☐

3. Remove the inner bearing.   ☐

4. Wash the bearing and hub cavity using a parts washer.   ☐

5. Use reduced-pressure-compressed air to dry the bearing. Use a catch basin to collect the cleaner and waste.   ☐

6. Place the bearing, small end down, into the bearing packer.

7. Force the grease into the bearing until clean ribbons of grease are visible at the top of the rollers.   ☐

8. Smear a light coat of grease on the outer sides of the rollers.   ☐

9. Install the bearing, small end first, into the cleaned and dried hub cavity.   ☐

10. Place the grease seal, lip first, over the cavity.   ☐

Task Completed

11. Hold the seal in place with the seal driver while using the hammer to drive the seal in place. ☐

   INSTRUCTOR CHECK POINT _____

12. Drive the seal in until it is flush with the edge of the hub. ☐

13. Use some fingers to rotate the bearing within the hub. If it moves freely, the task is complete. ☐

Instructor's Response

_____

_____

_____

CHAPTER 11

HEATING VENTILATION AND AIR-CONDITIONING SYSTEMS

Upon completion and review of this chapter, you should understand and be able to:

- Understand the safety precautions when working with heating, ventilation, and air-conditioning (HVAC) systems.
- Conduct a performance test of the HVAC system.
- Conduct a performance test of dash-mounted control panel and mode controls.
- Inspect HVAC blower motor and controls.
- Inspect engine cooling system and service.
- Inspect and test coolant strength and condition.
- Identify refrigerant and oil type.
- Identify proper manifold gauge set and record pressure readings.
- Inspect air-conditioning system, identify components, leak test, and determine action.
- Inspect and service the cabin air filter.

Terms To Know

Automatic climate control
Blend doors
Blower motor
Cabin air filter
Coolant
Fluorescent dye tracer

Heater control valve
High voltage
HVAC
Manifold gauge set
Performance test
Resistor block

Refractometer
Refrigerant
Thermal Limiter
Thermostat

INTRODUCTION

The heating, ventilation, and air-conditioning systems, also known as **HVAC** for short, are responsible for maintaining the passenger cabin temperature and controlling the engine coolant temperature. The air-conditioning system, also known as an A/C system, has now become a standard equipment. The A/C system and the controls have been unchanged for many years, but due to environmental concerns, control systems and **refrigerant** have changed. Heating and A/C systems have become more complex. Because many of the HVAC controls on today's vehicles require skills gained from many hours of experience and training, this chapter will cover some of the common tests and repairs that could be performed by entry-level technicians.

HVAC: Short for heating, ventilation and air-conditioning system, which provides heat and cooling to the passenger cabin.

Some **refrigerants** are flammable, and exposure to a spark or flame can cause the vapor to ignite.

HEATING AND AIR-CONDITIONING SAFETY

Covered in Shop Manual Chapter 2 "Automotive Shop and Safety Procedures," when servicing the cooling and air-conditioning systems, the coolant and refrigerant can be extremely dangerous. Coolant is the liquid that transfers heat away from the engine. During combustion, the engine temperature can reach 2,500°F. Use extreme caution when opening the cooling system, as it can be under high pressure and extremely high temperature. Just like coolant, refrigerant can also be extremely dangerous. Refrigerant also transfers heat and can be at extreme temperatures and pressures. Always use proper personal protection equipment (PPE) when working with coolant and refrigerant. Before servicing any system, follow service manual procedures and safety guidelines. Here are some of the general safety precautions that should be always followed to help prevent personal injury.

Caution

Contact with refrigerant vapor can cause frostbite and permanent eye damage.

- Always wear safety glasses, face shields, or goggles. When working with coolant or refrigerant, face shields and goggles can provide more protection than just safety glasses alone.
- Do not wear loose clothing or jewelry when working around moving components like belts, pulleys, and cooling fan blades.
- Never open a cooling system when the engine is hot. Coolant temperature can be over 200°F and under extreme pressure. Opening the radiator cap before the engine cools can result in an eruption of coolant causing severe burns and permanent eye damage.
- Never open an A/C system unless the refrigerant has been recovered from the vehicle. Release of refrigerant into the atmosphere is illegal and can be dangerous if inhaled by humans.
- Never attempt to disconnect or service any high-voltage components or wiring. Only a trained technician can perform service on a high-voltage vehicle.

Caution

High-voltage components are identified by bright orange conduit and connectors.

CLIMATE CONTROL SYSTEMS

Normally, an entry-level technician will not perform full repairs on climate control systems, but he or she could perform basic troubleshooting procedures and possibly correct a minor problem. Some shops will not allow noncertified technicians to do any air-conditioning (A/C) work. Because this system relies on the transfer of heat and the movement of air, usually the malfunction may be traced to a switch or component. The repair may not require the opening of the A/C system, which must be done by an Environmental Protection Agency (EPA)-certified technician. Some common complaints of manual heating and A/C systems that do not require the opening of the A/C system and a full systems performance test will be discussed.

Classroom Manual
pages 372–374

A **performance test** should be performed before and after the repair to verify the fault was repaired.

HVAC Control Panel Performance Tests

A **performance test** involves operating the system through every mode. The following tests are set up using certain terms based on a specified control system. The system being tested may have controls that are labeled a little differently, but the test procedures are the same. Before starting the engine, ensure that the controls for the heater, A/C, and blower are set to OFF.

Caution

Use eye and face protection while recovering or recharging A/C refrigerant. Refrigerant can cause eye and flesh injuries due to freezing.

⚙ **SERVICE TIP** Constant or full-flow heater cores do not have the water control valve. The cores on these heaters receive coolant at all times when the engine is operating. If there is no heat available, the problem is probably in the blend doors, their control circuits, or the engine coolant system.

Figure 11-1 Set the temperature to cool (cold), blower to LOWEST speed, and the mode control to dash or instrument panel. Note international pictorial symbols. COOL will usually be colored blue.

Figure 11-2 Set the temperature to mid-range, mode to HI/LO, and blower to mid-range or high speed for this test. Note that the AC and defrost buttons are shown just to highlight the possible differences between vehicles.

Start the engine and set the heater-A/C mode to VENT, the temperature to COOL, and then switch the blower to LOW or 1 (**Figure 11-1**). Air of about the same temperature as the outside air should be felt coming from the vents in the dash. Check each vent for airflow. If one vent does not have airflow, the ducting to that vent is disconnected or missing. Switch the blower to the other speeds, one at a time, and feel for an increase in airflow at each higher speed. Failure to obtain airflow in any speed normally indicates there is a problem with the blower or blower switch. Loss of flow at one or more speeds, but not all, indicates a problem with the blower switch. Leave the blower on HIGH for the next tests.

Move the mode lever (knob) to FLOOR and feel for airflow at the floor (**Figure 11-2**). Move the lever (knob or button) to DEFROST and feel for airflow at the bottom of the windshield. If there is a mode setting for HI/LO or mixed, move the lever to that position. The airflow should come from the floor and the vent or from the floor and the defrost. Failure to receive airflow at one of the modes indicates that the controlling cables or lever (knob) is broken or the blend door is broken. Move the mode to VENT for the following test. Keep the blower on HIGH.

> **Special Tools**
> Service manual
> Belt tension gauge

> 🛠️ **SERVICE TIP** Most of the newest vehicles do not use cables or vacuum to operate the blend doors. Small electric stepper motors are used to move the doors and are controlled by a computer. Units of this type require an experienced technician with good electronic expertise to perform the diagnosis and repair.

Set the temperature lever (knob) to HOT. Within a few seconds, there should be warm or hot air (depending on engine temperature) from the vents. If not, then the heater control valve or control mechanism is damaged. Move the temperature lever (knob) back to COOL and leave the blower on HIGH.

Select the MAX position with the A/C controls (**Figure 11-3**). The sound of the blower should increase immediately, and within a few seconds, the A/C compressor should switch on, sometimes noticeable by a loud under-the-hood click. A few seconds

> **Special Tools**
> Belt tension gauge
> Service manual

Figure 11-3 Set the temperature to cool (midrange may be selected for checking proper hot/cold air mix), switch on the AC by using the AC button. Note the MAX AC setting on the blend door. This actually causes the incoming air to be drawn from within the passenger compartment instead of outside air.

later, cool air should come from the vents. Switch the mode to NORMAL. The only noticeable difference will be less noise from the blower. Switch to DEFROST mode. There should be cold air at the bottom of the windshield. Failure to obtain cold air after a few minutes indicates that the A/C system is not functioning properly.

The following are some common faults found with climate control systems. In most cases, the suggested cause can occur on manual or automatic temperature control (ATC) systems. However, the typical sensors, actuator, and program commands in an ATC control module can cause what appears to be a mechanical problem. Experienced technicians should perform the final diagnosis and test of ATC systems.

Special Tools

Jumper wire
Multimeter or
voltmeter

⚙ **SERVICE TIP** Although most vehicles now use a serpentine belt to drive most accessory pulleys, some still require a separate belt to drive the A/C compressor. Most serpentine belts do not become loose unless there is a failure with the automatic tensioner; however, they may break from old age, excessive stress, or abuse.

HVAC BLOWER MOTOR ASSEMBLY

Blower motor: A component within a vehicles HVAC system that blows heated or cooled air through vents into the passenger cabin.

The passenger cabin fan is often called a **blower motor** (**Figure 11-4**). It circulates the air through the HVAC ductwork and the passenger cabin. The blower motor is controlled by a switch on the climate control panel and often has multiple fan speeds. The fan speeds are controlled by using a number of resistors connected in series, and the selected switch speed determines the combination of resistors used to control the motor's speed. The blower motor assembly spins the fan blades, commonly known as a squirrel cage, and the fan blades circulate air into the HVAC case and out the ductwork and selected vents.

Blower
housing

Blower and
motor assembly

Motor

Blower

Figure 11-4 A typical bower motor.

NO AIR FROM VENTS WHEN VEHICLE IS STOPPED

The most common problem here is a failed blower motor or motor circuit. If the customer states that air comes from the vents while driving but not when the vehicle stops, the blower motor is inoperative for some reason. Airflow during driving results from air entering the intake just below and forward of the windshield. Any time the vehicle stops or slows, there is no intake. Perform the following test on the blower motor circuit.

⚠ WARNING **Do not use a test light on automatic climate control or electronic systems. The electronic circuits can be damaged using test lights. Consult the service manual for proper test procedures and equipment.**

⚡ **WARNING** Do not use a test light on blower motors that are on automatic climate control systems unless directed to do so by the service manual. The motor may not be considered electronic, but its controls will be, and most do not have a resistor block for motor speed control. Damage to the circuit can result if test procedures are violated.

A **resistor block** is a series resistor with multiple inputs and a single output. It is usually used to control speed on an electric motor.

As usual, check the fuse first. If it is good, locate the feed wire to the blower motor and check for voltage with a multimeter or voltmeter. With the meter connected between the feed and a good ground, turn the ignition switch to ON and place the blower control switch in each position (**Figure 11-5**). If there is voltage in one or more speeds but not on all, check the service manual to locate any resistors. Most vehicles have a **resistor block** between the switch and the motor. The resistor block controls the amount of voltage reaching the motor, thus adjusting its speed (**Figure 11-6**). At the resistor block

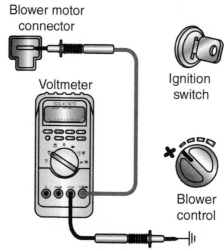

Figure 11-5 This motor has positive power even though the ignition key is on and the blower switch is off. The circuit uses a grounding switch that is typical of those found in electrical circuits.

Figure 11-6 Typical location of a resistor for manual heating and A/C systems.

Figure 11-7 Vacuum check valves are used to maintain system vacuum when the engine is under load or shut down.

connection, use a wiring diagram to determine which of the three or four wires are coming from the switch. Test for battery voltage in each switch position. Voltage on each or some of the conductors indicates that the resistor block is open. Voltage on none of them indicates there is a problem within the switch. The switch and block are not repairable and must be replaced individually.

No voltage at the resistor block output or at the motor indicates an open in the circuit. Some manufacturers use a **thermal limiter** in the resistor block. The limiter acts like a circuit breaker and opens because of excessive current or heat. The thermal limiter is usually part of the resistor block and the block must be replaced.

If there is voltage at the motor in all speeds, check the motor's ground before replacing it. Connect the meter between the feed wire and the ground wire on the motor. Move the blower switch to high and ignition to ON. A battery voltage reading indicates that the power and ground are good, and the motor must be replaced. No voltage at this point indicates the motor's ground is bad. Repair the ground wire or connection.

Air Comes from the Vent Only

The air is directed to selected vents by moving a lever to different modes: floor, panel, or defrost. The lever is attached to a vacuum switch or connected to the **blend doors** via cable. Before removing the climate control panel, consult the service manual to determine the type of controls. If vacuum operated, start your diagnosis by checking under the hood along the firewall. There should be one or two small vacuum hoses entering the passenger compartment. Locate the nearest vacuum connection in each hose found and check for vacuum with the engine running. It is best to use a vacuum gauge but placing a finger over the hose end on the engine side will be a quick check. Vacuum will be felt if it is present. If there is no vacuum, follow the hose to its connection with the engine's intake and then repair the problem. Observe the vacuum hose for any check valves that may be used (**Figure 11-7**). Many times, a loss of heater or air controls within the passenger compartment is a result of rotten or disconnected vacuum hoses or bad check valves.

Assuming vacuum is present on the hoses or the system is cable operated, it may be necessary to remove a portion of the dash or dash-mounted components to correct the problem. It is best that an experienced technician either performs the repairs or provides supervision.

A **thermal limiter** is a special circuit protection device that opens the circuit when overheated.

Automatic climate control systems use small electric motors to set any or all the various **blend doors** to almost any position needed to raise or lower the passenger compartment temperature.

 Special Tool
Vacuum gauge

 Caution
Do not remove the radiator cap from a warm or hot engine. Extreme burns could be caused by pressurized, hot coolant. Always allow the engine to cool before opening the cooling system.

THE COOLING SYSTEM

A basic cooling system includes the following components: engine, coolant, water pump, radiator, radiator fan, thermostat, heater core, heater hoses, and upper and lower radiator hoses.

All these components were discussed in Classroom Manual Chapter 7 "Engine Systems". For this chapter, we will focus on the coolant condition, thermostat operation, and heater control valve operation, which are important to the HVAC system. Testing the coolant condition and strength and keeping the coolant clean is an important part to a proper operating heating system. To inspect the coolant, first make sure that the cooling system is fully cooled down before removing the radiator cap or the pressure tank cap. Inspect the radiator cap and filler neck for signs of contamination or corrosion. If the coolant level and condition look acceptable, test the coolant strength with a **refractometer** or a coolant hydrometer. When using a refractometer, place a few drops of coolant on the sight glass or test window. Look through the eye piece and read the line between the dark and the light shaded area. Where the dark and light lines cross the scale shows the protection level of the coolant (**Figure 11-8**). With many types of coolant, you can use a traditional coolant protection level tester. Testing the thermostat and the radiator cooling fans are also an important part of a cooling system check to make sure that it is operating properly. Thermostat and cooling fans are covered in Shop Manual Chapter 7 "Engine System Inspection and Maintenance."

Refractometer: A tool used to measure fluid concentration or levels of strength and or charge for a battery.

1. Place a few drops of the sample fluid on the measuring prism and close the cover

2. Hold up to a light and read the scale

Figure 11-8 Using a refractometer to check coolant protection level.

PHOTO SEQUENCE 37
Cooling System Testing

P37-1 Visually inspect the coolant and use a coolant analysis strip to test coolant.

P37-2 Dip the strip in the coolant and compare the colors on the strip to the indications on the side of the bottle to determine if the coolant needs replacement.

P37-3 With many coolants, you can use a traditional antifreeze protection level tester.

P37-4 On some newer long-life coolants, a refractometer provides a more accurate analysis of the coolant.

P37-5 Test antifreeze protection level whenever cooling system maintenance is performed.

P37-6 Use a cooling system pressure tester to test the system for leaks and to test the function of the radiator cap.

P37-7 Pressure the system to the pressure stated on the cap, and make sure the system holds pressure while you look for leaks.

P37-8 Test the radiator cap to be sure it holds pressure. A weak cap can cause boiling over and overheating. The cap should also release pressure above its rating.

P37-9 Use a pyrometer to be sure that the engine is running at normal operating temperature.

P37-10 You can check your gauge accuracy by comparing your readings on the engine.

When performing a pressure test on the cooling system, don't forget to check the heater core for leaks. The only failure of a heater core is leaking or being plugged. If the heater core is leaking, you may notice that the carpet on the passenger floor may be wet. Another sign of a leaking heater core is fogging of the windshield or excess moisture in the passenger cabin. If you see signs of a heater core leaking, it should be replaced by an experienced technician.

⚡ **Caution**

Perform the following test with a cold or warm engine. Some valves are difficult to see easily and may be difficult to reach. A hot engine presents a probability of being burned. It will not make any difference if the engine is not hot because the valve is mechanically operated, not heat controlled.

HEATER DOES NOT WORK, A/C IS GOOD

In this instance, always check the coolant level and engine temperature. The coolant level on most vehicles can be checked through the translucent overflow reservoir. An engine that is still cool or cold after a mile or two of driving indicates a stuck-open or missing thermostat. Low coolant level will cause poor flow through the heater core, whereas an open thermostat will not allow the engine to warm. Both can cause heater problems.

Some older vehicles have a water control valve or heater valve located in the engine compartment. The valve is placed in the hose that delivers hot coolant to the heater core. The valve will be mechanically operated by a cable or vacuum motor. To check a cable control, have a coworker move the TEMPERATURE lever from hot to cold and back to hot. The cable and valve lever should move accordingly. If not, the cable or the control lever must be replaced or repaired (**Figure 11-9**).

Figure 11-9 Common types of heater control valves.

A vacuum-controlled valve is checked with the engine operating. Again, have the coworker move the TEMPERATURE control from hot to cold and back to hot. The valve lever and the arm of the vacuum motor should move accordingly. If they do not, the problem may be in the vacuum supply, temperature control, or the vacuum motor on the valve.

Disconnect the vacuum hose from the vacuum motor and place a finger over it or connect a vacuum gauge. With the engine operating, there should be vacuum in one direction and none in the other (e.g., 18 inches of vacuum in hot, zero in cold). If no vacuum is present in either position, use the procedures to check the vacuum supply outlined in the previous section on vent diagnosis. The entire heater control unit or head is usually replaced as a unit.

AIR-CONDITIONING SYSTEM

The basic air-conditioning system components are the following: refrigerant, compressor, condenser, evaporator, receiver/dryer or an accumulator, and an orifice tube or an expansion valve. The Classroom Manual Chapter 11 "Heating Ventilation and Air Conditioning" explains the components and operation. As stated before, HVAC repairs are typically performed by a more experienced technician. Understanding the A/C system can help with basic performance testing and leak testing. Performance testing is a term used to describe the basic air-conditioning process. You will need to look at all the components and compare how it is performing versus how it should perform. When testing the A/C system, make sure have the controls on the highest or maximum settings. An example would be: blower fan speed on the highest speed, the temperature setting on the coldest setting, and the air should be set to bi-level or dash vents. It is a good idea to have an external fan (**Figure 11-10**) to supply air across the condenser to simulate normal driving conditions. With all these conditions set, testing the engine idle should be held at 1200 rpms to help the compressor circulate the refrigerant through the A/C system. Once the engine stabilizes for a few minutes, the refrigerant should equalize and the temperature should reach the proper level. To test the proper temperature level, install a thermometer in the center dash vent and read and record the outlet temperature (**Figure 11-11**).

Figure 11-10 An external fan placed in front of the vehicle while performance testing.

Figure 11-11 Thermometer in dash vent to test outlet temperature. Heater works, but no A/C.

⚡ **WARNING** Do not open the A/C system before recovering the refrigerant using EPA equipment. Releasing refrigerant to the atmosphere could result in heavy fines, injury, or damage to the vehicle. Only EPA-certified technicians are allowed to recover refrigerant.

Most repairs on the A/C system are best left to EPA-certified and experienced technicians. However, there are a few checks that can be made quickly and safely by a new technician under the supervision of a certified person. Performing the following performance tests will give the technician a better idea of what is happening within the system.

Mechanical Diagnosing

Loose or worn engine drive belts tend to slip and produce a squeal when placed under load. The A/C compressor is one of the heaviest belt-driven loads. The A/C drive belt may drive other components such as the alternator, fan, and water pump. A seized compressor will cause the belt to slip or shed apart.

Check the tension of the belt by pressing down on it about halfway across its longest span or by using a belt tensioner tool. The belt should not deflect more than 1/2 inch on most systems. An oily or worn belt may slip even if the tension is correct. The belt must be replaced if it is oily or worn. The belt is replaced following similar procedures used on other drive belts. The most common problem with replacing A/C compressor drive belts is locating and accessing the fasteners. There may be up to four or more fasteners.

Loosen the pivot adjustment fasteners. Before forcing the component to move, check for a square 1/2-inch hole somewhere at the front. The hole is used for inserting the drive end of 1/2-inch-drive breaker bar (**Figure 11-12**). If there is no hole, then the component must be forced to move with a small pry bar because the belt tension cannot be set using hand and arm power alone. Use extreme caution in the placement of the bar. One end of the bar must be placed on the engine block without catching any wires or hoses. The portion of the bar touching the compressor must be placed against a solid metal portion. Ensure that the bar is not placed against the wires, hoses, or lines. In most cases when a pry bar must be used, the best place is beside the mounting bracket. This is the strongest part of the component outside the housing.

Force the component toward the engine and remove the belt. Install the new belt, ensuring that it is properly seated in each pulley. Apply outer force against the component until the correct belt tension is achieved. Hold the component in place while tightening

Special Tools

Small pry bar
1/2-inch breaker bar
Belt tension gauge

Special Tool

Service manual

Pulley

Wrench fittings

Pivot bolt

Adjusting slot

Lock bolt

Figure 11-12 A 1/2-inch-square breaker bar is used to make belt tension adjustments in this unit.

the adjustment or the easiest fastener to access. Remove the bar and tighten all other fasteners. Sometimes, and whenever possible, it is necessary to tighten the adjustment and pivot fasteners before removing the bar to prevent the compressor from twisting and loosening the belt. Measure the belt tension after all fasteners have been tightened.

MANIFOLD GAUGE SET

A service technician may need to connect a manifold gauge (**Figure 11-13**) set to determine if the A/C system has a full charge of refrigerant. Modern vehicles use high- and low-pressure sensors, which can disable a compressor if refrigerant pressure is too high

Special Tools

Safety glasses
Fender cover
Basic tool set
Manifold gauge set
  with hoses
Thermometer

Figure 11-13 A typical manifold gauge set.

or too low. Using a manifold gauge set can help diagnose the state of charge of the A/C system. A technician may need different manifold gauge sets depending on the type of refrigerant used in the vehicle. Remember the refrigerants are not compatible and different gauges sets must be used.

PHOTO SEQUENCE 38

Typical Procedure for Connecting a Manifold and Gauge Set to an R-134A or R1234YF Air-Conditioning System

P38-1 Locate the low side service valve and remove the cap.

P38-2 Locate the high side service valve and remove the cap.

P38-3 Position the gauge set so the hand valves can be easily reached, and the gauges can be easily viewed while performing the test.

P38-4 Close the low-side hand valve.

P38-5 Close the high-side hand valve.

P38-6 Connect the low-side service hose (blue) and the high-side service hose (red) to the vehicle service ports.

P38-7 The A/C system is ready to be tested.

Figure 11-14 Electronic leak detectors.

REFRIGERANT LEAK TESTING

Before connecting any equipment to the A/C system you should always perform a visual inspection. Inspect the hoses, connections, fittings, and the major components for signs of oil leakage and damage. Some service shops use a refrigerant oil with **fluorescent dye tracer** mixed into the system. It is generally believed that, if you see oil, that means the system may have a refrigerant leak. The dye tracer is generally yellow of green in color under an ultraviolet (UV) light. Another method of leak detection is using electronic leak detector (**Figure 11-14**). When using a leak detector, make sure the detector has been calibrated and is working properly. Read the instructions before using leak detection equipment. The detector is very sensitive, and if the tip gets dirty, it can give false readings and will also need to be replaced.

Fluorescent dye tracer: A method of tracing fluids by adding a bright colored dye which may show the component causing the leaking fluid.

PHOTO SEQUENCE 39
Typical Procedure for Checking for Leaks

P39-1 Power on the leak detector and zero out the sensitivity knob. Allow the detector to warm up for about 5 minutes.

P39-2 Place the probe at the reference test port. Adjust the control knob until the detector alarm sounds.

P39-3 Remove the probe from the test port and the alarm should stop.

P39-4 If the alarm does not stop the control knob is set to high. Repeat step 3 until the alarm stop when you remove the probe.

P39-5 Slowly move the test probe under and around the fittings and connections.

P39-6 Check all components and all screw in control devices.

P39-7 Check the vehicle service ports and remove the caps to test Schrader valves.

P39-8 Finally check the evaporator at the dash vent closest to the evaporator core. You can also remove the blower motor resistor to gain access.

CASE STUDY

This is a problem that was national in scope, but it only concerned one brand of vehicle and only for a model year or two. The customers complained that the vent air was not cool when the air conditioning was on MAX or NORMAL. The heater worked fine. This system used stepper motors on the blend doors. Routine electrical tests did not locate the problem. The doors could be heard moving back and forth as the controls were moved. Studying the service manual's system operation section revealed a possible cause. When the ignition key was moved to ON, the A/C computer ran a test on its circuits. Part of this test was the search for a 35-mA load on each of the motors as they move the doors through their entire range of movement. A DMM set to amps and connected between the heat blend door motor and the computer showed 35 mA at full cold and at full hot. Supposedly the system was functioning correctly. Further physical investigation found the problem on top of the air box. The arm from the motor was held in place to the door lever with a small plastic clip. The clip had broken and was hanging down. As the motor moved the door toward the cold position, the clip would hook a ridge of the air box. The motor would try to push the door further, but the 35-mA load was reached, and the computer assumed this was the limit of door travel and stopped the motor. The result was the door never reached its cold position even though the ampere reading was correct. Replacing the clip with a better connection solved the problem.

Figure 11-15 Location of the engine compartment cabin air filter.

Glove compartment

Figure 11-16 Location of the passenger compartment cabin air filter.

PASSENGER CABIN AIR FILTER

Most manufacturers have added **cabin air filters** to help purify the air coming into the passenger cabin. The filter is designed to catch dirt, dust, and other contaminates so they won't enter the passenger cabin. Most are paper-type filters and can be accessed from the engine compartment (**Figure 11-15**) or under the passenger glove box area (**Figure 11-16**). Cabin air filters should be replaced once a year or every 12,000 miles.

> **Cabin air filters:** A small pleated paper cotton material used to trap outside air contaminants before entering the passenger cabin.

ASE-STYLE REVIEW QUESTIONS

1. Climate control system diagnosis is being discussed.

    Technician A says an open blower motor fuse could cause a compressor clutch malfunction.

    Technician B says an inoperative blower motor may be caused by a poor ground connection.

    Who is correct?

    A. A only   C. Both A and B

    B. B only   D. Neither A nor B

2. Climate control is being discussed.

    Technician A says a noncertified technician can perform preliminary testing.

    Technician B says an EPA-certified technician must perform any repairs requiring removal and replacing of the A/C refrigerant.

    Who is correct?

    A. A only   C. Both A and B

    B. B only   D. Neither A nor B

3. Blower motor fan is being discussed.

    Technician A says a cabin air filter could cause low airflow to the passenger cabin.

    Technician B says a bad blower motor resistor could be the cause of no airflow to the passenger cabin.

    Who is correct?

    A. A only   C. Both A and B

    B. B only   D. Neither A nor B

4. HVAC system performance test is being discussed.

    Technician A says a vehicle inspection is necessary before starting any testing.

    Technician B says rinsing off the radiator and condenser should be performed before any test is performed.

    Who is correct?

    A. A only   C. Both A and B

    B. B only   D. Neither A nor B

5. A blower motor circuit is being discussed.

Technician A says that a blower motor resistor block controls the blower motor speed.

Technician B says the blower motor should be tested for power and ground before replacing.

Who is correct?

A. A only

B. B only

C. Both A and B

D. Neither A nor B

6. An serpentine belt is being discussed.

Technician A says worn serpentine belt and cause a belt to make noise

Technician B says an oily belt can be cleaned and reused if in good condition.

Who is correct?

A. A only

B. B only

C. Both A and B

D. Neither A nor B

7. Heater performance is being discussed.

Technician A says low coolant level can cause a no heat condition in the passenger cabin.

Technician B says low refrigerant can cause a no heat condition.

Who is correct?

A. A only

B. B only

C. Both A and B

D. Neither A nor B

8. A passenger cabin air filter is being discussed.

Technician A says the cabin air filter removes dirt and dust from the air coming into the cabin.

Technician B says the filter is usually paper and can be located under the dash.

Who is correct?

A. A only

B. B only

C. Both A and B

D. Neither A nor B

9. The cooling system is being discussed.

Technician A says the best way to test the coolant protection level is to use a refractometer.

Technician B says the heater core cools the coolant for the engine.

Who is correct?

A. A only

B. B only

C. Both A and B

D. Neither A nor B

10. An air-conditioning leak is being discussed.

Technician A says an electronic leak detector is the only way to find a leak.

Technician B says you should always perform a visual inspection of the A/C system before performing any test.

Who is correct?

A. A only

B. B only

C. Both A and B

D. Neither A nor B

ASE CHALLENGE QUESTIONS

1. Two technicians are discussing testing coolant.

Technician A says that you should use a refractometer to test coolant strength.

Technician B says that you can use a hydrometer to test coolant strength.

Who is correct?

A. A only

B. B only

C. Both A and B

D. Neither A nor B

2. Air-conditioning system is being discussed.

Technician A says that a loose AC belt can cause poor AC performance.

Technician B says an AC system should be checked for outlet temperature during a performance test.

Who is correct?

A. A only

B. B only

C. Both A and B

D. Neither A nor B

3. Low air flow from the dash vent is being discussed.

   Technician A says that a cabin air filter may be dirty and causing low air flow.

   Technician B says low freon in the system is causing low air flow from the vents.

   Who is correct?

   A. A only

   B. B only

   C. Both A and B

   D. Neither A nor B

4. Heating and air-conditioning safety is being discussed.

   Technician A says coolant and freon are extremely dangerous.

   Technician B says opening a cooling system or an AC system should be performed by a trained technician.

   Who is correct?

   A. A only

   B. B only

   C. Both A and B

   D. Neither A nor B

5. A climate control panel is being discussed.

   Technician A says Technician A says you should check every position on the panel and verify proper operation.

   Technician B says the performance test should be performed while the engine is running and at full operating temperature.

   Who is correct?

   A. A only

   B. B only

   C. Both A and B

   D. Neither A nor B

Name _____   Date _____

IDENTIFY A/C SYSTEM COMPONENTS

Upon completion of this activity, you will be able to understand how to identify the types of air-conditioning systems, identify the basic components, and research its operation, service precautions, and service history.

ASE Education Foundation Correlation

This job sheet addresses the following **MLR** tasks:

7.D.1   Inspect A/C-heater ducts, doors, hoses, cabin filters, and outlets; perform necessary action. **(P-1)**

7.D.2   Identify the source of A/C system odors. **(P-2)**

This job sheet addresses the following **AST/MAST** tasks:

7.D.6   Inspect HVAC ducts, doors, hoses, cabin filters, and outlets; determine needed action. (P-1)

7.D.7   Identify the source of HVAC system odors. **(P-2)**

We Support
ASE | Education Foundation

Tools and Materials

Goggles or safety glasses with side shields

Fender cover

Service manual

Thermometer

Refrigerant manifold gauge set

Describe the vehicle being worked on:

Year _____ Make _____ Model _____

VIN _____ Engine type and size _____

Procedure

Task Completed

1. Using service manuals, search for technical service bulletins related to A/C system issues. Record any bulletins.

   _____

   _____

2. Start the vehicle and let idle. Install an exhaust hose on the vehicle.   ☐

3. Operate the vehicle's air-conditioning system and determine if the A/C system is working. Is the system working? Explain.

   _____

4. Turn the A/C system on to its maximum output setting. Describe how to set the HVAC controls to maximum A/C.

   _____

   _____

   _____

5. Using a thermometer, insert it into the middle dash air vent and read the outlet temperature of the air coming out of the vents. Record _____. Turn off vehicle.

6. Using service manuals, determine what the outlet temperature specification is for this vehicle. Record.

   _____

7. View all the labels under the hood of the vehicle and find which one addresses the air-conditioning system. Record the location of the label.

   _____

8. Using the labels (or, if missing, use service manuals), fill in the chart below:

Type of Refrigerant	_____
Type of Refrigerant Oil	_____
Refrigerant Capacity	_____
Type of System (Orifice Tube, TXV)	_____

9. Locate and inspect the service fittings/service ports. Are they quick-disconnect style or threaded-flare fitting?

   _____

10. Based on the fitting style, can you determine the type of refrigerant this vehicle should use?

    _____

11. Using the service manuals as a guide, describe the components the vehicle uses to control the flow of refrigerant onto the evaporator.

    _____

    _____

    _____

12. Inspect the heater ducts, HVAC doors, vacuum hoses, cabin air filters, and outlet vents. Determine necessary action.

    _____

13. Inspect climate control head for proper functioning. Test all modes for proper airflow. Test temperature controls for proper function.   ☐

14. Visually inspect the components; **be prepared to identify to your instructor and show location.**   ☐

    INSTRUCTOR CHECK POINT _____

15. Describe the safety precautions that you should always use when working with refrigerants and air-conditioning systems.

    _____

    _____

16. The _____ system works with both the heating and air-conditioning systems to route airflow for the passenger compartment.

17. Describe what cabin air filters are used for and where they can typically be found.

_____

_____

18. The _____ separates the high- and low-pressure sides of the A/C system.

19. An accumulator is used on the _____ pressure side of the A/C system.

20. List three components found in the HVAC case.

a. _____

b. _____

c. _____

21. List six components of the air-conditioning system.

a. _____

b. _____

c. _____

d. _____

e. _____

f. _____

Instructor's Response

_____

_____

_____

Name _____   Date _____

PERFORMANCE TESTING THE A/C SYSTEM

Upon completion of this activity, you will be able to perform an A/C performance test and diagnose A/C malfunctions using principles of refrigeration.

ASE Education Foundation Correlation ————————————————————

This job sheet addresses the following **MLR** tasks:

7.A.2   Identify heating, ventilation, and air-conditioning (HVAC) components and configuration. **(P-1)**

7.B.1   Inspect and replace A/C compressor drive belts, pulleys, and tensioners; visually inspect A/C components for signs of leaks; determine necessary action. **(P-1)**

7.B.3   Inspect A/C condenser for airflow restrictions; determine necessary action. **(P-1)**

This job sheet addresses the following **AST/MAST** tasks:

7.A.1   Identify and interpret heating and air conditioning problems; determine needed action. **(P-1)**

7.B.1   Inspect, remove, and/or replace A/C compressor drive belts, pulleys, and tensioners; visually inspect A/C components for signs of leaks; determine needed action. **(P-1)**

7.B.7   Inspect A/C condenser for airflow restrictions; determine needed action. **(P-1)**

We Support

ASE | Education Foundation

Tools and Materials

Goggles or safety glasses with side shields

Fender cover

Service manual

Thermometer

Refrigerant manifold gauge set

Describe the vehicle being worked on:

Year _____   Make _____   Model _____

VIN _____   Engine type and size _____

Task Completed

Procedure

1. Locate the refrigerant manifold pressure gauge set. Record the make and model.

    _____

2. Perform a visual inspection of the pressure gauge set. Record any damage. _____
    _____ Note
    condition of the hoses. _____

3. Research available service information, vehicle history, any service precautions, technician service bulletins, and recalls. Record your findings.

    _____

    _____

4. Locate the high- and the low-side service ports on the vehicle and remove the caps.   ☐
    Put caps you will need to reinstall after servicing in a safe place.

Task Completed

5. Connect the appropriate pressure gauge set to the service ports.   ☐

   INSTRUCTOR CHECK POINT _____

6. After the hoses have been secured to the ports, open the valves on the manifold gauge   ☐
   set.

7. What is the static pressure on the gauges? High-side _____ Low-side
   _____

8. What is the current ambient air temperature? _____

9. Based on the service manual refrigerant chart, is the system undercharged,
   overcharged, or normal?

   _____

10. Insert a thermometer or temperature probe in the center vent of the dash.   ☐

11. Before starting the engine, inspect the service drive belt(s), pulley, and tensioners.
    Determine necessary action. Record.

    _____

    _____

12. Inspect the A/C condenser for proper airflow and that it is free from dirt and debris.   ☐
    Determine necessary action.

13. Start the engine, increase the engine speed to 1500 rpm, and turn on the A/C to   ☐
    maximum setting, blower on high.

14. With the compressor running, watch the gauge set. WARNING: If the high-side   ☐
    pressure gauge reaches 300 psi or higher, stop the test and see your instructor.

15. What happened to the high- and low-side gauge readings? Record findings.

    High-side gauge _____

    Low-side gauge _____

16. After letting the engine run at 1500 rpms, check the outlet temperature of the air
    coming out of the center of the dash. What is the lowest temperature reached?
    _____

17. If the system you are testing is using a cycling clutch, record the pressures the
    compressor started and stopped at:

    High-side compressor STARTS _____ **PSI**

    High-side compressor STOPS _____ **PSI**

    Low-side compressor STARTS _____ **PSI**

    Low-side compressor STOPS _____ **PSI**

    INSTRUCTOR CHECK POINT _____

18. Using service manuals as a guide, are the readings you took within specifications for
    performance testing?

    _____

19. The component used to remove heat from the refrigerant is the _____.

20. To reduce the pressure of the refrigerant entering the evaporator, either an _____ tube or _____ valve is used.

21. During A/C operation, _____ forms on the evaporator, which can cause mildew and odor from the HVAC system.

Instructor's Response

_____

_____

_____

Name _____   Date _____

COOLING SYSTEM INSPECTION

Upon completion of this activity, you will be able to understand how to check the cooling system for problems, and perform tests related to the cooling system's performance and the system's ability to seal. You will also determine what action is needed to keep the cooling system operating properly.

ASE Education Foundation Correlation ——————————————————————

This job sheet addresses the following **MLR** task:

7.C.1   Inspect engine cooling and heater systems hoses and pipes; perform necessary action. **(P-1)**

This job sheet addresses the following **AST/MAST** task:

7.C.1   Inspect engine cooling and heater systems hoses; determine needed action. **(P-1)**

We Support
ASE | **Education Foundation**

Tools and Materials

Goggles or safety glasses with side shields

Fender cover

Service manual

Thermometer

DMM

Coolant pressure tester

Describe the vehicle being worked on:

Year _____   Make _____   Model _____

VIN _____   Engine type and size _____

Procedure

Task Completed

1. Pull the vehicle into the service hoist area. Do not let the engine heat up. Turn vehicle off as soon as possible.   ☐

2. While letting the vehicle cool down (if needed), check for any related Technical Service Bulletins on cooling systems for this vehicle. List bulletins.

   _____

   _____

   _____

3. If engine is cooled down, remove the radiator cap or coolant recovery reservoir cap.   ☐

⚡ **WARNING Serious injury can occur if the cap is removed while the vehicle is still hot!!**

4. Check the coolant condition, type, and level. Record your findings.

   _____

5. Using service manuals, determine the type of coolant required by the manufacturer.   ☐

6. Check the coolant level in the radiator and the reservoir. Is either one low? Explain.

_____

7. If possible, perform a visual inspection of the inside walls of the radiator. Does it appear to have any buildup? Explain.

_____

8. Using Google, research and explain what "electrolysis" is.

_____

_____

9. What is a normal electrolysis reading for all cooling system?

_____

10. Using a DMM, set the meter to DC voltage. Place the negative meter lead on the battery negative terminal, and place the positive meter lead in the coolant. **NOTE: Only touch the coolant with the positive lead.** Record your reading and explain if it is normal.

_____

11. Using a hydrometer, measure the specific gravity of the coolant. Record your reading.

_____

12. Using a refractometer, measure the percentage of water and glycol in the coolant. Record your readings.

_____

13. Using your results, explain if the coolant needs to be flushed, refilled, or drained and replaced.

_____

14. Visually inspect the engine and cooling system for leaks. Inspect the following:

Upper radiator hose: _____

Lower radiator hose: _____

Heater hoses: _____

By-pass hoses: _____

All hose connections/clamps: _____

Water pump: _____

Thermostat housing: _____

Intake manifold gaskets: _____

Cylinder head gaskets: _____

Engine core plugs: _____

Engine oil dipstick: _____

Transmission oil dipstick: _____

INSTRUCTOR CHECK POINT _____

15. Using a coolant pressure tester, pressurize the cap to the manufacturer's recommended system pressure. At what pressure did the cap release the pressure?

_____

16. Did the cap open/release properly at the right pressure? Record pressure.

_____

17. Using the coolant pressure tester, pressurize the cooling system at the radiator cap opening. Pressurize the system to the manufacturer's recommended system pressure. ☐

18. The system should hold pressure for at least 15 minutes. During that time, visually inspect all the components that you inspected earlier in step 13. Did you find any leaks? Explain.

_____

_____

19. Does the system hold pressure for 15 minutes? _____

20. If the system does not hold pressure for 15 minutes and no external leaks were found, what areas/problems would you focus on?

_____

_____

21. Before disconnecting the pressure tester, release all the pressure then apply only 5 psi to the system. Start the engine, and read the gauge. Did the reading go up immediately? YES or NO Turn off the engine! Record the reading._____

22. If the pressure went up, what could cause this type of problem?

_____

23. Remove the pressure tester, and slowly release the pressure from the system. Leave the cap off the radiator. Start the engine and let the engine idle. ☐

24. Use a 4 or 5 gas analyzer to detect if there is any presence of combustion gases coming from the radiator opening. Record your results.

_____

25. What condition could cause combustion gases to be present in the cooling system? Explain.

_____

_____

26. Remove all test equipment, and make sure coolant is at the proper level. Reinstall the radiator cap. ☐

27. Rinse off engine with water. ☐

28. Return vehicle to parking spot. ☐

29. Clean shop floor WITH SOAP AND WATER. ☐

30. Describe what is meant by a liquid-cooled engine.

_____

_____

_____

31. What does the term HVAC stand for? _____ _____

_____

32. What is the purpose of the **RADIATOR**? Explain.

_____

_____

33. Draw a picture of the radiator tubes and the fins. Show how the coolant flows through the radiator in the picture you draw.

34. Describe how the coolant is cooled down as it flows through the radiator.

_____

_____

_____

35. List 7 basic cooling system components.

1) _____

2) _____

3) _____

4) _____

5) _____

6) _____

7) _____

36. Describe the purpose for a clutch fan and how it performs its task.

_____

_____

Instructor's Response

_____

_____

_____

CHAPTER 12

SAFETY INSPECTIONS AND PREVENTIVE MAINTENANCE

Upon completion and review of this chapter, you should understand and be able to:

■ Inspect the top of the engine for leaks.

■ Examine components under the vehicle and inspect.

■ Inspect under the vehicle for fluid leaks.

■ Inspect the exhaust system for leaks.

■ Inspect vehicle lighting and wipers for operation.

■ Describe how to use the five senses aid in diagnosis.

■ Understand how understanding theories of operation can help a technician diagnose a fault.

■ Understand how the air/fuel mixture can affect thermodynamics and engine performance.

■ Identify the location of components from under the hood.

■ Inspect basic components from under the hood.

Terms To Know

Baking soda
Carburetor
Chassis ear
Compressor
Control arm
Corrosion
Drive axle

Electronic fuel injection (EFI)
Fuel rail
Mass airflow sensor (MAF)
Positive crankcase ventilation (PCV)
Radiator
Ratio

Seepage
Subframe
Tachometer
Throttle body
Transaxle
Transverse
Water pump

INTRODUCTION

One major step in diagnosing a problem is to look at the simple things first. Understanding the theories of operation will assist the technician in determining the fault. In addition, an inspection of the vehicle will give the technician insight into how the vehicle was operated and maintained. The two should give the technician an idea of what to expect as the diagnosis and repairs are being performed.

INSPECTION ROUTINE USING THE FIVE SENSES

In Chapter 1 you learned how customer service plays an important role. Some of your best tools to do the job right are not always found in the toolbox. In this section, you will learn how to use those customer service skills along with your senses as a human to diagnose a problem.

Five of the best tools available for technicians are the five senses. Taste, touch, vision, smell, and hearing can speed up the diagnosis and repair process. Granted, the sense of taste will not apply in most cases. But consider that sometimes the atmospheric environment in the area will cause a bad (or good) taste on the tongue, so don't ignore taste entirely. A solid inspection gives a good indication of the vehicle overall and some systems specifically. In addition to the diagnosis inspection, most shops require their technicians to perform a vehicle safety check. This check has two results in mind: notifying the customer of potential problems and increasing business. Remember, if a problem is found and presented to the customer, he or she has the right to refuse that service. But, on the other hand, most customers appreciate the "heads up," and the repair may be done that day or scheduled later.

> ⚙ **SERVICE TIP** One of the simplest rules to remember is: Keep it simple, stupid, or KISS. If a starter doesn't work, check the battery first. A bad (dead) battery or corrosion on the terminals will cause very poor starter operation. Many times, the simple things will cost time and labor because they were not checked before replacing parts.

> ⚙ **SERVICE TIP** A good trick to try when learning how to do a proper visual inspection is to put one hand in your pocket and a pocket flashlight in the other, this helps to focus your attention on the components you are inspecting. People tend to touch what they see. Touching distracts from what you are supposed to be doing: looking and observing. Touching comes later after you have made your observations. Instead, concentrate on the situation in the engine compartment. Many times, your eyes will tell you almost everything needed to make an initial diagnosis.

Performance or operation checks are those simple tasks to see if something is working. To determine if a radio will receive a broadcast, turn it on. If sound comes out in some form, then it probably works. In many cases, the same scenario can be applied to other systems. Coupling a good inspection with performance checks determines quickly if the system should be working and whether it works correctly. For the discussion in this section, these procedures will be performed with the vehicle in the shop. To performance check the engine, transmissions, and other "moving vehicle" systems, it would be necessary to conduct a road test.

Diagnostic Steps

There are seven general and acceptable steps to a proper diagnosis and repair:

1. Verify the customer's complaint.
2. Collect vehicle data including maintenance history.
3. Collect vehicle Technical Service Bulletins (TSBs), recalls, or other information pertaining to this repair.
4. Test the system, including use of scan tools and other test instruments.
5. Determine the cause of the repair.
6. Complete the repair (correction).
7. Test the repair and verify that you took care of the original customer complaint.

Each vehicle manufacturer outlines and suggests these steps in some form for the dealership technicians to follow. For this discussion, "step 1, Verify customer's complaint," is the one that most involves the inspection and performance check.

Sometimes there is a communication disconnect between the customer, the service writer, and the technician. A customer says the left headlight is out, but he or she was facing the vehicle from the outside and really the right headlight is out. It can be embarrassing to replace the wrong headlight because the customer's concern was not confirmed.

AUTHOR'S NOTE One of the most embarrassing things I have ever done was when I was a young technician. I was installing a new belt on a Jeep because the old one was worn out—a simple and regular maintenance item. It was a tight fit with not much room for my hands to work. My arm had to touch the water pump pulley when installing the new belt. The water pump pulley felt hot and warmer than any other water pump pulley I have ever felt before. A week later the customer's vehicle was towed in and the water pump had blown the seal, coolant sprayed everywhere, and the engine overheated. In my head I thought that if I would have just thought this through, I would have been able to prevent the vehicle breakdown by going further in my inspection and notifying the customer. That would have saved them from overheating the engine and towing it in.

For this discussion, let's assume the customer's concern is engine overheating. Upon receiving the repair order, the technician approaches the vehicle. This is the first part of a good inspection. Observe the vehicle exterior during "The Walk Around." Assuming it has not been in a crash, damage to the body, tires, and overall appearance may point to a vehicle that is poorly maintained. Before starting any diagnosis on a single complaint, you should first perform a complete vehicle inspection. It's a good idea to evaluate the complete vehicle before spending time on the complaint to give the customer the overall condition of his or her vehicle. Once the vehicle inspection has been completed, you can start the diagnosis of the overheating complaint. The first clue to solving the overheating problem: When was the coolant system last serviced, and when was the drive belt last replaced?

Under Hood Inspection

Raising the hood and taking a good look at the engine compartment also gives an idea of the vehicle's previous maintenance. If the engine is fairly clean and shows signs of recent repairs, such as new hoses or spark plug wires, then the operator at least tried to keep it running (**Figure 12-1**). The presence of new radiator hoses may also provide a key to the diagnosis. There could be a leak at one of the connections, possibly a thermostat was installed incorrectly, or the water pump was replaced or needs replacing. Because we have an overheating problem, check the drive belts for a nonelectric cooling fan and, above all else, check the level of coolant in the system. Many times, this is checked in the recovery reservoir (**Figure 12-2**). Even if the reservoir is at the proper level, check the level in the radiator.

Continue the under hood inspection by checking for leaks. If none are found and no other obvious signs of faults are found, a quick performance check can be done. If not already cool, allow the engine to cool to the point where the upper radiator hose can be touched by your bare hand without discomfort.

Figure 12-1 Performing a good visual inspection is vital. A good inspection can detect signs of recent maintenance and signs of needed repairs.

Figure 12-2 This is for engine coolant only. There may be several other reservoirs nearby. Do not confuse reservoirs when topping them off.

⚠ **Caution**

Hybrid-powered vehicles have a high-voltage system. The positive cable for this system is colored orange. If this is your first hybrid or a different make or model than the ones you are used to repairing, stop and get someone with experience in this hybrid's electric power safety. The voltage on this cable can seriously injure or kill you.

⚙ **SERVICE TIP** If the coolant in the reservoir is below or above the *add* or *full* marks, it could point to a problem with the radiator cap. With the engine cooled to at least ambient temperature, remove the radiator cap and observe the coolant for discoloration and level. The coolant should cover the tubing and be near the bottom of the filler neck. It should be either green or reddish (sometimes orange) in color. Clear, clean liquid points to the operator adding water and the antifreeze is diluted or missing. Sometimes plain water will cause certain engines under certain conditions to overheat to a small extent.

⚡ **WARNING** Wear a face shield when checking coolant flow on an operating engine. A blown cylinder head gasket could cause the coolant to be ejected through the cap opening. Injury to the eyes, face, or body could occur.

This step requires the technician to pay attention and not allow the engine to operate for more than a minute once the upper radiator hose begins to warm. Start the engine and allow it to idle with the radiator cap off. Touch the upper hose repeatedly until heat is felt. It will become warmer closer to the engine first and then heat up rapidly to the radiator connection. This indicates the thermostat has opened. Use caution while observing the flow of coolant across the internal radiator tubes. Most times the flow is easily seen. Look for air bubbles in the coolant. They could mean outside air is in the system or could reveal a cylinder head problem. Shut down the engine and allow it to cool enough to install the radiator cap.

A **compressor** is used to pressurize a vapor.

If the vehicle is equipped with an electric-operated radiator cooling fan, there is one test that may help in the diagnosis. With the engine operating, turn on the air conditioning. At least one cooling fan should switch on as the air-conditioning **compressor** engages. Although this doesn't completely eliminate the cooling fan system as the problem, it at least tells the technician that the fan itself is operating.

Based on the visual observations and heat felt in the upper radiator hoses, most of the following can be determined to be correct or not:

a. Belts are present, tight, and serviceable.
b. Coolant is at proper level and appears to be the correct mixture of water and antifreeze.
c. Electric cooling fan is on with air conditioning on.
d. Thermostat is functioning.
e. No leaks are observed.
f. No air or combustion gases are present in coolant.

This diagnosis inspection may point directly to the problem or eliminate some components as the possible fault. But before continuing with the repair, make a good visual inspection of the other components under the hood. Items like corroded or old batteries, power-steering or transaxle leaks, damaged cables, and add-on electrical or mechanical components may reveal other possible or future problems. If the inspection finds other things that the technician honestly believes may be a problem, tell the customer. As stated before, it is the customer's vehicle and he or she has the right to refuse or accept the technician's advice.

IDENTIFYING THE AREA OF CONCERN

There are many parts on the vehicle that wear out, break, fall off, make noise, get dirty, require adjustment, and need repair or replacement. It is your job as a technician to understand almost all of them and determine which area and component need repair.

To do this, you must be able to focus on one area of the vehicle. Often, components like the engine will have had problems for a period of time before you see the vehicle. The customer may not have noticed any problems at the time they started. When asking customers what is wrong with their vehicle, be sure that you understand them clearly. If language is an issue, get an assistant to help with communications. I always say "A PROBLEM WELL DEFINED IS HALF SOLVED."

Being able to communicate with a customer in technical terms proves to be the most challenging of all tasks. When trying to explain something to or communicating with a customer, you should avoid using technical terms and use clear, everyday terms. Avoid using very general terms, like "that thing," as these may make the customer feel very inferior or substandard. Some customers will understand the word "piston" and be able to relate it to an engine component, but they may not know what one looks like or what it does. You will have to be the judge of your actions when you are trying to understand what the customer needs.

When focusing on one area, try to ask as many questions as possible. Make sure to get a phone number where the customer can be reached quickly. Having a good working relationship helps here because you can feel comfortable talking with the customer. If you are still trying to establish the area to focus on with the customer, try test-driving the vehicle with the customer. Ask them to drive the vehicle and attempt to repeat the problem. Sometimes this may be difficult but remember to be patient and calm (**Figure 12-3**).

Once you have determined the area to focus on, try to zone out other areas that are not related. This is tricky, especially because some other areas may be related to the problem. An example of trying to zone out other nonrelated area is when the customer has an engine noise, but it only makes noise when you accelerate hard from a dead stop. If the customer has a lot of papers and things lying on the dash or center console, those papers may make a lot of annoying noises when accelerating hard. You definitely don't want to clean it up or move it for the customer, but a phone call or simply asking the customer to move them so that you can hear the engine better is polite.

 Caution

Never remove the cap on a hot radiator. Always allow the engine to cool before opening the radiator. The coolant temperature may be as high as 212°F (100°C) or higher. Serious burn injury could result.

Caution

Use care when touching or moving anything under the hood. The engine and its components may be hot, or there will be moving belts and fans if the engine is running. Injuries could result from heat or moving components touching or grabbing fingers or hands.

Figure 12-3 Sometimes a test-drive with the customer is necessary.

AUTHOR'S NOTE Just to prove the value of the test-drive with the customer, I once had a neighbor ask if I could help her out and fix a noise that was driving her crazy. She knew that I was a GM technician but didn't want to pay the dealership prices. I decided to help her out. She described the problem over the phone, and I was very familiar with the problem. She described the noise as a soft rattling noise at highway speeds on light acceleration while cruising at 60 mph. She asked if she could drop off the car at my house late that evening and she would leave the keys under the floor mat. The next day when I took the car for a road test and verified the complaint. I heard the noise, and it was a light pinging sound coming from the engine. I repaired the problem and verified my repair. The noise was gone. The neighbor came and paid the bill and picked up the car. The next day I got an angry phone call saying that nothing had changed the noise was still there. I asked if she could stop by, so I could go for a ride with her. We went for a ride and after a long time we finally heard the noise she was talking about. The noise sounded like it was coming from the middle of the dash right by the right wiper blade. I crawled around the car trying to locate that noise while she was driving. (Which is not safe! I do not recommend this for anyone to do!) I opened the glove box to find a box of Tic-Tacs rattling. I picked up the Tic-Tacs, and her noise disappeared. Even though I fixed a problem her car had, it was not the problem she wanted fixed. I could have saved myself and the neighbor time and money if I had just gone for a road test with her the first time.

A **chassis ear** machine uses several small microphones that can be placed under the hood and then routed to the passenger compartment where the driver can use a set of headphones.

Sometimes using a **chassis ear** machine (**Figure 12-4**) works well when trying to figure out where the noises are coming from. You can attach the ends of the chassis ear to components on the engine and then bring the wires into the passenger compartment and listen to them while driving. The chassis ear is designed for listening to suspension, steering, and chassis problems but can be used for engines as well. Be careful when using

Figure 12-4 A chassis ear machine can help you figure out where the noises are coming from.

this tool because you don't want to hear suspension or steering problems and assume they are engine problems. Make sure to fully read the directions before using the chassis ear.

Once you have focused on the problem area, ask the customer about the vehicle's past service history. If they have had other major problems like a head gasket leak, overheating, or oil pressure problems in the past, ask what repairs have been completed. It is good practice to ask when and where the previous repairs have been done. They may have the old receipts and repair orders from other shops. These are all clues to the puzzle. If things are not adding up, try asking the owner if the vehicle has been in an accident or has a salvage or flood title. Although this seems strange, many flood and salvage vehicles can have preexisting damage from a previous accident that was never fully repaired. The body and exterior may be in great shape, but the engine may have suffered damage and have never received the proper repairs.

UNDER THE VEHICLE INSPECTION

 WARNING Before lifting a vehicle, ensure that the vehicle's lift points are identified and the lift pads are properly placed. Failure to do so may result in damage to the vehicle and the creation of a work hazard. If necessary, review Chapter 3 of the Shop Manual, "Using Automotive Tool and Equipment," on using a lift.

Under-vehicle inspection and work can be done using a floor jack and jack stands. However, for this discussion, it is assumed that the vehicle will be raised on a double-post, above-ground lift.

Position the vehicle between the posts and align the lift's pads with the vehicle's lift points. Raise the vehicle slightly above the desired work height and then lower it until the lift's locks engage and support the vehicle's weight. Start at the front of the vehicle and work rearward observing for components and leaks not visible from the top of the engine. First, inspect the front of the engine for components not found or not readily visible from the top of the engine.

⚠ Caution

Never work under a vehicle that is not supported by jack stands or the locks of a lift. The vehicle could slip off a jack and cause severe injury or death.

Figure 12-5 The air-conditioning compressor is capable of producing up to 600 psi and draws horsepower from the engine during operation. Note the engine oil dipstick.

Probably the most visible major component at this point is the air-conditioning (A/C) compressor, which is a type of pump (**Figure 12-5**). A/C compressors come in many different sizes, but modern ones usually have a round body, which is slightly less in diameter than the length of the compressor. At the rear or on one side of the compressor will be two large aluminum lines. This is the only component that will have this type of line. One end of the compressor will have the pulley and drive belt. Just behind the compressor pulley is a magnetic locking device with a small electrical plug. Make sure the plug is connected securely to the compressor and there are no broken or damaged wires visible.

On some vehicles, it may be easier to see the power-steering pump from this viewpoint. This is not necessarily the position from which to remove the pump, but the technician may be able to find a hidden fastener that is invisible from the top. It may also be easier to see the bottom of the alternator from this view, but its removal will probably be from the top.

Other components visible or at least partially visible include those of the engine, transmission, steering, suspension, and brake systems. (Note the corrosion and grime in Figures 12-5 and 12-6. This is not unusual on older vehicles.) Each system is discussed in earlier chapters. Directly under the lower portion of the engine is the oil pan. The oil filter is usually mounted low on the engine near the top edge of the pan. The filter may be found on either side or near either end of the engine. A plug will be located at the lowest part of the oil pan for easy removal of the engine oil (Figure 12-6).

Check around the oil filter, oil pan, and the engine in general for evidence of oil leaks. Engine oil will be black. Because liquids will flow to the lowest point before dropping off, do not assume that oil on the pan means the leak originated there. The oil could be coming from the top of the engine. Sometimes it is necessary to completely clean the entire engine compartment, test-drive the vehicle, and then attempt to find the source of the leak.

Behind the engine on the opposite end from the drive belts is the **transaxle** (**Figure 12-7**). The transaxle allows the driver to select gears for movement of the vehicle. There is a pan under this unit as well. It is usually flat and parallel to the floor, but it may be angled because of the space available for the transaxle. Usually, there is no place to drain the oil. The pan must be removed for service. Most transaxles have the filter inside the pan. Inspect the transaxle at its pan and anywhere else that appears to be a connection between the two parts. Most of the external parts of a transaxle have some type of seal

A rear-wheel drive vehicle uses a transmission instead of the transaxle. They basically work in the same manner.

Figure 12-6 This is a drain plug for the engine's oil pan. Some transaxles have a similar plug. Ensure that the correct one is removed, and replacement oil is added to the correct component.

Figure 12-7 The transaxle is normally mounted to the rear of the engine and could be an area of leaks.

between its mating parts. Transaxle oil or fluid is usually red but may be very dark because of dirt and overheating. One common leak area is the connection between the engine and the transaxle. Fluid found there might be engine oil or transaxle fluid, and both could be dark in color and difficult to identify. However, burned transaxle fluid will have a very burned smell.

The engine and transaxle are mounted on a **subframe** that reaches from the passenger floorboard to the front bumper (**Figure 12-8**). It also extends from side to side under the vehicle. Extending from the subframe to an area near the bottom of each wheel assembly

Rear-wheel drive vehicles usually do not use a **subframe.**

Figure 12-8 The subframe supports all the power components in a FWD vehicle.

Figure 12-9 The lower control arm provides the bottom mounting for the suspension components. There is one for each front wheel and on some rear wheels.

Control arm Also known as an "A-arm" is a hinged suspension link between the chassis and the steering knuckle that carries the wheel.

Drive axles transaxle the engine's power from the transaxle to the drive wheels.

Classroom Manual pages 305–306

is a lower **control arm** (**Figure 12-9**). This provides a pivot and support point for the wheel.

Looking upward past the arm and into the fender area, the lower part of the upper suspension components is visible. Also, on both sides of this area is a long, slender rod extending from the center of the vehicle to an attachment near the wheel. These are the outer parts of the steering system (**Figure 12-10**). The braking devices for each wheel may also be visible. If this is a late-model Chrysler car, notice the fasteners that hold the inner panel for the left fender. The vehicle's battery is located behind that panel.

Extending from each side of the transaxle out to the wheels is a large, round bar with bellowed rubber boots at each end (**Figure 12-11**). These are the **drive axles**. They transfer power from the transaxle to the wheels.

Figure 12-10 Shown is the outer tie rod end of one side of the steering linkage. It is often used to align the front wheels.

Figure 12-11 A FWD drive axle. One critical item to check during an inspection is the condition of the rubber boots at each end of the axle. Some RWD vehicles have similar axles at the rear.

Usually, no fluid leaks are found in these areas unless the brake system is leaking, but grease may be slung around the inside of the fender and any components mounted there. This is a result of a torn rubber boot. The boot is designed to hold special-purpose grease. A tear in the boot, combined with the high-speed rotation of the drive axle, slings the grease out in a vertical, circular pattern.

Moving rearward from the engine, the technician will find the floor pan or floorboard for the passenger compartment. Along one or both sides of the vehicle are the brake and fuel lines and the electrical conductors. On some vehicles, the lines and conductors are installed inside a tubular section of the frame and are not visible from this view. There should be no leaks or **seepage** along any of the lines.

At the rear of the vehicle is the most obvious component: the fuel tank. The suspension for each wheel is also visible. There can be several different components visible depending

Seepage is seen as a damp spot with no visual of fluid drops. Seepage is considered to be normal. Many seals will not function correctly without some seepage.

Figure 12-12 The rear suspension allows the wheel to move in relation to the road surface and keeps the wheel attached and in line with the body.

Two lighting systems are considered emergency lighting. They are the brake lights and the hazard warning lights. They should operate in almost any condition if electrical power is available.

on the design of the suspension system. Almost everything visible between the body and the wheel is some part of either the suspension or the brake system (**Figure 12-12**). Each of the systems mentioned in this section is covered in detail in earlier chapters.

The inside of the wheel assemblies and the mounting components should be clean and free of any oil. Fluid or oil at this point usually indicates a leaking brake system. Naturally, there should be no leakage or seepage from the fuel tank or any of the lines and hoses under the vehicle. On the side of the vehicle where the fuel cap is located, check the area between the fuel tank and the fender or body for a large and small hose (**Figure 12-13**).

Figure 12-13 There should never be leaks around the fuel filler hose, cap, or air breather. The rubber should be flexible without signs of deterioration.

Common leak areas

Figure 12-14 There should be no exhaust leaks anywhere along the exhaust system. Leakage is commonly seen as soot.

The larger one is the fuel inlet hose. The second and smaller hose allows air to escape the tank during filling. Each is held in place with clamps. Like all fuel hoses, there should be no leaks or seepage.

Inspect the exhaust system that runs under the vehicle (**Figure 12-14**). There should be no soot at the connections. The larger parts of the exhaust are silencers called *mufflers* or *resonators,* and they may have small holes near one of the welded seams. Ideally, there should be no soot around the holes either, but they should be examined closely, and corrective action taken as needed.

A large unit in the exhaust system near the engine is part of the emission control system. It operates at a temperature near or exceeding 1,600°F. It can be uncomfortable working near this system, so it should be allowed to cool before repairs are made. There should be no kinks, holes, or cracks anywhere else in the exhaust system. If such defects exist, the sections in which they appear must be repaired or replaced.

INSPECTING BODY-MOUNTED COMPONENTS

The entry-level technician performs only basic repairs to body-mounted components. He or she can, however, inspect and test many body-mounted components.

Classroom Manual
pages 130–133

Lights

Inspecting the lighting system is a relatively straightforward process, but it should be done every time the vehicle is brought in for servicing or repairs. An inoperative lamp may not be apparent to the operator until it is too late. Repairs to the lighting system are covered in Chapter 6, "Accessory Lighting and Electrical System Inspection, Maintenance, and Testing."

Start your inspection with the lights that must be connected directly to the battery. In other words, the ignition switch does not need to be in the RUN position.

Turn on the parking lights and walk around the vehicle (**Figure 12-15**). All parking lights should be on. Switch on the headlights. Some imported vehicles require the ignition switch to be in the RUN position before the headlights will work.

Figure 12-15 A typical headlight switch. Note the dimmer control.

Depress button to turn on

Pull out collar to turn off

Figure 12-16 This switch is used to flash all four (or more) marker (parking) lights in emergencies.

Even though light may be visible on the wall in front of the vehicle, do not assume everything is correct and that the lights are in working order. Move to the front of the vehicle and make sure both headlamps are operating. Check each headlight in its high- and low-beam position and remember to check the high-beam (bright) indicator to be sure it is working in the dash.

Check the hazard warning lights next. The headlights and parking lights may be left on or off but ensure that the ignition switch is off. The hazard lights are on a circuit by themselves and must operate regardless of the position of any other switches. The hazard switch will be on the right side of the steering column near the ignition key slot, on the top center of the steering column just forward of the steering wheel, or on the instrument panel (**Figure 12-16**). It will be marked with a red marking in most cases.

Move the hazard switch to the ON position and observe the direction or turn signal indicators in the dash. Both should be flashing. The two front parking lights should be flashing, as should the brake lights in the rear. This system will work even if three lamps are burned out. The fourth light will still flash.

With the hazards working, apply the brakes. The front lights should still flash, but the rear lights should shine solidly without flashing. Releasing the brakes will allow the rear lights to flash again. Switch off the hazard lights and retest the brakes.

The ignition switch must be in the RUN position to test most turn or signal lights. Switch on the ignition and select the left turn signal. The turn signal indicator should blink. Select the right turn signal and observe the right indicator. With the right turn signal on, check the rear and front lights on the right side of the vehicle. Reselect the left turn signal and check the left side.

Leave the left turn signal on and apply the brakes. Depending on the design of the lighting circuits, one of several things should happen. If this is a typical two-lamp system, the rear left turn signal will flash, and the right brake light will glow solidly. On a three-lamp system, the brake and turn signal are on separate lamps, so there will be brake lights on both sides in the rear and a flashing light for the left turn. After checking the brake and left turn signal, do the same test for the right turn signal.

With the ignition still in the RUN position, operate the headlights and hazard lights again in the same manner noted earlier. Headlights that require the ignition switch to be in the RUN position can be checked now. One other headlight system that requires checking is the daytime running lights. In this system, the headlights are automatically switched on when the ignition switch is on. The lights will be dimmer than normal because of the low voltage and current supplied. If in doubt, turn on the ignition switch and observe the headlight illumination. Switch on the headlight switch and the illumination should brighten considerably.

Figure 12-17 Some internal lights are switched on or off by pressing one end of the light lens. Other types may use push-button or flip-type switches.

Check the lights in the dash while the ignition switch is on. There will be several red lights visible. The system at this time is operating in a test mode of the warning lights. The warning lights and gauges were covered in Chapter 6, "Accessory Lighting and Electrical System Inspection, Maintenance, and Testing."

There should be one or more interior lights that work when a door is opened. In addition, the interior lights usually have a switch that can be used for control. Observe the interior lights as each door is opened and closed. Remember to close the other doors as one is checked. Close all doors and locate the switch for the interior lights. Check the lights using this switch. Some interior lights will only work with the doors while others may have independent switches for each light (**Figure 12-17**). Check all of them.

Windshield Wipers

Windshield wipers tend to be left out of the maintenance process. A badly damaged wiper blade can scratch the windshield glass and will not sufficiently clear it of rain or debris. Repairs on the wiper system are covered in Chapter 6, "Accessory Lighting and Electrical System Inspection, Maintenance, and Testing."

⚡ WARNING **Do not allow the wiper arm to snap against the glass. The glass may chip or crack, resulting in a high cost of replacement.**

⚡ WARNING **Do not operate the windshield wipers on dry glass. Using the windshield wipers on dry glass could scratch the glass and/or damage the new wipers.**

Inspect the rubber insert portion of the wiper that lies on the windshield (**Figure 12-18**). The rubber should not be torn or ripped. If so, replace the insert before proceeding. With good inserts on each side, turn the ignition to RUN. The wiper control is on a lever on the left or right side of the steering column or on the dash. Most vehicles have interval wipers, meaning that a position on the switch allows the wiper to sweep,

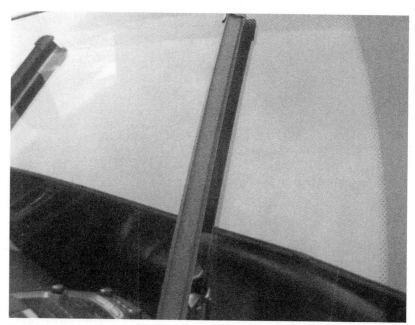

Figure 12-18 Only the rubber blade portion of this wiper should touch the glass. The metal arm could scratch the glass.

pause for a set amount of time, and then make another sweep. Most wiper systems have two speed positions in addition to the interval. The system includes a washer position on the wiper lever or a separate switch for the washer. Before operating the wipers, use the washer to wet the glass.

Operate the wiper in each switch position and ensure that the glass is wet by using the washer. In the interval position, adjust the time by operating the interval switch (**Figure 12-19**). Some vehicles, however, do not have the option of adjusting the time. Also check the rear windshield. Some vehicles have a wiper and washer for this glass, and it should be checked in the same manner as the front.

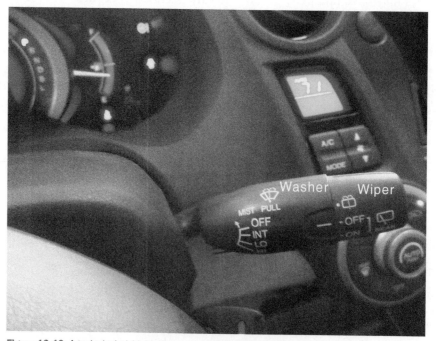

Figure 12-19 A typical windshield wiper and washer switch mounted on the turn signal (multifunction) switch arm.

Doors

While checking the interior lights with the door, observe the movement of the door. It should be smooth and quiet. A grinding or scraping noise indicates a lack of lubricant. The door should shut solidly and swing in an even arc through its entire range. A sagging door will not close quietly and may have to be slammed shut to latch properly. This usually indicates that the door hinges are worn, allowing the door to drop out of alignment. Repairs to the door should be performed by an experienced technician and in some cases by a body shop technician.

DIAGNOSING BY THEORY

When a system fails, it is because some part of it did not fulfill the requirements of a theory. Often, a lamp does not work because the electricity did not flow from the power source, through the lamp, and back to the power source. A lamp may be "burned out," but the electricity failed to return to the power source, hence, no light. The same diagnostic procedure can be applied to any system on the vehicle. Every technician will encounter a fault that fails to respond to typical diagnostic tests. By thinking about the theory or theories that make the system work, the technician can track through the system and find the point where the theory breaks. In this area the fault will be found. It may be caused by the lack of electricity, fuel, or air. It may be a broken part or any other item that is needed to make the system work.

Classroom Manual
pages 115–118

Engine Operations

In the Classroom Manual, we discussed the need for fuel, oxygen, and spark to be present in the combustion chamber to produce power. Two of those items, fuel and oxygen, require pressure and vacuum to be drawn into the chamber. The air and fuel must also be mixed in the correct **ratio**.

The laws of motion also apply to the movement of air and fuel. The engine is designed to make maximum use of physics to move, mix, and fire the fuel and transfer the produced power to the vehicle's driveline.

Understanding the theory of operation of vehicle systems will give the technician a head start on diagnosing. Let us discuss a possible situation. The technician does not have any high-tech diagnostic tools available but has a vehicle that will not start. As discussed, the engine must receive air and fuel in the correct mixture. The mixture is then fired with a spark. The easiest item to check is the airflow. Remove the air filter housing and try to start the vehicle. It is not often that an air intake is so plugged that no air can pass. But a dirty air filter could reduce the amount of air and cause a rich mixture or too much fuel for the air available. This could cause a no-start situation in the engine but more likely will allow it to start but run poorly.

A broken or missing vacuum line will cause the opposite condition or a lean mixture, which is too much air for the fuel available. Both conditions reduce the power produced and cause more harmful emissions. Most vacuum line connections are on the intake manifold or on the throttle body.

Let us consider a second situation concerning a poorly operating engine. The fact that it is running indicates that fuel, air, and spark are present and being used. Apparently, something is wrong with one of the three. Conduct a quick test to see if the problem is the same at idle and higher engine speeds. After allowing the engine to warm, speed up the engine until it reaches 1,200 to 1,500 rpm. If a meter (**tachometer**) is not available, hold the accelerator at about one-quarter throttle.

If the engine smoothes out and runs evenly, then the problem exists only at idle. The engine intakes little air and fuel at idle. The air valve, sometimes called a throttle valve, is

A **ratio** is the proportional amount of one element to another. An ideal air-to-fuel ratio is 14.7 parts of air to 1 part of fuel (14.7 to 1).

A **tachometer** is a meter used to measure the speed of a device in revolutions per minute (rpm).

Figure 12-20 This throttle valve or air valve is controlled mechanically by the driver using the accelerator pedal. Newer vehicles may be driven by an electrical motor based on accelerator pedal movement input to a computer.

A **carburetor** is an old air/fuel mixture device.

⚠ **Caution**

Never look into a carburetor when the engine is running. The fuel in a cylinder could ignite early and cause a flame to run back through the intake system and cause injury to the eyes and face.

When the first electronically controlled engines began to show up for repairs, many good sensors were replaced due to a lack of understanding about system and test equipment operation.

closed and air enters through an idle air control valve (**Figure 12-20**). Carbon buildup can partially block the valve and reduce the oxygen available to the engine. The air-to-fuel ratio is rich, and combustion cannot be supported. Because the valve has no effect when the engine is off idle, the engine will run smoothly at speed.

A second problem that affects idle speed is a vacuum (air) leak around the intake manifold. As mentioned, a vacuum leak will cause the air-to-fuel ratio to be lean, but unless the leak is very large, the engine will run smooth off idle. A fuel or spark problem will probably be present at idle and may increase at higher engine speeds. In this example, we have found that a small valve or a leak can upset the application of thermodynamic theory by changing the air-to-fuel ratio, so combustion is not completed.

The fuel delivery cannot be checked easily on a fuel injection system. It is best left to technicians with the proper test equipment. On a **carburetor** system, operating the throttle once or twice without the engine running will check the fuel delivery. Observe the inside of the carburetor while moving the throttle. There should be a squirt of fuel sprayed into the carburetor each time the throttle is moved. Once or twice is sufficient.

Remember that for the engine to work properly, the laws of physics must be followed. If one of the checks fails, then the faulty system has been found. From this point, the technician can inspect and test that system until the root cause is found.

This was a simple diagnostic test using the theory of operation. Similar diagnosing occurs every day, except we now rely on electronic devices to do the initial testing for us. An electronic sensor "stuck on lean" indicates that there is not enough fuel or that there is too much air in the mixture. Fuel injection relies on constant pressure and volume. If a sensor indicates the mixture is lean, then the fuel is insufficient, or the air volume may be too large. On most engines using a fuel injection system, the fuel delivery is the most common cause of a lean condition. The fuel may not be present in either pressure, volume, or both. A little common sense tells us that if there is no volume, then there can be no pressure, so in this instance we can assume there is insufficient fuel available. The most common cause of this problem is a partially clogged fuel filter. A $20 part could restore engine performance and ensure that thermodynamics is accomplished.

A point to remember as technicians advance in their careers is to avoid replacing a sensor until it is determined that the sensor is bad. In most cases, the sensor is doing exactly what it was designed to do. In the previous case, the oxygen sensor is telling the technician that the fuel flow is insufficient for the amount of air available.

Engine Noises and Smells

Much can be determined about an engine's operation based on its sound and the odor or color of the exhaust gases. If the engine sounds like metal is ripping apart during operation, it should be obvious that something is broken inside the engine. The only true way to determine what is wrong is to disassemble the engine. Most experienced technicians can give a fairly accurate diagnosis based on the sound(s) alone. Many internal mechanical failures produce sounds that point to the failed component. The only way to prevent further damage is engine shutdown as soon as the noise is first heard. Further operation will only increase the damage.

The smell of an engine, particularly when it is at operating temperature, can lead the technician to a possible diagnosis that can be confirmed with follow-up testing. The smell of burnt oil in and around the engine compartment usually means serious oil leaks. The smell of antifreeze may indicate a coolant leak. The smell of fuel in the same area requires immediate engine shut down and a nearby fire extinguisher. Leaking fuel on a hot surface can start a fire quickly. Another quick check can also narrow down some engine problems.

Observe the color of the exhaust gases at the rear of the vehicle. Blue smoke indicates bad piston rings, allowing engine oil into the combustion chamber. Black smoke usually indicates excessive fuel or a rich air/fuel mixture. This could be caused by several things but is most often caused by bad engine sensors or malfunctioning fuel injectors. With a diesel engine, black exhaust smoke is almost a sure sign that the air filter needs changing. Also, on a diesel-fueled engine, white smoke could be caused by a rich air/fuel mixture. Gray smoke that may be accompanied by drips of water from the tail pipe is a sign of cylinder head or head gasket failure.

Smell and feel senses can also be applied while observing the exhaust gases. A rotten egg smell is associated with a faulty catalytic converter, possibly caused by a long-term rich air/fuel mixture. If the eyes and nasal passage become irritated while behind an operating engine, it could be nitrogenous oxide or some version of smog in the exhaust gases. This could be caused by one of several subsystems, including the exhaust gas recirculating system, lean air/fuel mixture, or very high combustion chamber temperature. A clogged fuel filter will also contribute to this condition.

> ⚠ **Caution**
> Do not attempt to perform an inspection or repairs on hot engines. Serious burns may result, or injury could occur if a temperature-sensitive cooling fan switches on.

Classroom Manual
pages 397–398

COMPONENT LOCATION AND INSPECTION

The various components of the vehicle's systems are located in different areas of the body and frame. Belt-driven components are mounted on and driven by the engine. Steering, brake, and suspension systems have components mounted on the engine, body, and frame. It is also sometimes difficult to locate and view a component.

Location and Identification

With front-wheel-drive (FWD) and **transverse** engines there is not much room under the hood. Most engines have the alternator, power-steering pumps, **water pump**, and air conditioning compressor at thevhe engine where the drive belts are located (**Figure 12-21**). On an FWD vehicle, this may be located on the left or right side of the vehicle. When dealing with a vehicle's left or right side, always look at it from the vehicle's point of view. The left side is always the driver's side in most countries. One or all components may require the vehicle to be lifted, the engine mounts removed, or both for access and repair. In addition, the battery may be located within the wheel well or

> **Transverse** means the engine is mounted crossway to the length of the vehicle.

> Some **water pumps** are driven by the timing belt positioned behind a cover on the front of the engine.

Figure 12-21 FWD vehicles have the engine and transaxle mounted crossways (transverse) within the engine compartment. The front of this one is to the right of the vehicle (to left in figure); others may be reversed.

elsewhere in the vehicle and require the removal of other components for access and service. In most cases, the battery has remote terminals mounted in the engine compartment for jump starting and electrical tests (**Figure 12-22**).

Rear-wheel-drive vehicles usually have sufficient room under the hood and elsewhere to repair or replace components (**Figure 12-23**). In addition, the major components are

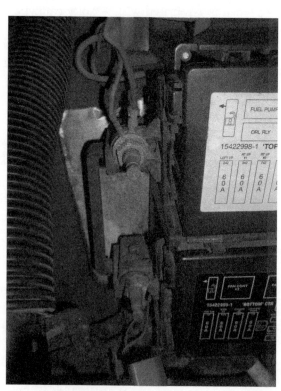

Figure 12-22 This battery is out of sight and sometimes out of mind. Remote terminals provide a means to recharge, jump off the battery, and make electrical measurements.

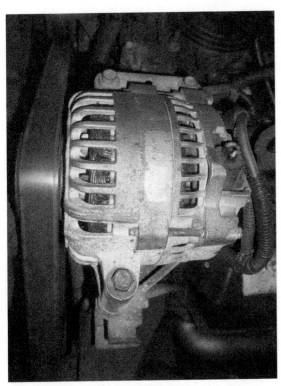

Figure 12-23 The alternator may be mounted at the top or to one side of the engine. Note the drive belt.

Figure 12-24 This alternator is mounted at the top of the engine. Note the drive belt and pulley.

easy to identify. For these reasons, we will discuss component location and inspection using a typical FWD vehicle. Left and right directions apply to any vehicle.

Before starting the inspection, study the engine compartment and locate the alternator (**Figure 12-24**). It is usually at the top of the engine but may be mounted low on the engine. It will usually have a single groove pulley with a fan behind it. The alternator is short, round, and will appear to be made of aluminum. It will have slots running across the rear cover next to the area where the wires are positioned (**Figure 12-25**). The end

Figure 12-25 Slots at the rear of the alternator allow the airflow produced by the alternator fan to exhaust.

opposite the pulley will have a two-wire electrical connection and usually a heavy red wire attached to another terminal. In most cases, the alternator will be the easiest belt-driven device to access.

While checking for the alternator, locate any other belt-driven components. The water pump may not be located on the outside of the engine, but there may be several pulleys, counting the crankshaft pulley at the bottom center of the engine. There may be up to eight pulleys depending on idler and tension pulleys used and the number of belt-driven components available.

The power-steering pump and reservoir are oddly shaped (**Figure 12-26**). The easiest way to identify these devices is to locate a drive pulley that has a small cap behind the metal bracket between the pulley and the pump. The cap is used for checking the fluid level. However, some vehicles have a remote-mounted reservoir with two hoses, one of which routes down to the pump. In either case, the filler cap will be marked in some way to indicate that this is the power-steering reservoir. Usually, the marking will indicate that only a certain type of power-steering fluid should be used in that system.

Many times, the power-steering pump is located to the lower left side of the engine in rear-wheel-drive vehicles and the upper right side of the engine in front-wheel drive. Again, remember that the directions are based on the vehicle or engine, *not* the position the technician is facing. The front of an engine is almost always the end with the drive belts, and the left and right of an engine may not be the left and right of the vehicle (**Figure 12-27**).

Basic Wiring Inspection

⚠ WARNING **Before performing any electrical repairs, disconnect the negative cable from the battery. Damage to electronic components or injury to the technician may result.**

Before we start discussing the electrical system, we need to quickly survey the electrical system of a hybrid vehicle. All of the systems normally associated with non-hybrid vehicles

⚠ **Caution**

Hybrid-powered vehicles have a high-voltage system. The positive cable for this system is colored orange. If this is your first hybrid, or a make or model that you have not worked on before, then stop and consult someone with experience in this hybrid's electric power safety. The voltage on this cable can seriously injure or kill you.

Figure 12-26 Layout for a power-steering pump.

Belt-driven power-steering pump

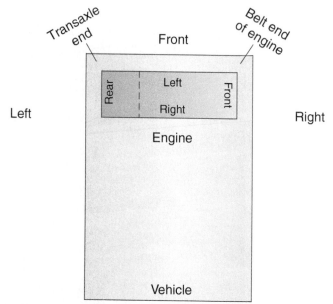

Figure 12-27 Directions are based on the front of each major component.

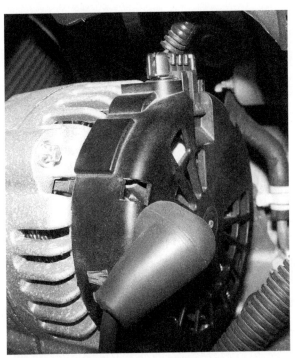

Figure 12-28 Alternator electrical connectors should be clean and fit tightly to each other.

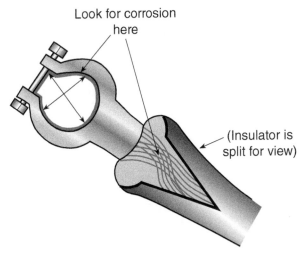

Look for corrosion here

(Insulator is split for view)

Figure 12-29 Once corrosion gets under the cable's insulation, the entire cable should be replaced.

still operate on 12 VDC except for the alternator, drive wheels, and sometimes the starter. The hybrid drive batteries provide high voltage that is converted or stepped down to 14 volts, whereas high voltage is used to operate the vehicle's drive motor(s). The drive batteries are charged by a large alternating-current generator (alternator) usually mounted between the engine and transaxle or transmission.

There are two things that quickly damage electrical wiring and batteries: lack of maintenance and **corrosion**. If preventive maintenance is performed, corrosion will not be a problem. But along with antifreeze changes, battery preventive maintenance does not appear to be of great concern for some drivers until the vehicle quits working.

Inspect the wiring at the back of the alternator for tightness and routing (**Figure 12-28**). The battery end of the cable is the most probable place for massive corrosion. This is partially due to the heavy current needed during cranking. Sometimes the corrosion can be removed, but if it extends under the cable's insulation it is best to replace the cable (**Figure 12-29**). The cable's terminals can also become a point of corrosion. Terminals can be replaced without changing the complete cable. There are temporary terminals that can be clamped to the bare cable strands, but they tend to increase resistance and lose voltage between the clamp and cable. The best method is to use the type of terminal that must be soldered to the cable (**Figure 12-30**).

Other starting and charging system inspections include checking the security of the battery and corrosion on the battery tray or box. The tray should be cleaned and the battery held in place by proper hold-downs. **Baking soda** and water work well to remove corrosionvolyte leakage. Use plenty of water to wash all of the soda from the vehicle. A final check includes the routing of all starting and charging system wires and harnesses. Ensure that there are no contacts between the wire and engine hot spots.

Corrosion can be caused by oxidation (oxygen acting on metal) or by acid or acid fumes.

Baking soda is a common household product.

Side

End

Fill open end
with heated solder
and insert cable

Figure 12-30 The permanent type of battery terminals are best suited for preventing corrosion.

CHECKING FOR LEAKS IN THE ENGINE COMPARTMENT

Classroom Manual
pages 5, 214–219

There are many places for leaks to occur under the hood. Among them are fuel clamps and hoses, air intake hoses, cover gaskets, and engine seals. The origin of some leaks may not be easy to locate, but they are evident based on the amount of liquid on the outside of engine and engine mounted components. Smell and color can identify leaking fluids and their possible source.

1. Pink or red—Power-steering fluid; automatic transmission fluid; some manual drivelines.
2. Green—Engine coolant, antifreeze; has distinct odor and very greasy to touch; newer fluids may be pinkish.
3. Black or very dark—Engine oil; manual drivelines; driveline fluid has distinct odor.
4. Clear—Brake fluid; some power-steering fluid; brake fluid is very slick compared to other fluids; fluid may look dirty.
5. Clear with distinct odor—Gasoline; will feel cold even in warm weather.
6. Dirty clear with distinct odor—Diesel fuel; oily to the touch.

The oil will be dirty and the color not readily apparent sometimes, but a little investigation by the technician can pinpoint the type and system. Automatic transmission fluid contains a lot of detergent, so if the leak is similarly colored and relatively clean, it may be transmission fluid. Engine oil tends to flow slowly and collect road dust and dirt, so if the leak looks like it has a lot of grease and dirt on it, it is likely an engine oil leak. Always remember that leaks will be conditioned by gravity and wind. A leak coming from a part of the valve cover where it is difficult to see may be first seen from under the vehicle and on the back side. Always trace the source of the leak.

Fuel System

A **throttle body** may mount one or two injectors and regulate the fuel pressure.

The fuel system ends at the intake manifold. It may be just one fuel line connected to the carburetor or multiple lines to a **throttle body**.

Figure 12-31 There is a fuel pressure line, pump to throttle body, and a fuel return line, throttle body to tank.

Electronic fuel injection (EFI) systems use a pressure line and a return line. On an EFI system, there are two fuel lines—pressure and return. Both lines are close to each other. One leaves the fuel tank and goes to the **fuel rail**, the other leaves the fuel rail and goes back into the fuel tank (**Figure 12-31**). Some EFI systems are returnless and only have a pressure fuel link. EFI systems using individual injectors and fuel rails may have both lines close to each other or the lines may be at opposite ends of the fuel rail (**Figure 12-32**). The checks are the same on either of the systems. Visually check each connection in the lines and each connection to the fuel rail, throttle body, or carburetor. Feel under each connection with a finger.

⚡ **WARNING** **Use a flashlight for illumination if a fuel leak is suspected. An electric light may create an electrical spark when it is switched on and off, thereby causing a fire.**

EFI A computer controlled fuel system. Combines sophisticated computer controls and high pressure fuel delivery system to provide optimum power and fuel efficiency.

Fuel rails provide a common channel to maintain constant fuel pressure and volume at each injector in a multi-injector system.

Figure 12-32 A fuel rail connects each fuel injector to the pressurized supplied fuel from the pump.

There must be no leaks or any type of seepage in any fuel system. In other words, if there is dampness at a fuel connection or anywhere on the line, find the problem before releasing the vehicle. This is especially true if fuel lines were disconnected during repair work. While checking the lines, also inspect the fuel rail connection to the injectors, the fuel injectors, and the carburetor or throttle body to the extent possible. Some of the fittings, hoses, lines, and components may not be visible. Older carburetors tend to seep at their various seals and gaskets. The carburetor should be rebuilt or replaced.

Air Intake

The air system is fairly easy to check. Air usually enters the intake system behind one of the headlights and travels through some flexible or rigid tubing to the air cleaner housing (**Figure 12-33**). Many times, there is a **mass airflow (MAF) sensor** located in the air tubing (**Figure 12-34**). Any air entering the engine that does not flow through the sensor will result in poor engine performance. The sensor measures the amount of air entering the engine and the powertrain control module (PCM) selects the amount of fuel based on that measurement. Inspect air ducts for cracks or improper mounting and fastening.

The **MAF** measures the amount of air entering the engine's intake system.

> **SERVICE TIP** Customers typically purchase the cheapest air filter available and later complain that the engine is operating poorly. This may be caused by the air filter. Some of the cheap air filters coat the filter fibers with oil. The oil works well in collecting dust and dirt entering the filter. However, this oil will also dissipate over time only to collect on the sensing grid of a MAF sensor. The oil coating now reduces the sensitivity of the grid and transmits incorrect airflow data to the PCM. If a customer complains of a poorly running engine, check the air filter condition and the sensing grid of the MAF. Should the feel or visual inspection of the grid indicate an oil film, then the condition may be resolved with a gentle cleaning of the grid and installing the correct air filter.

Figure 12-33 Cracked or missing air ducts can allow unfiltered air into the engine and may bypass the airflow sensor.

Figure 12-34 Mass airflow sensors measure the volume of air and sometimes the temperature and the density of the air entering the engine.

Figure 12-35 The clogged air filter on the left restricts the volume of air entering the engine. This is very critical on fuel-injected engines, both gasoline and diesel.

Vacuum leaks like the ones mentioned earlier also allow air to enter the system without going through the sensor. The end result is a lean mixture because of the extra air plus a wrong air volume measurement being sent to the PCM.

The air cleaner in the housing cleans the air before it enters the intake manifold (**Figure 12-35**). A dirty air filter restricts airflow and results in poor engine performance by causing a rich mixture. While checking the air filter, look for any oil or oil residue in the filter housing. The presence of oil there indicates a problem with the **positive crankcase ventilation (PCV) system**, which is covered in Chapter 7, "Engine Systems Inspection and Maintenance."

A further note concerns industrial and equipment engines. Many of the old ones are still running strong. They use what is called an "oil bath" air cleaner. There are no paper fibers here, only a half-quart of engine oil and a metal mesh similar to steel wool. The incoming air passes through the mesh drawing some oil into the mesh. The oil captures the dust and debris in the air. If working on this type of engine, replace the oil after cleaning the bath and wash out the steel mesh. The best method I have found is to use carburetor cleaner, rinse in the hottest running water available, and then blow dry with compressed air.

> The **PCV system** is used to vent the pressure created inside the engine by heat and pressure. Failure to vent the engine will cause oil leaks.

Oil

Oil leaks are noticeable by black, oily dirt around the engine covers and on the sides of the engine. Each cover on the engine has some type of seal or sealing material between it and its mating components. Leaks result when the seal is broken or damaged. The PCV system may also cause an engine oil leak. The system is supposed to vent the pressure in the oil sump of a running engine (**Figure 12-36**). If the system is inoperative, pressure will build, and oil will be forced through or around gaskets and seals or through the PCV valve.

1. Fresh air in from air cleaner

Filter

Throttle body

6. PCV valve controls the flow rate into the intake manifold

2. Flows through hose to valve cover

5. To PCV valve

3. And into crankcase

4. Fresh air and blowby gases

Figure 12-36 The PCV system vents the crankcase vapors into the engine for burning. Clogged PCV systems can lead to serious engine oil leakage.

A **radiator** is connected to a vehicle's engine by two hoses. Damage or rot to these hoses can cause coolant leaks.

⚠ **Caution**

The exhaust system gets hot very quickly after the engine is started. Extreme care must be taken when checking or working around the exhaust system. Serious burn injury could result.

Oil will drain to the lowest portion of a component before dropping on the ground. Leaks around seals on the front of the engine will cause the drive belts to be oily. The belts may also sling oil onto the underside of the hood and body panels in the area of the belts.

Coolant

Coolant leaks can be identified by the green or reddish color of the liquid and may occur because of loose clamps or water pump seals. The **radiator**, radiator cap, and hoses may also develop leaks. The radiator is mounted at the front of the vehicle and is connected to the engine by two large hoses, one at or near the top and one at the bottom (**Figure 12-37**). While checking for coolant leaks, inspect the two radiator hoses for dry rot or other damage. If a coolant leak is suspected but not visible, a coolant pressure tester can be used (**Figure 12-38**). The cooling system can be checked with a hand pump and adapter equipment. This system will be covered in Chapter 7, "Engine Systems Inspection and Maintenance."

FRONT

Upper radiator hose

Lower radiator hose

Figure 12-37 The two radiator hoses transfer engine coolant from the engine to the radiator and back.

Figure 12-38 This is a common coolant system pressure tester.

Exhaust manifold
gasket

Exhaust
manifold

Exhaust
doughnut seal

Figure 12-39 A common exhaust leak point is between the exhaust manifold and the cylinder head. It is not uncommon for the exhaust manifold gasket to fail as vehicles age.

Exhaust

Sound is the best and safest method to check the exhaust system for leaks. The sound of a leaking exhaust is different from almost any other sounds from the vehicle. If the sound is coming from the engine area, it usually means the leak is at the connection between the exhaust manifold and engine or between the manifold and exhaust pipe (**Figure 12-39**).

CASE STUDY

A customer complained of adding coolant every week or so. He stated that he could not find any coolant on the garage floor or inside the engine compartment. The technician performing the diagnosis noticed that the overflow reservoir was completely full, but the engine was only warm. Further testing with a pressure tester revealed no leaks, including the radiator cap. The technician located the end of the overflow reservoir drip hose. There was an indication of coolant in the end of the hose. It was decided that the cap was allowing coolant out of the system as designed but not drawing it back into the system as it was supposed to do. As a result, the overflow would fill to capacity as the car was running and the excess would drain into the air stream flowing under the car. Because the driver checked the coolant regularly, the vehicle never overheated. But the driver did not check under the vehicle immediately after stopping. A replacement radiator cap solved the problem.

ASE-STYLE REVIEW QUESTIONS

1. *Technician A* says pressure and vacuum can be directly influenced by a leak.
   Technician B says pressure and vacuum are used to move the air and fuel into the engine.
   Who is correct?
   A. A only
   B. B only
   C. Both A and B
   D. Neither A nor B

2. Theories of operation are being discussed.
   Technician A says Newton's Laws affect the movement of air and fuel.
   Technician B says the laws of motion can affect thermodynamics.
   Who is correct?
   A. A only
   B. B only
   C. Both A and B
   D. Neither A nor B

3. *Technician A* says a dirty air filter can affect thermodynamic theory.
   Technician B says thermodynamics will be affected by fuel flow through the fuel filter.
   Who is correct?

   A. A only
   B. B only
   C. Both A and B
   D. Neither A nor B

4. Leaks are being discussed.
   Technician A says a leak could cause problems within the system if hydraulic theory cannot be maintained.
   Technician B says leaks may cause problems because the proper application of pressure is not maintained.
   Who is correct?

   A. A only
   B. B only
   C. Both A and B
   D. Neither A nor B

5. The air/fuel mixture is being discussed.
   Technician A says a leak in a vacuum line could cause the PCM to set the amount of fuel injected incorrectly.
   Technician B says a crack in the air intake ducting will result in a rich mixture.
   Who is correct?

   A. A only
   B. B only
   C. Both A and B
   D. Neither A nor B

6. All of the following statements concerning leaks are false EXCEPT:

   A. Leaks will not cause a system pressure drop.
   B. Seepage is not considered a leak.
   C. Leaks will not cause damage to flexible components of the vehicle.
   D. Leaks are never internal.

7. Application of electrical theory will be affected by each of the following EXCEPT:

   A. incorrect conductors.
   B. damaged controls.
   C. overcharged battery.
   D. undercharged battery.

8. When inspecting a leak under the vehicle, all of the following steps are important to keep in mind, EXCEPT:

   A. Oil leaks will look very dirty.
   B. Automatic transmission leaks will look very clean.
   C. Gravity may cause you to think that the oil pan is leaking when in reality it may be the valve cover leaking.
   D. Coolant is often similar in color to engine oil.

9. A test-drive with the customer:

   A. allows a technician to always try driving different vehicles.
   B. can be very useful when diagnosing a problem.
   C. should always be short.
   D. is not important.

10. *Technician A* says fluid seepage on a rear wheel is acceptable.
    Technician B says power-steering fluid may be the same color as transaxle fluid.
    Who is correct?

    A. A only
    B. B only
    C. Both A and B
    D. Neither A nor B

ASE CHALLENGE QUESTIONS

1. *Technician A* says that a good visual inspection using your five senses is the best way to understand the overall condition of the vehicle.
   Technician B says when you give your customer a good inspection they usually appreciate the heads up.
   Who is correct?

   A. A only
   B. B only
   C. Both A and B
   D. Neither A nor B

2. Testing a vehicle is being discussed.
   Technician A says the first step in diagnosing a complaint is the test the system.
   Technician B says the first step is to search for recalls and bulletins.
   Who is correct?

   A. A only
   B. B only
   C. Both A and B
   D. Neither A nor B

3. *Technician A* says when starting a new repair on a vehicle you should perform a "walk around" to identify any body damage.

   Technician B says you should always verify the customers complaint before an service is performed.

   Who is correct?

   A. A only

   B. B only

   C. Both A and B

   D. Neither A nor B

4. Two technicians are discussing a vehicle inspection.

   Technician A says that the wiper blades are often overlooked during an inspection.

   Technician B says a badly damaged wiper blade can scratch the windshield.

   Who is correct?

   A. A only

   B. B only

   C. Both A and B

   D. Neither A nor B

5. *Technician A* says that when closing a door it makes a snapping sound and thinks that it is normal wear.

   Technician B says door alignment should be performed by a body shop technician.

   Who is correct?

   A. A only

   B. B only

   C. Both A and B

   D. Neither A nor B

Name _____   Date _____

LOCATING THE SOURCE OF A LEAK

Upon completion and review of this job sheet, you should be able to locate the source of a leak.

ASE Education Foundation Correlation ————————————————————

This job sheet addresses the following **MLR** task:

1.A.3   Inspect engine assembly for fuel, oil, coolant, and other leaks; determine necessary action. **(P-1)**

This job sheet addresses the following **AST/MAST** task:

1.A.4   Inspect engine assembly for fuel, oil, coolant, and other leaks; determine needed action. **(P-1)**

We Support
ASE | Education Foundation

Tools and Materials

Goggles or safety glasses with side shields

Basic tool set

Flashlight

Mechanic mirror

Service manual

⚡ **Caution**

The engine, engine components, or other vehicle components may be hot. Do not touch any component until it has cooled. Injury can result from touching a hot component.

Describe the vehicle being worked on:

Year _____ Make _____ Model _____

VIN _____ Engine type and size _____

Procedure

Task Completed

☐

1. Start a repair order on the vehicle being inspected.

2. Identify the type of liquid and system leaking.

    Color and smell of liquid _____

    Suspected system _____

3. Raise the hood and inspect any visible components of the suspected system. Use the flashlight and mirror as needed.

    Remarks _____

    _____

    _____

4. Inspect the sides and covers of the engine and other components on or near the suspected system components.

    Remarks _____

    _____

    _____

5. Is the leak source visible from this view? If not, prepare the vehicle for lifting.

    Remarks _____

    _____

6. For the purpose of this job sheet, assume that the vehicle has to be lifted. Lift the vehicle to a good working height.   ☐

7. Observe the point at which the liquid drops from the vehicle's components.   ☐

8. Use the light and mirror to trace the liquid up or over the components until a dry area is found. A dry area is defined as any component that does not show signs of the leaking liquid.   ☐

NOTE: It may take some effort and time to locate the highest point of the leak. Practice with the flashlight and mirror to become accustomed to the different angles of sight with the mirror.

9. Identify the source of the leak. Students may want the assistance of the instructor or an experienced technician for the proper terminology at this point in their training.

    Remarks _____

    _____

    _____

10. Lower the vehicle and complete the repair order. A recommendation for repair or service can be made if desired. Use a service manual for information as needed.

    Remarks _____

    _____

    _____

    INSTRUCTOR CHECK POINT _____

Instructor's Response

_____

_____

_____

Name _____   Date _____

UNDER HOOD INSPECTION

Upon completion and review of this job sheet, you should be able to visually inspect the engine compartment, determine the general condition of the engine, and make maintenance recommendations.

ASE Education Foundation Correlation

This job sheet addresses the following **MLR** task:

1.A.3   Inspect engine assembly for fuel, oil, coolant, and other leaks; determine necessary action. **(P-1)**

This job sheet addresses the following **AST/MAST** task:

1.A.4   Inspect engine assembly for fuel, oil, coolant, and other leaks; determine needed action. **(P-1)**

We Support
ASE | Education Foundation

Tools and Materials

Goggles or safety glasses with side shields

Assigned vehicle

Flashlight or drop light

Mirror

Service manual

Describe the vehicle being worked on:

Year _____ Make _____ Model _____

VIN _____ Engine type and size _____

⚠ Caution

The engine, engine components, or other vehicle components may be hot. Do not touch any component until it has cooled. Injury can result from touching a hot component.

⚠ Caution

Exercise caution when working around a hot engine. Serious burn injuries could result.

⚠ Caution

Keep the fingers and hands clear of electrical cooling fans. They could switch on even with the engine and ignition key set to off. Injuries to the fingers or hands could result when the fan is running.

⚠ Caution

Do not remove the radiator cap on a warm or hot engine. Hot coolant could be expelled rapidly and cause serious burn and/or eye injury.

Procedure   **Task Completed**

1. Start a repair order on the vehicle. Also use a vehicle inspection form if one is   ☐
    available.

2. Move the vehicle to a well-lit area and allow the engine to cool if necessary.   ☐

3. Raise the hood and record your general impression of the engine, its mounted
    components, and the overall sense of the engine compartment.

    _____

    _____

4. Locate the battery and record the visual condition of the battery and its cables.

    _____

    _____

5. Use the light if necessary to inspect and record the condition of all drive belts. Check
    each belt for the proper tension.

    _____

    _____

6. Inspect the battery and its cables. Record your overall impression of the battery and its
    connections. Record the cold-cranking amps of the battery

    _____

    _____

7. Inspect the parts of the spark plugs and other ignition components that are visual and
    record your findings.

    _____

    _____

8. Inspect the top and sides of the engine (and transaxle if equipped) that are visible. Are
    there signs of leakage of any type? If so, what liquid is leaking? Record your overall
    impression of the engine.

    _____

    _____

9. Check the following fluid levels: engine oil, power-steering and brake fluids,
    transmission and transaxle fluid level (if automatic), and coolant level. Are any fluids
    low and which one(s)? Do the respective fluids have the correct color and are not
    burned, discolored, or dirty?

    _____

    _____

10. Remove the radiator cap and inspect the coolant in the radiator. Does it appear to be the right color for the antifreeze used? Is it at the proper level?

_____

_____

11. Ensure the ignition key is off and the engine is cold. Spin the cooling fan by hand. Is it freewheeling without binding or dragging noises?

_____

_____

12. Inspect the front and rear of the radiator as much as possible. On air-conditioning equipped vehicles, the front side of the condenser core blocks a view of the radiator. Are there bent or missing radiator or condenser fins? Is there excessive trash or road debris on the front side of the radiator or condenser? Are there signs of leakage from either the radiator or condenser?

_____

_____

13. Based on your visual inspection, what repairs, if any, are recommended? Consult the service manual for guidance. If available, complete a vehicle inspection (21-point safety check) form. If a form is not available, record your recommendation on the repair order with reason(s) why the service is recommended.

_____

_____

INSTRUCTOR CHECK POINT _____

Instructor's Response

_____

_____

_____

Name _____   Date _____

PERFORMANCE CHECK OF THE ENGINE

Upon completion and review of this job sheet, you should be able to perform a performance check of the engine at idle speed, determine the general condition of the engine, and make maintenance recommendations.

ASE Education Foundation Correlation

This job sheet addresses the following **MLR** tasks:

8.A.2   Perform engine absolute manifold pressure tests (vacuum/boost); document results. **(P-2)**

8.A.3   Perform cylinder power balance test; document results. **(P-2)**

8.A.4   Perform cylinder cranking and running compression tests; document results. **(P-2)**

This job sheet addresses the following **AST/MAST** tasks:

8.A.5   Perform engine absolute manifold pressure tests (vacuum/boost); determine needed action. **(P-1)**

8.A.6   Perform cylinder power balance test; determine needed action. **(P-2)**

8.A.7   Perform cylinder cranking and running compression tests; determine needed action. **(P-1)**

⚠ **Caution**

The engine, engine components, or other vehicle components may be hot. Do not touch any component until it has cooled. Injury can result from touching a hot component.

We Support

ASE | **Education Foundation**

Tools and Materials

Goggles or safety glasses with side shields

Assigned vehicle

Service manual

⚠ **Caution**

Exercise caution when working around a hot engine. Serious burn injuries could result.

Describe the vehicle being worked on:

Year _____ Make _____ Model _____

VIN _____ Engine type and size _____

INSTRUCTOR NOTE: An instructor, senior student, or an experienced technician should be on hand to observe and supervise a student's first attempt at performance checking an engine or vehicle.

Procedure

Task Completed

1. Start a repair order on the vehicle.   ☐

2. Move the vehicle to a well-lit area and allow the engine to cool if necessary. Chock the wheels, place an automatic transmission in PARK (manual in NEUTRAL), set the parking brakes.   ☐

3. Raise the hood and check the fluid levels in the engine and the cooling system. Are they correct?

   _____

NOTE: All tasks in this job sheet are done with the engine at idle speed, transmission in PARK (NEUTRAL).

4. Turn the ignition switch to RUN, ENGINE OFF. Record which warning, or caution light is illuminated. Are all lights correct? If not, stop and consult the instructor and the service manual.

   _____

   _____

Task Completed

5. Start the engine and allow it to idle. ☐

6. Record your impression of the engine and starter when the ignition switch was in the start position. Was the cranking and starting noise "normal?" Did the starter respond correctly? Did the engine crank over easily and start up within 15 seconds? Did all of the warning or cautions lights go out when the engine started (within 15 seconds)? If the answer to any question is NO, shut down the engine and consult the instructor and service manual. Record your observations.

_____

_____

7. If the answer to all questions in step 4 was YES, proceed with this step. Does the engine operate quietly? If equipped with a tachometer, what is the registered engine speed? According to the service manual, what is the desired engine speed? If too high or too low, what do you recommend with regard to repairs?

_____

_____

8. Is there any excessive vibration that can be felt when the engine is running?

_____

_____

9. Check around the vehicle for clearance of objects and personnel. Ensure the parking ☐ brakes are on and the transmission is in PARK (NEUTRAL).

NOTE: Step 9 is only performed on vehicles equipped with an automatic transmission.

10. Apply the service brakes and shift to REVERSE. Did the engine speed remain stable? Did you feel the transmission engage? Shift back to PARK. Record your observation and recommend any repairs you feel are needed.

_____

_____

11. Turn the ignition key to OFF. Record your overall impression of the engine idle operation. Complete the repair order and recommend any maintenance that should be performed and why.

_____

_____

INSTRUCTOR CHECK POINT _____

Instructor's Response

_____

_____

_____

Name _____   Date _____

UNDERCARRIAGE INSPECTION

Upon completion and review of this job sheet, you will be able to perform an undercarriage (under car) inspection and make maintenance recommendations.

ASE Education Foundation Correlation

There are no specific task(s) directly related to an undercarriage inspection, but there are related tasks in each task area that may be used as a guide.

We Support

ASE | Education Foundation

Tools and Materials

Goggles or safety glasses with side shields

Assigned vehicle

Vehicle inspection sheet if available

Service manual

Describe the vehicle being worked on:

Year _____ Make _____ Model _____

VIN _____ Engine type and size _____

⚡ Caution

The engine, engine components, or other vehicle components may be hot. Do not touch any component until it has cooled. Injury can result from touching a hot component.

Procedure

1. Start a repair order on the vehicle.

2. Move the vehicle to a lift. An above-ground, two-post lift is best suited for this job sheet. Set the parking brakes, chock the wheels, and place the transmission in PARK (NEUTRAL).

3. Set the lift and raise the vehicle until wheels are clear of the floor. Shake the vehicle to ensure it is solidly supported.

4. Reach (enter) into the vehicle and release the parking brakes. Shift the transmission to NEUTRAL if necessary. It may be necessary to switch the ignition to RUN and apply the service brakes. DO NOT start the engine.

5. Lift the vehicle until easy access is gained to the undercarriage.

6. Lower the lift until the manual safety locks engage.

NOTE: For the purpose of clarity, this job sheet will start at the front of the vehicle and proceed to the rear, performing checks on the left then the right as appropriate.

7. Observe the bottom of the engine and transaxle, if equipped, for leaks of any types. Observe the lower radiator and hoses for leaks and serviceability. Were any leaks or apparent damage to either component inspected? If yes, what do you recommend with regard to repairs?

_____

_____

8. Observe the components near or attached to the left front wheel including the tire tread and hydraulic lines and hoses. It may be necessary to pull or push accessible steering components to feel for any looseness or damage. Spin the wheel and listen for any noise or observation of wobble. If this is an FWD vehicle, inspect the drive axles and the CV joint boots. Note any damaged or apparently worn or broken parts.

_____

_____

Task Completed

☐

☐

☐

☐

☐

☐

9. Repeat step 7 on the right front wheel.   ☐

10. If this is a RWD vehicle, perform an inspection of the transmission and its external components for leaks or damage. Record your results.

_____

_____

11. Beginning at the backside (rear) of the engine, inspect the exhaust system from front to rear. Is the exhaust suspended on serviceable hanger? Is there any black soot around the connections or anywhere on the pipes, converter, and muffler? Attempt to move the exhaust side to side and up and down. Were there any indications that any portion of the exhaust system was hitting the frame, body, or other vehicle component?

_____

_____

12. Once at the rear of the vehicle, inspect the filler and breathing tubes from the filler cap to the fuel tank. Are there any signs of leakage or deterioration?

_____

13. Follow the brake lines and fuel lines back to the area of the engine. Are there any signs of leakage or deterioration?

_____

14. Return to the left rear wheel and repeat step 7.   ☐

15. Repeat step 7 on the right rear wheel.   ☐

16. If this is RWD vehicle, inspect the seals on the axle housing and around the wheels for signs of lubricant leaks.   ☐

17. Lower the vehicle until the wheels are just clear of the floor. Shift the transmission to PARK. Lower the vehicle to the floor, ensuring the wheel chocks are properly placed.   ☐

18. Based on your observations, what repairs are recommended and why? Complete the repair order.

_____

_____

_____

_____

INSTRUCTOR CHECK POINT _____

Instructor's Response

_____

_____

_____

Name _____   **Date** _____

PERFORM AN EXPRESS SERVICE AND COMPLETE VEHICLE INSPECTION (45 MINUTES)

Upon completion and review of this job sheet, you should be able to perform an express service.

NOTE: The ASE tasks involved here are numerous as are the various tasks to be performed. This job sheet will list the general tasks typically performed by express lube technicians along with preparation of a repair order. Many of the individual tasks will include a reference (Job Sheet 1, for example) to a job sheet in this Shop Manual for the student to use as necessary. ASE tasks are listed on the individual job sheets for reference. It is expected that the instructor will have locally produced repair orders and inspection sheets pertinent to local businesses. A below-ground lube pit will be used in this job sheet as the service area. Additional instructions and procedures will be required if an above-ground lift is used.

NOTE: The following procedures are general and may require adjustment based on shop policy and equipment available.

Tools and Materials

Goggles or safety glasses with side shields

Service information

Repair order

Inspection sheet

Battery tester

Tire depth gauge

Tire pressure gauge

Oil, oil filter, oil filter wrenches

Grease gun

Describe the vehicle being worked on:

Year _____ Make _____ Model _____

VIN _____ Engine type and size _____

Procedure   **Task Completed**

1. Greet the customer and determine type of service requested.   ☐

2. Initialize the repair order. Secure an inspection sheet and record the service tag   ☐
   number, mileage, and make of vehicle.

3. Attach the service tags on rear-view mirror and ignition key. Perform the first phase of   ☐
   the inspection in the service lane and note results on the inspection sheet

   Interior lights; horn; parking brakes; lit warning lights with engine running; exterior
   lighting; tire pressure, tread condition, and tread depth; and exterior condition of the
   vehicle.

NOTE: Some shops require that certain inspection findings also be recorded on the repair order.

4. Inform instructor of any defects found.   ☐

5. Move vehicle to express service bay and drive onto express lube pit. Note and record   ☐
   operating condition of engine and drivetrain as vehicle is moved. If an above-ground
   lift is used.

Task Completed

6. Determine type and amount of oil and oil filter designation.   ☐

NOTE: The following is based on two-person team: top and pit.

7. Pit person: Drain the engine oil and replace the oil filter.   ☐

8. Top person: Upon notification that the drain plug and oil filter are in place, install correct amount and type of engine oil.   ☐

9. Pit person: Inspect the complete undercarriage for leaks, damage, and other defects. Record findings on inspection sheet.   ☐

10. If possible, measure brake pad wear. This may have to be done from top and with wheel assemblies removed. Record findings on inspection sheet if measurement is made.   ☐

11. Lubricate steering and suspension components.   ☐

12. Top person: Inspect each of the following and record findings on inspection sheet.   ☐

    Belt(s) condition and tension; battery performance check; fluid levels (master cylinder, coolant, power steering, automatic transmission, windshield washer); leaks and type of leak; and overall visual impression of engine compartment.

NOTE: Some shops offer free top off of some fluids.

13. Top person: If necessary, lift vehicle and remove wheel assemblies to measure brake pad or shoe wear. Record measurement and make estimate, in percentage, of wear remaining on inspection sheet.   ☐

14. Either person: If vehicle passes inspection or customer refuses additional services, complete technician portion of repair order, inspection sheet, and return vehicle to service lane.   ☐

15. Either person: Remove the disposable mat and seat covers, complete repair order and turn in repair order, inspection sheet, and ignition key to instructor.   ☐

   INSTRUCTOR CHECK POINT _____

Instructor's Response

_____

_____

_____

Name _____   Date _____

CONNECTING A SCAN TOOL

Upon completion of this activity, you will be able to understand how to properly connect the scan tool and communicate with the vehicles control systems.

ASE Education Foundation Correlation

This job sheet addresses the following **MLR/AST/MAST** task:

8.B.1   Retrieve and record diagnostic trouble codes (DTC), OBD monitor status, and freeze frame data; clear codes when applicable. **(P-1)**

This job sheet addresses the following **MLR/** task:

8.B.2   Describe the use of the OBD monitors for repair verification. **(P-1)**

This job sheet addresses the following **AST/MAST** task:

8.B.4   Describe the use of the OBD monitors for repair verification. **(P-1)**

We Support
ASE | **Education Foundation**

> **⚠ Caution**
>
> The engine, engine components, or other vehicle components may be hot. Do not touch any component until it has cooled. Injury can result from touching a hot component.

> **⚠ Caution**
>
> Exercise caution when working around a hot engine. Serious burn injuries could result.

Tools and Materials

Goggles or safety glasses with side shields

Assigned vehicle

Service manual

Scan tool

Describe the vehicle being worked on:

Year _____ Make _____ Model _____

VIN _____ Engine type and size _____

Procedure

Task Completed

1. Begin by locating the data link connector. (DLC)   ☐

2. Where did you find the DLC? Please describe the location.

   _____

   _____

3. Draw a picture of the DLC and label the pin locations.

4. Connect a scan tool to the data link connector (DLC).   ☐

5. Enter vehicle information into the scan tool. You will need to use the VIN #.   ☐

6. With the scan tool powered application manger screen: select Scan Diagnostic.   ☐

   You must press enter on the keypad.

7. Enter vehicle information. Follow screen prompts.   ☐

8. List the ECUs (computers) the scan tool detects.

_____	_____
_____	_____
_____	_____
_____	_____
_____	_____
_____	_____
_____	_____
_____	_____

9. Select Global OBD II and begin communication. List the communication protocol displayed on the scan tool.

_____	_____
_____	_____
_____	_____
_____	_____

10. Check and record any stored DTCs. Check all the ECUs.

_____

_____

_____

_____

11. Select Engine data. Locate the following fields. Start the engine and idle. Enter the data at idle.

IDLE

a. Throttle position sensor (TPS) _____
b. Engine temperature sensor (ECT) _____
c. Idle air control valve (IAC) _____
d. Mass air flow sensor (MAF) _____
e. Engine speed (RPM) _____

12. Increase RPM's to 2000 and enter data.

2000 rpm's

f. Throttle position sensor (TPS) _____
g. Engine temperature sensor (ECT) _____
h. Idle air control valve (IAC) _____
i. Mass air flow sensor (MAF) _____
j. Engine speed (RPM) _____

Instructor's Response

_____

_____

Name _____   Date _____

REPLACE FUEL FILTER

Upon completion of this activity, you will be able to understand how to properly relieve fuel pressure and remove and replace a fuel filter.

ASE Education Foundation Correlation

This job sheet addresses the following **MLR** tasks:

8.C.1   Replace fuel filter(s) where applicable. **(P-1)**

8.C.2   Inspect, service, or replace air filters, filter housings, and intake duct work. **(P-1)**

This job sheet addresses the following **AST/MAST** tasks:

8.D.3   Replace fuel filter(s) where applicable. **(P-2)**

8.D.4   Inspect, service, or replace air filters, filter housings, and intake duct work. **(P-1)**

We Support

ASE | **Education Foundation**

> ⚠ **Caution:**
>
> The engine, engine components, or other vehicle components may be hot. Do not touch any component until it has cooled. Injury can result from touching a hot component.

Tools and Materials

Goggles or safety glasses with side shields

Assigned vehicle

Service manual

Basic hand tools

Describe the vehicle being worked on:

Year _____ Make _____ Model _____

VIN _____ Engine type and size _____

> ⚠ **Caution:**
>
> Exercise caution when working around a hot engine. Serious burn injuries could result.

Procedure

Task Completed

1. Using service manuals, **review** the service procedure for removing and replacing a fuel filter.   ☐

2. If the vehicle you are working on has a fuel service port located on the fuel rail, connect a fuel pressure gauge.   ☐

3. Start the engine and read the fuel pressure. Record your findings. _____

4. Using service manuals, look up the vehicles fuel pressure specification. Record manufacturer's specification _____

5. Using service manuals, locate the fuel filter, raise the vehicle and locate the filter on the vehicle. Describe the location of the filter.

   _____

6. Describe the manufacturer's procedure for depressurizing the fuel system.

   _____

   _____

⚠ **WARNING ALWAYS USE CAUTION WHEN WORKING WITH FUEL!**

7. Depressurize the fuel system by following the manufacturer's procedure.   ☐

Task Completed

8. Make sure the fuel pressure is released. Use the gauge if installed. Record pressure.
   _____

9. With the fuel feed line depressurized, raise the vehicle and describe the general appearance of the fuel filter.

   _____

   _____

10. Describe the type of fuel fittings used on the fuel filter, will you need any special tools to remove the lines from the filter? Describe them.

    _____

    INSTRUCTOR CHECK POINT _____

11. Remove the fuel filter and mounting bracket. Inspect the fuel filter. Record any findings.

    _____

    NOTE: FUEL FILTERS ARE DIRECTIONAL. MAKE SURE YOU REINSTALL IN THE SAME DIRECTION.

12. Write down any information printed on the fuel filter.

    _____

    _____

13. Reinstall the fuel filter and any brackets that may have been removed earlier.   ☐

14. Start the vehicle and make sure the fuel filter is not leaking. Record the fuel pressure reading. _____   ☐

15. The engine may run rough or stall at first until the air is purged from the fuel lines.   ☐

16. Check for fuel leaks a second time. Clean up any fuel that may have leaked onto the floor with soap and water.   ☐

    NOTE: IF YOU SPILL A LARGE AMOUNT OF FUEL ON THE FLOOR, DILUTE WITH WATER AS SOON AS POSSIBLE!!

17. Remove the fuel pressure gauge and install the service port cap on the rail.   ☐

18. Rinse off any fuel that may have leaked while removing the fuel pressure gauge.   ☐

19. Using service manuals, what is the recommended mileage interval for fuel filter replacement on this vehicle? _____

20. Using service manuals, look up the flat rate labor for fuel filter replacement.

    _____

Instructor's Response

_____

_____

_____

Name _____   Date _____

REPLACE SPARK PLUGS

Upon completion of this activity, you will be able to understand how to properly remove and inspect all spark plugs from the engine and test the spark plug wires.

ASE Education Foundation Correlation ——————————————————

This job sheet addresses the following **MLR** task:

8.A.7   Remove and replace spark plugs; inspect secondary ignition components for wear and damage. **(P-1)**

This job sheet addresses the following **AST/MAST** task:

8.C.4   Remove and replace spark plugs; inspect secondary ignition components for wear and damage. **(P-1)**

We Support

ASE | **Education Foundation**

Tools and Materials

Goggles or safety glasses with side shields

Assigned vehicle

Service manual

Basic hand tools

Describe the vehicle being worked on:

Year _____ Make _____ Model _____

VIN _____ Engine type and size _____

Procedure

1. Using service manuals, **review** the service procedure for removing and installing spark plugs.

2. Using service manuals locate the firing order for this vehicle. Record your findings.

   _____

3. Locate the engine's cylinder arrangement and draw it here.

4. Before beginning, make sure the engine is cool to the touch. Do not attempt to remove spark plugs from a hot engine.   ☐

5. Carefully remove the spark plug wire or coil from the spark plug. Record any difficulties.

   _____

   _____

6. Record the size of the **spark plug socket** needed to remove the spark plug.

   _____

⚠ Caution

The engine, engine components, or other vehicle components may be hot. Do not touch any component until it has cooled. Injury can result from touching a hot component.

⚠ Caution

Exercise caution when working around a hot engine. Serious burn injuries could result.

Task Completed

7. Remove the spark plugs from the engine. Note you may need an extension for some of the plugs. Record any difficulties. _____

   KEEP ALL THE PLUGS IN ORDER SO YOU CAN TELL WHICH PLUG CAME FROM WHICH CLYINDER.

8. Identify the following parts of the spark plug below:

a. _____

b. _____

c. _____

d. _____

e. _____

f. _____

g. _____

9. With all the plugs removed, examine all the plugs one at a time. Record the condition of the following:

   Spark plug part # _____

   a. Plug wire terminal _____

   b. Insulation _____

   c. Steel shell and hex _____

   d. Threads _____

   e. Center and ground electrodes _____

   f. Color of the electrodes and insulator _____

10. Using a spark plug gapping tool, carefully measure and record the plug gap.

   SPARK PLUG GAP SPECIFICATIONS _____

Measured gap plug #1	_____
Plug #2	_____
Plug #3	_____
Plug #4	_____
Plug #5	_____
Plug #6	_____
Plug #7	_____
Plug #8	_____

☐

11. Adjust all spark plugs to the manufacturer's specification.

12. Does the manufacturer recommend using antiseize on the threads? ☐ YES or ☐ NO?

13. Locate the spark plug torque specification. **RECORD TORQUE SPECIFICATION**
_____

14. Measure the resistance of each plug wire. Use a DMM on OHM's setting. Record the resistance for each wire.

   NOTE: A good plug wire reading is less than 10,000 ohms per foot in length.

Plug wire	Length	OHM Reading
#1	_____	_____
#2	_____	_____
#3	_____	_____
#4	_____	_____
#5	_____	_____
#6	_____	_____
#7	_____	_____
#8	_____	_____

   INSTRUCTOR CHECK POINT _____

15. **CAREFULLY THREAD THE SPARK PLUGS BACK INTO THE HEAD USING YOUR FINGERS ONLY. MAKE SURE THE SPARK PLUG IS THREADING IN EASILY! SEE INSTRUCTOR IF YOU ARE HAVING**

Task Completed

DIFFICULITY! ☐

16. Once the plugs have been started, torque the plug to specifications. ☐

17. Reinstall the spark plug wires. ☐

18. Start the engine and make sure you do not have an engine misfire.

INSTRUCTOR CHECK POINT _____

19. Using service manuals, what is the recommended mileage interval for spark plug replacement on this vehicle? _____

20. Describe why it is important to let the engine cool down before removing the spark plugs.

_____

_____

21. Using service manuals, look up the flat rate labor for spark plug replacement.

_____

Instructor's Response

_____

_____

_____

GLOSSARY

Note: **Terms are highlighted in bold**, followed by Spanish translation in color.

A **bulb test** or proof-out test of the warning light is done every time the ignition key is switched to ON. Other warning lights are tested while the key is in the START position.

Se realiza una **prueba de bombilla** o prueba de prueba de la luz de advertencia cada vez que se enciende la llave de encendido. Se prueban otras luces de advertencia mientras la llave está en la posición de ARRANQUE.

A **burned area** will be bluish in color.

Un **área quemada** será de color azulado.

A **capacitor** is a small electrical storage device.

Un **capacitor** es un pequeño dispositivo de almacenamiento eléctrico.

A **capacity (load) test** uses a carbon pile to draw energy out of the battery. The voltage level of the battery should not drop below 9.6 volts during the test.

Una **prueba de capacidad (carga)** usa una pila de carbón para extraer energía de la batería. El nivel de voltaje de la batería no debe caer por debajo de 9.6 voltios durante la prueba.

A **carburetor** is an old air/fuel mixture device. When the first electronically controlled engines began to show up for repairs, many good sensors were replaced due to a lack of understanding about system and test equipment operation.

Un **carburador** es un viejo dispositivo de mezcla de aire/combustible. Cuando los primeros motores controlados electrónicamente comenzaron a aparecer para reparaciones, muchos sensores buenos fueron reemplazados debido a la falta de comprensión sobre el funcionamiento del sistema y del equipo de prueba.

A **castellated nut** is so-named because of the turret-type projections on one side.

Una **tuerca almenada** se llama así debido a las proyecciones de tipo torreta en un lado.

A **C-clamp** is so-named because of its shape and function.

Una **abrazadera en C** se llama así debido a su forma y función.

A **chassis ear** machine uses several small microphones that can be placed under the hood and then routed to the passenger compartment where the driver can use a set of headphones. A rear-wheel drive vehicle uses a transmission instead of the **transaxle**. They basically work in the same manner. Rear-wheel drive vehicles usually do not use a **subframe**.

Un aparato **auricular de chasis** utiliza varios micrófonos pequeños que se pueden colocar debajo del capó y luego dirigirse al compartimiento de pasajeros donde el conductor puede usar un conjunto de auriculares. Un vehículo con tracción trasera usa una transmisión en lugar del **transeje**. Básicamente funcionan de la misma manera. Los vehículos con tracción trasera generalmente no usan un **bastidor auxiliar**.

A clutch friction disc can get overheated and have hot spots like those found in brake friction material. It usually results in chattering or grabbing during clutch engagement. Very few hydraulic control clutch systems have pedal **free travel**. Almost all manual clutch linkage systems have a specific free travel.

Un disco de fricción del embrague puede sobrecalentarse y tener puntos calientes como los que se encuentran en el material de fricción del freno. Por lo general, produce charlas o agarres durante el acoplamiento del embrague. Muy pocos sistemas de embrague de control hidráulico tienen un **recorrido libre** en los pedales. Casi todos los sistemas de embrague manual tienen un recorrido libre específico.

A **compressor** is used to pressurize a vapor.

Se usa un **compresor** para presurizar un vapor.

A **conductance test** sends a low-current AC signal into the battery. The return signal is then analyzed.

Una **prueba de conductancia** envía una señal de CA de baja corriente a la batería. Luego se analiza la señal de retorno.

A **cotter key** is a piece of metal doubled back on itself. One use is to keep a nut from loosening.

Una **llave de chaveta** es una pieza de metal con doble respaldo sobre sí misma. Un uso es evitar que una tuerca se afloje.

A **cylinder leakage test** places air pressure into a combustion chamber so you can listen for leaks from possible faulty components.

Una **prueba de fugas en el cilindro** coloca presión de aire en una cámara de combustión para que pueda escuchar las fugas de posibles componentes defectuosos.

A **diesel particulate filter** is used to capture soot and incinerate it during regeneration cycles.

Se usa **un filtro de partículas diésel** para capturar hollín e incinerarlo durante los ciclos de regeneración.

A **female coupling** houses the locks that hold the male coupling in place after hookup. This coupling also has the release mechanism.

Un **acoplamiento hembra** aloja las cerraduras que mantienen el acoplamiento macho en su lugar después de la conexión. Este acoplamiento también tiene el mecanismo de liberación.

A **hard-pedal** condition is when high driver input braking force does not sufficiently slow the vehicle.

Una condición de **pedal duro** es cuando la alta fuerza de frenado de entrada del conductor no frena suficientemente el vehículo.

A **helicoil** is also called a *thread insert*.

A un **helicoide** también se le llama *inserto de hilo*.

A **journal** is a finely machined portion of a shaft fitted through a bearing or housing. This machined surface reduces the friction and heat as the shaft operates.

Un **bulón** es una porción finamente maquinada de un eje encajado a través de un rodamiento o carcasa. Esta superficie mecanizada reduce la fricción y el calor a medida que funciona el eje.

A **labor guide** is a book or guide that the service writer uses to estimate the amount of time a repair will take. This rate is then used to calculate the amount that the customer is charged.

Una **guía laboral** es un libro o guía que el escritor de servicios utiliza para estimar la cantidad de tiempo que llevará una reparación. Esta tarifa luego se usa para calcular la cantidad que se le cobra al cliente.

A **limited slip** differential is designed to reduce wheel spin using some type of clutch mechanism. Using incorrect differential fluid will damage the clutches.

Un diferencial de **deslizamiento limitado** está diseñado para reducir el giro de la rueda utilizando algún tipo de mecanismo de embrague. El uso de fluido diferencial incorrecto dañará los embragues.

A **logo** is artwork used by a business to promote its name and product more easily. The "Golden Arches" is a visual logo for McDonald's restaurants. The "blue seal" (ASE) is a visual logo for technicians who are certified.

Un **logotipo** es una obra de arte utilizada por una empresa para promocionar su nombre y producto más fácilmente. Los "Arcos Dorados" son un logotipo visual para los restaurantes McDonald's. El "sello azul" (ASE) es un logotipo visual para técnicos certificados.

A **male coupling** slides into a mated female unit similar to the manner in which a ratchet's drive stud fits into a socket. The male is fitted with grooves and ridges for locking.

Un **acoplamiento macho se** desliza dentro de una unidad hembra acoplada similar a la forma en que el espárrago de accionamiento de un trinquete encaja en un receptáculo. El macho está equipado con ranuras y crestas para bloquear.

A **memory saver** is a small battery-powered device used to save the vehicle's various memories, such as radio settings.

Un **protector de memoria** es un pequeño dispositivo alimentado por batería que se utiliza para guardar las diversas memorias del vehículo, como la configuración de la radio.

A **micro-vee** belt is a belt composed of multiple small vee grooves in it.

Un cinturón **micro-vee** es un cinturón compuesto por múltiples ranuras pequeñas en forma de v.

A multifunction **switch** is a device with multiple switches opened or closed by one mechanical lever, arm, or button. A turn signal switch is a multifunction switch that connects two different contacts to switch on a front and rear marker lamp.Some vehicles, especially imports, have more than four terminals on the relay. Relays of this sort have more than one switch within them.

Un **interruptor** multifunción es un dispositivo con múltiples interruptores abiertos o cerrados por una palanca mecánica, brazo o botón. Un interruptor de señal de giro es un interruptor multifunción que conecta dos contactos diferentes para encender una luz de señalización delantera y trasera.Algunos vehículos, especialmente los importados, tienen más de cuatro terminales en el relé. Los relés de este tipo tienen más de un interruptor dentro de ellos.

A **neutral flame** is an oxyacetylene flame with a small, sharp inner cone used for heating and welding.

Una **llama neutral** es una llama de oxiacetileno con un cono interno pequeño y afilado que se utiliza para calentar y soldar.

A **performance test** should be performed before and after the repair to verify the fault was repaired.

Se debe realizar una **prueba de rendimiento** antes y después de la reparación para verificar que se reparó la falla.

A **petcock** is a drain plug for the coolant system.

Una **llave de paso** es un tapón de drenaje para el sistema de refrigerante.

A **pigtail** is a replacement connector with a short length of the wiring for connection to the vehicle wiring.

Una **coleta** es un conector de reemplazo con una longitud corta del cableado para la conexión al cableado del vehículo.

A **positive camber** is the inward tilt of the wheel at the top.

Una **inclinación positiva** es la inclinación hacia adentro de la rueda en la parte superior.

A **power balance test** evaluates approximately how much each cylinder is contributing to the overall engine power. It is used to help identify a weak cylinder.

Una **prueba de equilibrio de potencia** evalúa aproximadamente cuánto contribuye cada cilindro a la potencia general del motor. Se usa para ayudar a identificar un cilindro débil.

A **pressure bleeder** is a small holding tank with a diaphragm between a compressed air chamber and a fluid chamber.

Un **purgador de presión** es un pequeño tanque de retención con un diafragma entre una cámara de aire comprimido y una cámara de fluido.

A **radiator** is connected to a vehicle's engine by two hoses. Damage or rot to these hoses can cause coolant leaks.

Un **radiador** está conectado al motor de un vehículo mediante dos mangueras. El daño o la pudrición de estas mangueras puede causar fugas de refrigerante.

A **ratio** is the proportional amount of one element to another. An ideal air-to-fuel ratio is 14.7 parts of air to 1 part of fuel (14.7 to 1).

Una **razón** es la cantidad proporcional de un elemento a otro. Una relación ideal de aire a combustible es 14.7 partes de aire por 1 parte de combustible (14.7 a 1).

A **resistor block** is a series resistor with multiple inputs and a single output. It is usually used to control speed on an electric motor.

Un **bloque de resistencia** es una resistencia en serie con múltiples entradas y una sola salida. Por lo general, se usa para controlar la velocidad en un motor eléctrico.

A **sealing tool** is a pliable cover shaped like the slip joint. The plug slides over the output shaft to prevent oil from leaking out during servicing.

Una **herramienta de sellado** es una cubierta flexible con forma de junta deslizante. El tapón se desliza sobre el eje de salida para evitar que el aceite se escape durante el servicio.

A **spongy pedal** is the result of air in the hydraulic system. It feels like stepping on an inflated balloon.

Un **pedal esponjoso** es el resultado del aire en el sistema hidráulico. Se siente como pisar un globo inflado.

A **tachometer** is a meter used to measure the speed of a device in revolutions per minute (rpm).

Un **tacómetro** es un medidor utilizado para medir la velocidad de un dispositivo en revoluciones por minuto (rpm).

A **test light** is connected between a positive conductor and ground. It will only show voltage presence, not amount. A **self-powered** test light is connected across a portion of the circuit. If the lamp lights, then there is continuity between the test points.

Se conecta una **luz de prueba** entre un conductor positivo y tierra. Solo mostrará la presencia de voltaje, no la cantidad. Una luz de prueba **autoalimentada** está conectada a través de una parte del circuito. Si la lámpara se enciende, entonces hay continuidad entre los puntos de prueba.

A **thermal limiter** is a special circuit protection device that opens the circuit when overheated.

Un **limitador térmico** es un dispositivo especial de protección de circuito que abre el circuito cuando se sobrecalienta.

A **throttle body** may mount one or two injectors and regulate the fuel pressure.

Un **cuerpo del acelerador** puede montar uno o dos inyectores y regular la presión del combustible.

A **vacuum pump** is a belt- or gear-driven pump to remove air from a system.

Una **bomba de vacío** es una bomba accionada por correa o engranaje para eliminar el aire de un sistema.

A **welding tip** is the part attached to an oxyacetylene torch handle where the gases are mixed to create the flame.

Una **punta de soldadura** es la parte unida a un mango de antorcha de oxiacetileno donde los gases se mezclan para crear la llama.

Accountant A person whose job is to keep track or inspect financial account records.

Contador Una persona cuyo trabajo es realizar un seguimiento o inspeccionar los registros de cuentas financieras.

All the internal components of an automatic transmission or transaxle are heavy, continuously lubed, and tend not to fail unless abused.

Todos los componentes internos de una transmisión automática o transmisión son pesados, continuamente lubricados, y tienden a no fallar a menos que se abusa de ellos.

American Wire Gauge A standardized wire gauge system to determine the diameter or a round wire.

Calibre de Alambre Estadounidense Un sistema estandarizado de calibre de alambre para determinar el diámetro o un alambre redondo.

An **instrument voltage regulator (IVR)** limits the voltage supplied to the gauges to 12 volt (DC) so the gauges are more accurate.

Un **regulador de voltaje del instrumento (IVR)** limita el voltaje suministrado a los medidores a 12 voltios (CC) para que los medidores sean más precisos.

An **on-vehicle brake lathe** is designed to machine a brake rotor on the vehicle.

Un **torno de freno en el vehículo** está diseñado para mecanizar un rotor de freno en el vehículo.

Anaerobic means without air. When you drop liquid sealant onto the threads of a bolt, it will remain liquid until you install it. After installing it and removing the air, it starts to harden and become plastic.

Anaeróbico significa sin aire. Cuando deja caer sellador líquido sobre las roscas de un perno, permanecerá líquido hasta que lo instale. Después de instalarlo y eliminar el aire, comienza a endurecerse y convertirse en plástico.

Anti-seize A lubricant to prevent mechanical seizing of two components.

Anti-agarrotamiento Un lubricante para evitar el agarrotamiento mecánico de dos componentes.

Automatic climate control systems use small electric motors to set any or all the various **blend doors** to almost any position needed to raise or lower the passenger compartment temperature.

Los sistemas automáticos de control de clima usan motores eléctricos pequeños para configurar cualquiera o todas las **puertas de mezcla** en casi cualquier posición necesaria para aumentar o disminuir la temperatura del compartimento de pasajeros.

Axle stubs are the outer ends of the axle.

Los **trozos del eje** son los extremos exteriores del eje.

Backlash is the small amount of clearance between the gear teeth that allows room for expansion as the gears heat up.

El **contragolpe** es la pequeña cantidad de espacio libre entre los dientes del engranaje que deja espacio para la expansión a medida que los engranajes se calientan.

Baking soda is a common household product.

El **bicarbonato de sodio es** un producto doméstico común.

Ball joint stud on ball joints and steering linkages are tapered and are interference fitted into a tapered hole.

Los pernos de las juntas de **rótula** y los enlaces de dirección son cónicos y son una interferencia encajada en un agujero cónico.

Bench Bleeding Removing air from a master cylinder on a workbench remove from the vehicle.

Sangrado de banco Eliminación del aire de un cilindro maestro en un banco de trabajo, eliminar del vehículo.

Blind hole A hole that is drilled but does not break through the part.

Agujero ciego Un agujero que se perfora pero que no atraviesa la pieza.

Blower motor A component within a vehicles HVAC system that blows heated or cooled air through vents into the passenger cabin.

Motor de soplador Un componente dentro del sistema HVAC de un vehículo que sopla aire calentado o enfriado a través de los respiraderos hacia la cabina de pasajeros.

Brinelling, from the French, is a type of bearing failure where dents appear due to the roller "hammering" against the race.

Brinelling, del francés, es un tipo de falla del rodamiento donde aparecen abolladuras debido al "martilleo" del rodillo contra la carrera.

Bulb check is a program where certain dash indicator lights will illuminate for only a few seconds so that the technician can check their operation.

La **verificación de la bombilla** es un programa en el que ciertas luces indicadoras del tablero se iluminan durante unos segundos para que el técnico pueda verificar su funcionamiento.

Burnishing is an operation used to heat and seat the brake pads to the rotor.

El **bruñido** es una operación utilizada para calentar y asentar las pastillas de freno en el rotor.

Cabin air filters A small pleated paper cotton material used to trap outside air contaminants before entering the passenger cabin.

Filtros de aire de la cabina un pequeño material de algodón de papel plisado utilizado para atrapar contaminantes del aire exterior antes de ingresar a la cabina de pasajeros.

Camber Camber angle is the inward or outward tilt of the wheel from the top view.

Comba El ángulo de inclinación es la inclinación hacia adentro o hacia afuera de la rueda desde la vista superior.

Camber wear and toe wear can be confusing. Remember that toe wear will create a sawtooth pattern across all or most of the tread.

El **desgaste de la comba** y la puntera pueden ser confusos. Recuerde que el desgaste de los dedos creará un patrón de diente de sierra en toda o la mayor parte de la banda de rodadura.

Charge rate is the rate of current that a battery can be charged at.

La **tasa de carga** es la tasa de corriente a la que se puede cargar una batería.

Commission pay is when the technician receives a percentage of the labor and part sales as part or all of his or her pay.

El pago de **comisión** es cuando el técnico recibe un porcentaje de la mano de obra y venta de piezas como parte o la totalidad de su pago.

Control arm Also known as an "A-arm" is a hinged suspension link between the chassis and the steering knuckle that carries the wheel.

Brazo de control También conocido como "brazo en A" es un enlace de suspensión con bisagras entre el chasis y el muñón de la dirección que lleva el volante.

Cooling fans force airflow draw the radiator to help in the transfer of heat from the coolant to the air.

Los **ventiladores de enfriamiento** obligan al flujo de aire a extraer el radiador para ayudar en la transferencia de calor del refrigerante al aire.

Corrosion can be caused by oxidation (oxygen acting on metal) or by acid or acid fumes.

La **corrosión** puede ser causada por la oxidación (oxígeno que actúa sobre el metal) o por vapores ácidos o ácidos.

Crimping is a procedure used to join a wire to a solderless connector.

El **crimpado** es un procedimiento utilizado para unir un cable a un conector sin soldadura.

Cupping wear caused by balance or rotation is not as clearly distinguishable as other types of tire wear.

El **desgaste de las ventosas** causado por el equilibrio o la rotación no se distingue tan claramente como otros tipos de desgaste de los neumáticos.

Diagnostic trouble codes (DTCs) are alphanumeric fault codes that represent a circuit failure in a monitored system.

Los **códigos de diagnóstico de fallas (DTC)** son códigos de falla alfanuméricos que representan una falla de circuito en un sistema monitoreado.

Disabled airbag To remove the power source from the airbag system preventing an unwanted deployment.

Bolsa de aire desactivada Para eliminar la fuente de alimentación del sistema de bolsa de aire evitando un despliegue no deseado.

Distributor To route secondary, or high voltage, current from the ignition coil to the spark plugs in the correct firing order.

Distribuidor Para dirigir la corriente secundaria o de alto voltaje desde la bobina de encendido a las bujías en el orden de encendido correcto.

Diverter valve A valve designed to supply air to the exhaust system to help control emissions.

Válvula de desvío Una válvula diseñada para suministrar aire al sistema de escape para ayudar a controlar las emisiones.

Drifting usually refers to a slow movement of the vehicle toward one side if the steering wheel is released.

A la deriva generalmente se refiere a un movimiento lento del vehículo hacia un lado si se suelta el volante.

Drive axles transaxle the engine's power from the transaxle to the drive wheels.

Los **ejes de transmisión** transmiten la potencia del motor desde la transmisión a las ruedas de transmisión.

Dry-park check Diagnostic procedure for checking loose steering components.

Verificación de estacionamiento en seco: Procedimiento de diagnóstico para verificar componentes sueltos de la dirección.

EFI A computer-controlled fuel system. Combines sophisticated computer controls and high pressure fuel delivery system to provide optimum power and fuel efficiency.

EFI Un sistema de combustible controlado por computadora. Combina controles informáticos sofisticados y un sistema de suministro de combustible a alta presión para proporcionar potencia y eficiencia de combustible óptimas.

Entry-level technicians are usually recent graduates or may still be attending school. This is their first official job as automotive technicians. An entry-level technician may have different titles when he or she is employed. Such titles often are lube technician, MLR technician, express service technician, or general service technician.

Los **técnicos de nivel de entrada** suelen ser graduados recientes o que aún pueden ser asistir a la escuela. Este es su primer trabajo oficial como técnicos automotrices. Un técnico de nivel de entrada puede tener diferentes títulos cuando está empleado. Dichos títulos a menudo son técnicos de lubricación, técnicos de MLR, técnicos de servicio rápido o técnicos de servicio general.

Experience is the best method to learn how to set up and use an **angle indicator**. The manual one is awkward to use in tight places.

La experiencia es el mejor método para aprender a configurar y usar un **indicador de ángulo**. El manual es incómodo de usar en lugares estrechos.

Fade The reduction of stopping power usually due to extreme heat build up during extended high load braking.

Desvanecimiento La reducción de la potencia de frenado generalmente debido a la acumulación extrema de calor durante el frenado prolongado a alta carga.

Fading is a condition where the brakes initially work but lose their stopping power because of heat or other conditions affecting the friction material. The most common cause of brake fade is heat, but it should be noted that water and hydraulic fluid can also cause brake fade.

La **fatiga de frenado** es una condición en la que los frenos funcionan inicialmente, pero pierden su poder de frenado debido al calor u otras condiciones que afectan el material de fricción. La causa más común del desvanecimiento del freno es el calor, pero debe tenerse en cuenta que el agua y el fluido hidráulico también pueden causar el desvanecimiento del freno.

Filter wrenches are specifically designed to remove oil filters.

Las **llaves para filtros** están diseñadas específicamente para eliminar los filtros de aceite.

Flat rate is the time it takes to perform a job. It is an average based on research and practice. Some technicians are paid by flat rate.

La **tarifa plana** es el tiempo que lleva realizar un trabajo. Es un promedio basado en la investigación y la práctica. Algunos técnicos reciben un pago de tarifa plana.

Fluorescent dye tracer A method of tracing fluids by adding a bright colored dye which may show the component causing the leaking fluid.

Trazador de tinte fluorescente: un método para rastrear fluidos agregando un tinte de color brillante que puede mostrar el componente que causa la fuga de líquido.

Flux A substance added to molten metals to clean and bond the surface together better.

Flujo Una sustancia añadida a los metales fundidos para limpiar y unir mejor la superficie.

Fuel rails provide a common channel to maintain constant fuel pressure and volume at each injector in a multi-injector system.

Los **rieles de combustible** proporcionan un canal común para mantener la presión y el volumen constantes de combustible en cada inyector en un sistema multi inyector.

Gas-charged shocks are a hydraulic shock absorber with nitrogen gas installed to pressurize the hydraulic fluid for quicker and more responsive shock action.

Los **amortiguadores cargados de gas** son amortiguadores hidráulicos con gas nitrógeno instalado para presurizar el fluido hidráulico para una acción de absorción más rápida y para capacidad de respuesta.

Gas-filled shocks have some nitrogen gas in them to increase the shock's action.

Los **amortiguadores llenos de gas** tienen algo de nitrógeno gaseoso para aumentar la acción de amortiguación.

Glazing is a condition when the friction material and rotor/drum develops a slick, nearly non-friction surface. This is usually caused by heat.

El **acristalamiento** es una condición cuando el material de fricción y el rotor/tambor desarrollan una superficie resbaladiza, casi sin fricción. Esto generalmente es causado por el calor.

Glycol is a member of a group of alcohols. It should not be confused with prefix "glyco," which is a combined form of glycerol, a sugar.

El **glicol** es miembro de un grupo de alcoholes. No debe confundirse con el prefijo "glico", que es una forma combinada de glicerol, un azúcar.

Grabbing brakes are usually caused by oil or brake fluid on the friction material.

Los **frenos de agarre** generalmente son causados por aceite o líquido de frenos en el material de fricción.

Ground reference voltage testing is used to determine the voltage within a circuit.

La **prueba de voltaje de referencia a tierra** se usa para determinar el voltaje dentro de un circuito.

Heat-resistant gloves and forearm sleeves are usually made of woven Kevlar. If the Kevlar becomes wet, the heat will be absorbed by the liquid and transferred to the skin.

Los **guantes resistentes al calor** y las mangas de antebrazo generalmente están hechos de Kevlar tejido. Si el Kevlar se moja, el calor será absorbido por el líquido y transferido a la piel.

Hot spots are formed by excessive heat on the weakest area of the disc or drum.

Los **puntos calientes** se forman por el calor excesivo en el área más débil del disco o tambor.

Hourly rate is the amount paid by clock hour to the technician regardless of the amount or type of work done in that hour. Entry level technicians are usually paid by the hour.

La **tarifa por hora** es la cantidad pagada por hora de reloj al técnico, independientemente de la cantidad o tipo de trabajo realizado en esa hora. Los técnicos de nivel de entrada generalmente se les paga por hora.

HVAC Short for heating, ventilation and air-conditioning system, which provides heat and cooling to the passenger cabin. Some **refrigerants** are flammable, and exposure to a spark or flame can cause the vapor to ignite.

HVAC Abreviatura de sistema de calefacción, ventilación y aire acondicionado, que proporciona calor y refrigeración a la cabina de pasajeros. Algunos **refrigerantes** son inflamables, y la exposición a una chispa o llama puede hacer que el vapor se encienda.

Hydrostatic lock is the result of attempting to compress a liquid in the cylinder. Because liquid is not compressible, the piston is not able to travel in the cylinder.

El **bloqueo hidrostático** es el resultado de intentar comprimir un líquido en el cilindro. Debido a que el líquido no es compresible, el pistón no puede viajar en el cilindro.

Hygroscopic A fluid that absorbs moisture from the air.

Higroscópico Un fluido que absorbe la humedad del aire.

Ignition timing refers to the timing of spark plug firing in relation to position of the piston.

El **tiempo de ignición** se refiere a la sincronización del encendido de la bujía en relación con la posición del pistón.

In automotive applications, a **dipstick** is a thin flexible rod used to check fluid levels in certain components.

En aplicaciones automotrices, una **varilla medidora** es una varilla delgada y flexible que se usa para verificar los niveles de líquido en ciertos componentes.

Infinite means forever or too large. In automobile circuit testing, it means the value being measured is too great for the test device being used.

Infinito significa para siempre o demasiado grande. En las pruebas de circuitos de automóviles, significa que el valor que se está midiendo es demasiado grande para el dispositivo de prueba que se está utilizando.

Isolators are made of rubber or rubber-like material to reduce noise.

Los **aisladores** están hechos de caucho o similar al caucho de material para reducir el ruido.

Jump starting a battery uses another battery source to quickly charge a dead battery.

Arranque de salto de una batería utiliza otra fuente de batería para cargar rápidamente una batería agotada.

KOEO means ignition key on, engine off, a test condition required to performed certain tests.

KOEO significa llave de encendido encendida, motor apagado, una condición de prueba requerida para realizar ciertas pruebas.

KOER means ignition key on, engine running, another test condition required for certain tests.

KOER significa llave de encendido encendida, motor en marcha, otra condición de prueba requerida para ciertas pruebas.

The **filter gasket** is usually an O-ring. A cartridge-style oil filter requires a new O-ring as well (Figure 7-6). There is a specific amount of torque from the container to the filter housing.The offset position of the container's **spout** allows air to enter the container. It is suggested that the container not be squeezed to speed the process. This will block the airflow.

La **junta del filtro** suele ser una junta tórica. Un filtro de aceite tipo cartucho también requiere una nueva junta tórica (Figura 7-6). Hay una cantidad específica de torque desde el contenedor a la carcasa del filtro.La posición desplazada de la **boquilla** del contenedor permite que el aire ingrese al contenedor. Se sugiere que el contenedor no se apriete para acelerar el proceso. Esto bloqueará el flujo de aire.

Lineman's gloves are highly insulated rubber gloves that protect the technician from lethal high voltages.

Los **guantes de Lineman** son guantes de goma altamente aislados que protegen al técnico de altos voltajes letales.

Locking brakes may be caused by misadjusted shoes, blockage in the brake line, or slick tires/road surface.

El **bloqueo de los frenos** puede ser causado por zapatos mal ajustados, bloqueo en la línea de freno o llantas resbaladizas/superficie de la carretera.

Lug nuts are the hex-sided fasteners used to hold the wheel rim on the studs of the brake rotor or drum.

Las **tuercas de rueda** son los sujetadores de lado hexagonal que se utilizan para sostener la llanta de la rueda en los pernos del rotor o tambor de freno.

MIL (malfunction indicator lamp) is a dash-mounted warning light commonly referred to as the check engine light.

MIL (**lámpara indicadora de mal funcionamiento**) es una luz de advertencia montada en el tablero, comúnmente conocida como la luz de verificación del motor.

Monitor The vehicles computer observes engine data and adjusts the controls to keep emissions as low as possible.

Monitor La computadora del vehículo observa los datos del motor y ajusta los controles para mantener las emisiones lo más bajas posible.

Most **front-wheel-drive (FWD)** axle bearings are one-piece, double ball bearings that are permanently lubed and sealed.

La mayoría de los rodamientos de eje de **tracción delantera (FWD)** son cojinetes de una sola pieza, de doble bola, lubricados y sellados de forma permanente.

Most original equipment U-joints do not have **grease fittings**. Replacement U-joints will include a fitting with the joint.

La mayoría de las juntas en U del equipo original no tienen **accesorios de engrase**. Las juntas en U de reemplazo incluirán un accesorio con la junta.

Negative camber is the outward tilt of the wheel.

Una **inclinación negativa** es la inclinación hacia afuera de la rueda.

Panic button A button on the key remote used to alert people in a potential emergency.

Botón de pánico Un botón en el control remoto de llave usada para alertar a las personas en una posible emergencia.

PASS A method to extinguish a fire using a pull, aim, squeeze, sweep (PASS) technique.

PASS Un método para extinguir un incendio usando una técnica de tirar, apuntar, apretar, barrer (PASS).

Pedal pulsation is a rapid up-and-down movement of the brake pedal during brake application. The driver may complain that the vehicle shakes when coming to a stop.

La **pulsación del pedal** es un movimiento rápido hacia arriba y hacia abajo del pedal del freno durante la aplicación del freno. El conductor puede quejarse de las sacudidas del vehículo cuando se detiene.

Preloaded Axial or thrust load applied to a bearing that removes excess play.

Carga axial o de empuje precargada aplicada a un rodamiento que elimina el exceso de juego.

Professionalism A person who shows consideration and respect for others and demonstrates loyalty and exceeds expectations.

Profesionalismo Una persona que muestra consideración y respeto por los demás y demuestra lealtad y supera las expectativas.

Pulling usually refers to a sudden movement of the vehicle to one side if the steering wheel is released.

Tirar generalmente se refiere a un movimiento repentino del vehículo hacia un lado si se suelta el volante.

Quick-disconnect coupling Used to provide a fast and easy connection and disconnection to air hoses for air-operated tools.

Acoplamiento de desconexión rápida Se utiliza para proporcionar una conexión y desconexión rápida y fácil a las mangueras de aire para herramientas accionadas por aire.

Recalls A safety-related defect that does not comply with federal safety standards. The automaker must alert and repair the problem at no charge to the vehicle owner.

Retiros Un defecto relacionado con la seguridad que no cumple con los estándares federales de seguridad. El fabricante de automóviles debe alertar y reparar el problema sin cargo para el propietario del vehículo.

Refractometer A tool used to measure fluid concentration or levels of strength and or charge for a battery.

Refractómetro herramienta utilizada para medir la concentración de fluidos o los niveles de resistencia y / o carga de una batería.

Rotor runout When a rotor deviates from its axial plane, also known as wobble.

Desgaste del rotor Cuando un rotor se desvía de su plano axial, también conocido como oscilación.

Safety Data Sheet (SDS) A document that contains information on the potential health effects of exposure to chemicals or other potentially dangerous substances.The extinguisher's **tag** is usually a small card tied directly to the extinguisher. The inspector may enter his or her initials, date, and status of the extinguisher.

Hoja de datos de seguridad (SDS) Un documento que contiene información sobre los posibles efectos en la salud de la exposición a productos químicos u otras sustancias potencialmente peligrosasLa **etiqueta** del extintor suele ser una pequeña tarjeta atada directamente al extintor. El inspector puede ingresar sus iniciales, fecha y estado del extintor.

Seepage is seen as a damp spot with no visual of fluid drops. Seepage is considered to be normal. Many seals will not function correctly without some seepage.

La **filtración** se ve como un lugar húmedo sin visuales de gotas de líquido. La filtración se considera normal. Muchos sellos no funcionarán correctamente sin alguna filtración.

Seepage is the very small amount of fluid that seeps past the seal. Most seals are designed to allow some lubrication seepage around the seal's contact point with the shaft.

La **filtración** es la cantidad muy pequeña de líquido que se filtra más allá del sello. La mayoría de los sellos están diseñados para permitir cierta filtración de lubricación alrededor del punto de contacto del sello con el eje.

Soldering A process in which two or more items are joined together by melting solder into the joint.

Soldadura Un proceso en el que dos o más elementos se unen fundiendo soldadura en la unión.

Soldering gun or a **soldering iron** A tool used to melt solder. A person whose job is to keep track or inspect financial account records.

Pistola de soldar o un **soldador** Una herramienta utilizada para fundir la soldadura. Una persona cuyo trabajo es realizar un seguimiento o inspeccionar los registros de cuentas financieras.

Some **water pumps** are driven by the timing belt positioned behind a cover on the front of the engine.

Algunas **bombas de agua** son accionadas por la correa de distribución colocada detrás de una cubierta en la parte delantera del motor.

Spalling, from the German, means "to break or split off."

Desconchados, del alemán, significa "romper o separarse".

Special tools Tools that are needed to perform a special operation the complete a repair of task.

Herramientas especiales: Herramientas que se necesitan para realizar una operación especial para completar una reparación de la tarea.

Splice Joining two or more electrical wires together using solder.

Empalmar Unir dos o más cables eléctricos juntos mediante soldadura.

Split system Brake fluid is applied to the wheels on separate lines one is usually the front and rear.

Sistema dividido El líquido de frenos se aplica a las ruedas en líneas separadas, una generalmente es la delantera y la trasera.

Steering knuckle The part that contains the spindle or wheel hub which attaches to the steering components.

Nudillo de dirección La parte que contiene el eje o el cubo de la rueda que se une a los componentes de la dirección.

Steering wheel free play is the amount of steering wheel movement before the steering gear or linkage moves.

El **juego libre del volante** es la cantidad de movimiento del volante antes de que se mueva el mecanismo de dirección o la articulación.

Stripped threads A fastener failure were the threads have been damaged beyond use.

Hilos dañados Una falla en el sujetador donde los hilos se han dañado más allá del uso.

Stripping and crimping pliers are used to strip insulation from wire and crimp a solderless terminal on wire.

Los **alicates de pelado y engarzado** se utilizan para pelar el aislamiento del cable y engarzar un terminal sin soldadura en el cable.

Stripping is the procedure used to remove insulation from an electrical wire.

El **pelado** es el procedimiento utilizado para eliminar el aislamiento de un cable eléctrico.

Stud remover A special tool that can remove a stud with damaged thread by gripping on the damaged stud.

Removedor de pernos Una herramienta especial que puede quitar un perno con rosca dañada al sujetar el perno dañado.

Technical Service Bulletins (TSBs) Recommended procedures for an unanticipated problem that the automaker has had several occurrences. They are a form of diagnosis.

Boletines de servicio técnico (TSB) Procedimientos recomendados para un problema imprevisto en el que el fabricante de automóviles ha tenido varios sucesos. Son una forma de diagnóstico.

The **3 Cs** of automotive repair are concern, cause, and correction. The correction (actual repair the technician does) needs to match the concern (the reason it was brought in for service in the first place).

Las **3 C** de la reparación automotriz son preocupación, causa y corrección. La corrección (reparación real que realiza el técnico) debe coincidir con la inquietud (la razón por la que se solicitó el servicio en primer lugar).

The **battery terminal voltage drop** test checks for poor electrical connections between the battery cables and terminals.

La prueba de **caída de voltaje del terminal de la batería** verifica si hay malas conexiones eléctricas entre los cables de la batería y los terminales.

The **center bearing** is used on two-piece drive shafts.

El **cojinete central** se utiliza en ejes de transmisión de dos piezas.

The **coupler** is a flexible connection between the steering column and the gear or the upper and lower shafts.

El **acoplador** es una conexión flexible entre la columna de dirección y el engranaje o los ejes superior e inferior.

The **current draw test** measures the amount of current the starter draws when actuated. It determines the electrical and mechanical condition of the starting system.

La **prueba de consumo actual** mide la cantidad de corriente que consume el arrancador cuando se activa. Determina la condición eléctrica y mecánica del sistema de arranque.

The engine computer uses a strategy to spray a fine mist of **diesel exhaust fluid (DEF)** in the exhaust during certain periods to help clean up the exhaust emissions and burn the soot in the filter.

La computadora del motor utiliza una estrategia para rociar una fina niebla de **fluido de escape diésel (DEF)** en el escape durante ciertos períodos para ayudar a limpiar las emisiones de escape y quemar el hollín en el filtro.

The **front pipe**, **muffler**, and **tailpipe** are the three main components used to channel exhaust gases from the engine to the rear bumper and reduce exhaust noises.

El **tubo delantero**, el **silenciador** y el **tubo de escape** son los tres componentes principales utilizados para canalizar los gases de escape desde el motor hacia el parachoques trasero y reducir los ruidos de escape.

The **grease fitting** on some aftermarket U-joints are placed in the bearing cup instead of the joint cross.

El **engrasador** en algunas juntas en U del mercado de accesorios se coloca en la copa del rodamiento en lugar de la cruz de la junta.

The **MAF** measures the amount of air entering the engine's intake system.

El **MAF** mide la cantidad de aire que ingresa al sistema de admisión del motor.

The **open circuit voltage test** is used to determine the battery's state of charge.

La **prueba de voltaje de circuito abierto** se utiliza para determinar el estado de carga de la batería.

The **owner** or **manager** is responsible for the daily operation of a segment of a business and reports to the general manager.

El **propietario** o **gerente** es responsable de la operación diaria de un segmento de un negocio e informa al gerente general.

The **PCV system** is used to vent the pressure created inside the engine by heat and pressure. Failure to vent the engine will cause oil leaks.

El **sistema PCV** se usa para ventear la presión creada dentro del motor por el calor y la presión. Si no se ventila el motor, se producirán fugas de aceite.

The **repair order** is a document that tracks the vehicle from its arrival at the shop for repairs to the time the repairs are paid by the customer. This is considered a legal document. This document helps protect the customer and the business from any problems that may arise during or after the repairs.

El **número de identificación del vehículo (VIN)** es un número asignado a un vehículo específico. La información sobre el tamaño y tipo de motor, la transmisión, el año de producción y otros datos del vehículo están codificados en el VIN.

The **serpentine belt** is so named because of its winding route among the various pulleys at the front of the engine.

La **correa serpentina** se llama así debido a su ruta sinuosa entre las diversas poleas en la parte delantera del motor.

The **service manager** controls the operation of the service department, including parts, technicians, support staff, and the training of all department employees.

El **gerente de servicio** controla la operación del departamento de servicio, incluidas las piezas, los técnicos, el personal de soporte y la capacitación de todos los empleados del departamento.

The **service writer** meets the customer, writes the repair order, and assigns a technician based on company policy.

El **redactor del servicio** se reúne con el cliente, escribe la orden de reparación y asigna un técnico según la política de la empresa.

The **shaft** with the type bearings discussed here is called a *spindle*. The shaft (or spindle) is part of a steering knuckle, which is replaced as a unit. A rough cloth will remove most of the **grease** from a technician's hands after the bearing has been repacked.

El **eje** con los cojinetes de tipo discutido aquí se llama *husillo*. El eje (o eje) es parte de un muñón de dirección, que se reemplaza como una unidad.Una tela áspera eliminará la mayor parte de la **grasa** de las manos de un técnico después de que el rodamiento haya sido reempaquetado.

The **shims** are used to provide proper pinion-to-ring gear clearance.

Las **cuñas** se utilizan para proporcionar una separación adecuada del piñón al anillo.

The **starter quick test** will isolate the problem area; that is, if the starter motor, solenoid, or control circuit is at fault.

La **prueba rápida de inicio** aislará el área del problema; es decir, si el motor de arranque, el solenoide o el circuito de control están en falla.

The **striker** has a flint that creates a spark to ignite an oxyacetylene flame.

El **percutor** tiene un pedernal que crea una chispa para encender una llama de oxiacetileno.

The **suction pump** is used to draw fluid from a container and force the fluid into the transmission or transaxle by hand.

La **bomba de succión** se usa para extraer fluido de un recipiente y forzar el fluido hacia la transmisión o la transmisión manualmente.

The term **out-of-round** only applies to a journal or shaft that is supposed to be perfectly round.

El término **fuera de ronda** solo se aplica a un bulón o eje que se supone que es perfectamente redondo.

The term **race** refers to the machined area(s) of the bearing, shaft, or gear where the bearing's rollers or balls contact and roll on.

El término **carrera** se refiere al área o áreas mecanizadas del rodamiento, eje o engranaje donde los rodillos o bolas del rodamiento entran en contacto y ruedan.

The **toothed wheel** or **ring** is a metal ring or plate with teeth at a specific distance apart. As each tooth passes the sensor, it will cause the sensor to generate a voltage.

La **rueda dentada** o **anillo** es un anillo de metal o placa con dientes a una distancia específica. A medida que cada diente pasa por el sensor, hará que el sensor genere un voltaje.

The **vehicle identificaton number (VIN)** is a number assigned to a specific vehicle. Information on the engine size and type, transmission, year of production, and other vehicle data is encoded in the VIN.

El **número de identificación del vehículo (VIN)** es un número asignado a un vehículo específico. La información sobre el tamaño y tipo de motor, la transmisión, el año de producción y otros datos del vehículo están codificados en el VIN.

The **yoke** of a light drive shaft (small car) can be easily distorted. The correct use of the proper tool will help prevent this distortion.

El **yugo** de un eje de transmisión ligero (automóvil pequeño) se puede distorsionar fácilmente. El uso correcto de la herramienta adecuada ayudará a prevenir esta distorsión.

This type of **spring-loaded switch** is sometimes referred to as a "dead-man switch."

Este tipo de **interruptor de resorte a** veces se denomina "interruptor de hombre muerto".

Thread depth gauge A tool used to measure the tire thread left on the tire.

Medidor de profundidad de hilo Una herramienta utilizada para medir el hilo del neumático que queda en el neumático.

Thread-locking compound an adhesive applied to threads to prevent a bolt from loosening.

Compuesto de bloqueo de roscas Un adhesivo aplicado a las roscas para evitar que un perno se afloje.

Thread-sealing compound an adhesive applied to threads to prevent leaking or corrosion.

Compuesto de sellado de roscas un adhesivo aplicado a las roscas para evitar fugas o corrosión.

Through hole also known as a ("thru-hole") threads that go all the way through the part.

A través de un agujero también conocido como un ("thru-hole") hilos que van todo el camino a través de la pieza.

Tinning Application of a thin layer of soft solder to the ends of wires before soldering them.

Estañado Aplicación de una capa delgada de soldadura blanda a los extremos de los cables antes de soldarlos.

Toe wear Feather edge on a tire caused by the toe-in adjustment being out of specifications.

Desgaste del dedo del pie El borde de la pluma en un neumático causado por el ajuste del dedo del pie está fuera de las especificaciones.

Top off is a term that means fluid is added to a system to bring it to the correct level.Antifreeze is normally slick and oily. Metal dissolving in the coolant, however, causes it to feel like **slime**. You can also tell a lot about coolant by its color. Green is the color of traditional coolant. This type of coolant will last about 3 years. Extended life coolants are very common on modern vehicles. They will last anywhere from 3 to 6 years. Extended life coolants are available in many colors: red, pink, orange, yellow, and so on. But they are not green.

Relleno es un término que significa que se agrega fluido a un sistema para llevarlo al nivel correcto.El anticongelante es normalmente liso y aceitoso. Sin embargo, la disolución de metal en el refrigerante hace que se sienta como **limo**. También puede decir mucho sobre el refrigerante por su color. El verde es el color del refrigerante tradicional. Este tipo de refrigerante durará aproximadamente 3 años. Los refrigerantes de larga duración son muy comunes en los vehículos modernos. Durarán entre 3 y 6 años. Los refrigerantes de larga duración están disponibles en muchos colores: rojo, rosa, naranja, amarillo, etc. Pero no son verdes.

TPMS relearn procedure Steps used to identity the proper location of the tire pressure sensors after a rotation or tire service.

Procedimiento de reaprendizaje de TPMS Pasos utilizados para identificar la ubicación adecuada de los sensores de presión de los neumáticos después de una rotación o servicio de neumáticos.

Transverse means the engine is mounted crossway to the length of the vehicle.

Transversal significa que el motor está montado en forma transversal a la longitud del vehículo.

Two lighting systems are considered emergency lighting. They are the brake lights and the hazard warning lights. They should operate in almost any condition if electrical power is available.

Dos sistemas de iluminación se consideran iluminación de emergencia. Son las luces de freno y las luces de advertencia de peligro. Deben funcionar en casi cualquier condición si hay energía eléctrica disponible.

Upwind means the wind is blowing on the person's back. If the wind is in the firefighter's face (downwind), the flame and the firefighting material can be blown toward the fighter.

Viento arriba significa que el viento sopla sobre la espalda de la persona. Si el viento está en la cara del bombero (a favor del viento), la llama y el material de extinción de incendios pueden volar hacia el bombero.

Vehicle ride height The amount of space between the base of the vehicle and the road surface. As the vehicle **waves** from poor shocks or struts, the alignment angle changes on the front wheels and can cause steering control problems.

Altura de conducción del vehículo La cantidad de espacio entre la base del vehículo y la superficie de la carretera. A medida que el vehículo se **agita** por golpes o puntales deficientes, el ángulo de alineación cambia en las ruedas delanteras y puede causar problemas de control de la dirección.

Voltage drop is a measurement of the voltage across a portion of the circuit.

La **caída de voltaje** es una medida del voltaje en una parte del circuito.

Warranties A written guarantee, issued to the purchaser a repair by the business, promising to repair or replace if necessary within a specified period of time.

Garantías Una garantía por escrito, emitida al comprador por una reparación de la empresa, que promete reparar o reemplazar si es necesario dentro de un período de tiempo especificado.

Wheel bearings A wheel bearing, Also known as a hub assembly is a component that helps decrease friction when the wheel is turning by the use of ball bearing.

Cojinetes de rueda Un rodamiento de rueda, también conocido como conjunto de cubo, es un componente que ayuda a disminuir la fricción cuando la rueda gira mediante el uso de un rodamiento de bolas.

INDEX

Note: Page numbers followed by "f" indicate material in a figure.